T0260836

FIAT
124 Sport
Automotive
Repair Manual

by John H Haynes
Member of the Guild of Motoring Writers
and Adrian Sharp

Models covered:

UK
124 Sport Coupe
USA
124 Sport Coupe
124 Sport Spider
1438 cc (87.8 cu in), 1608 cc (98.1 cu in), 1592 cc (97.2 cu in) and 1756 cc (107 cu in) engine capacities

Does not cover Special T (UK) and Spcial TC (USA)

ISBN 0 90050 94 5

© **Haynes North America, Inc. 1974, 1976, 1978, 1986, 1987**
With permission from Haynes Group Limited

(34010-8U5)

Haynes Group Limited
Haynes North America, Inc.
www.haynes.com

Acknowledgements

First, thanks are due to Fiat (England) Limited for their co-operation and use of official manuals and to their Public Relations agency Woolf, Laing and Christie and Partners.

Secondly, more particular thanks are due to those firms who know about the Fiat 124 Sport and were so forthcoming with information and help.

The Champion Sparking Plug Company Limited provided the illustrations showing the various spark plug conditions and the bodywork repair photographs used in this manual were provided by Lloyds Industries Limited who supply 'Turtle Wax', 'Dupli-color Holts', and other Holts range products.

Special acknowledgement must also go to Ted Frenchum. His advice on the text structure greatly helped the author. His conscientious and detailed attention to the total layout is particularly appreciated.

Last, but not least thanks must go to all those people at Sparkford who assisted in the production of this manual, particularly Brian Horsfall, Les Brazier, Colin Barge, Lee Saunders and John Rose.

About this manual

Its aim

The aim of this Manual is to help you get the best value from your car. It can do so in several ways. It can help you decide what work must be done (even should you choose to get it done by a garage), provide information on routine maintenance and servicing, and give a logical course of action and diagnosis when random faults occur. However, it is hoped that you will make full use of the Manual by tackling the work yourself. On simpler jobs it may even be quicker than booking the car into a garage, and having to go there twice, to leave and collect it. Perhaps most important, a lot of money can be saved by avoiding the costs the garage must charge to cover its labour and overheads.

The Manual has drawings and descriptions to show the function of the various components so that their layout can be understood. Then the tasks are described and photographed in a step-by-step sequence so that even a novice can do the work.

Its arrangement

The manual is divided into thirteen Chapters, each covering a logical sub-division of the vehicle. The Chapters are each divided into consecutively numbered Sections and the Sections into paragraphs (or sub-sections), with decimal numbers following on from the Section they are in, eg 5.1, 5.2, 5.3 etc.

It is freely illustrated, especially in those parts where there is a detailed sequence of operations to be carried out. There are two forms of illustration: figures and photographs. The figures are numbered in sequence with decimal numbers, according to their position in the Chapter: eg Fig. 6.4 is the 4th drawing/illustration in Chapter 6. Photographs are numbered (either individually or in related groups) the same as the Section or sub-section of the text where the operation they show is described.

There is an alphabetical index at the back of the manual as well as a contents list at the front.

References to the 'left' or 'right' of the vehicle are in the sense of a person facing forwards in the driver's seat.

Whilst every care is taken to ensure that the information in this manual is correct no liability can be accepted by the authors or publishers for loss, damage or injury caused by any errors in, or omissions from, the information given.

Modifications to the Fiat 124 Sport

The policy of the manufacturer of these vehicles is one of continuous development and designs and specifications are frequently being changed as a result. It follows naturally that spares may sometimes be purchased which differ both from the original part removed and that referred to in this manual. However, suppliers of genuine FIAT spare parts can usually settle queries about interchangeability by reference to their latest technical information from FIAT.

Wherever possible a summary of modifications incorporated by FIAT, up to the publishing date of this manual, has been included near the end of the relevant Chapters in this manual.

The cars: FIAT 124 Sport and Spider

These cars were first manufactured in 1967 and were designed around the floor structure of the 124 saloon and a 1438 cc twin overhead camshaft engine originating from the FIAT 125.

In 1970 the coachwork was restyled to include twin headlights on the Sport, and dual braking systems were added to both Sport and Spider.

The cars were also offered with a 1608 cc engine and five speed gearbox. The 1438 cc engine option was discontinued in 1972 and then, at the beginning of 1973, the latest 124 Sport and Spider were announced.

The most recent Sport and Spider are offered with either the 1592 cc or 1755 cc twin ohc engines as fitted to the FIAT 132 Saloon, and a 4- or 5-speed gearbox, and have a further re-styling of front and rear.

On any count these are fabulous cars and although still basically FIAT they can be favourably compared with the BMW, Lancia and Alfa Romeo.

It is confusing - the coupe version is known as the FIAT 124 Sport whilst the open version is called the FIAT 124 Spider. The Spider is not available in right-hand drive (the Abarth version of the Spider is not covered by this Manual).

Vehicles exported to the USA have the statutory pollution control devices on the engine and several more electrical circuits, the primary purpose of which is to visually warn the driver of failure in major mechanical systems in the car, brakes in particular.

Contents

Fiat 124 Sport (1st Series) 1968 (UK)

Fiat 124 Sport (2nd Series) 1971 (UK)

Fiat 124 Sport (3rd Series) 1973 (UK)

Fiat 124 Sport Spider 1973 (North America)

Buying spare parts
and vehicle identification numbers

Buying spare parts

Replacement parts are available from many sources, which generally fall into one of two categories – authorized dealer parts departments and independent retail auto parts stores. Our advice concerning these parts is as follows:

Retail auto parts stores: Good auto parts stores will stock frequently needed components which wear out relatively fast, such as clutch components, exhaust systems, brake parts, tune-up parts, etc. These stores often supply new or reconditioned parts on an exchange basis, which can save a considerable amount of money. Discount auto parts stores are often very good places to buy materials and parts needed for general vehicle maintenance such as oil, grease, filters, spark plugs, belts, touch-up paint, bulbs, etc. They also usually sell tools and general accessories, have convenient hours, charge lower prices and can often be found not far from home.

Authorized dealer parts department: This is the best source for parts which are unique to the vehicle and not generally available elsewhere (such as major engine parts, transmission parts, trim pieces, etc.).

Warranty information: If the vehicle is still covered under warranty, be sure that any replacement parts purchased – regardless of the source – do not invalidate the warranty!

To be sure of obtaining the correct parts, have engine and chassis numbers available and, if possible, take the old parts along for positive identification.

Vehicle identification numbers

The *body shell number* is on the top of the firewall directly behind the engine.

The *engine number* is just above the cartridge oil filter fitting, beside the dipstick hole in the crankcase.

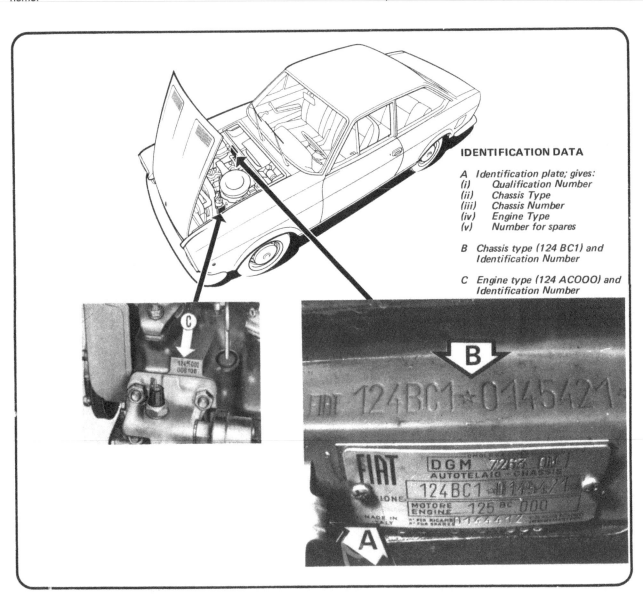

IDENTIFICATION DATA

A *Identification plate; gives:*
(i)   *Qualification Number*
(ii)   *Chassis Type*
(iii)   *Chassis Number*
(iv)   *Engine Type*
(v)   *Number for spares*

B   *Chassis type (124 BC1) and Identification Number*

C   *Engine type (124 ACOOO) and Identification Number*

Routine maintenance

Regular maintenance is essential to ensure the safety of the car, and to retain the economy and performance of the car.

The maintenance tasks and instructions listed are those recommended by the manufacturer and the majority are visual checks; they are supplemented by additional tasks which, from practical experience, need to be carried out. The additional tasks are primarily of a preventative nature - they will assist in eliminating the unexpected failure of a component.

Weekly, before a long journey or every 250 miles (400 km)

---

1 Remove the dipstick and check the engine oil level which should be up to the 'Max' mark. Top up the oil in the sump with Castrol GTX. On no account allow the oil to fall below the 'Min' mark on the dipstick.

2 Check the battery electrolyte level and top up as necessary with distilled water. Make sure the top of the battery is kept clean and dry. If the battery requires regular topping up refer to Chapter 11.

3 When the engine is cold inspect the level of coolant in the translucent plastic system expansion tank. The level must always be between 2.3/8 inch and 2¾ inches above the 'Min' mark on the tank. Top up the tank with water and anti-freeze solution as necessary.

Should the tank require topping up on consecutive checks, a week or less apart, refer to Chapter 2. Should the reservoir be found empty, remove the radiator filler cap, completely fill the radiator and replace the cap. Remove the reservoir cap and fill the reservoir to the aforementioned level. Refit the cap and inspect the system hoses for leaks - see Chapter 2.

4 Check the tyre pressures with an accurate gauge, and alter as necessary. Inspect the tyre walls and tread for damage such as cuts and blisters. Regular inspection of tyres is essential to safety since damage can quickly develop to a catastrophic degree if not attended to. Remember that the law requires at least 1 mm deep tread across three quarters of the width of the tread, around the whole circumference of the tyre.

5 Refill the windscreen washer container with water. Add an anti-freeze solution satchet in cold weather. Do not use ordinary engine anti-freeze as it corrodes paint work. Check that the washer jets are aligned effectively.

6 Remove the wheel trims and check all wheel bolts for tightness - do not overtighten; the correct torque on the wheel bolts is 50 lbs ft.

7 Ensure that all lights and electrical systems are working properly. Chapter 11 includes fault diagnosis information.

Every month or 1,500 miles

---

1 Carry out those checks listed for completing weekly.

2 Lubricate the electropneumatic horn compressor by pouring a few drops of FIAT OCT oil, in the oiler after removing the cap

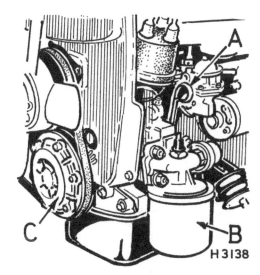

RM1. ENGINE LUBRICATION

A Engine oil dip stick
B Oil filter. Cartridge replaced at 6,000 mile intervals
C Centrifugal by-pass filter. Cleaned at major engine overhauls only. 124 ACOOO engines

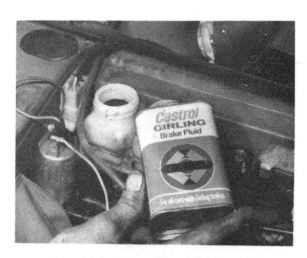

RM2. TOPPING UP BRAKE HYDRAULIC FLUID RESERVOIRS
One for the front brakes and one for the rear on later models

on the top of the compressor.

3 Check the level of hydraulic fluid in the brake fluid reservoir(s) and top up as necessary. Use Castrol Girling.

Every 3 months or 3,000 miles

1 Carry out those tasks listed in the sections for weekly and monthly maintenance.

2 Gearbox

Wipe the area around the gearbox oil level/filler plug. Unscrew the plug and check the level of the oil; it should be up to the bottom of the threads. Top up if necessary with Castrol GTX Oil and refit the plug. Wipe away any spilled oil.

3 Rear axle

Wipe the area around the rear axle oil level/filler plug. Remove the plug and check the oil level which should be to the bottom of the threads in the hole. Top up if necessary with Castrol EP90 Gear Oil; refit the plug and wipe away any spilled oil.

4 Steering box

Remove the oil filler plug from the top of the steering box. Check the level of the oil with a clean metal rod, it should be 1 inch from the top of the plug hole. Top up with Castrol EP90 Gear Oil as necessary and refit the plug.

5 Braking system

As well as checking the hydraulic oil level in the fluid reservoir, the hoses, system pipes and joints should be brushed clean and inspected for signs of corrosion, cracks or leaks. The metal pipes may be sprayed with a de-watering wax (WD40) to maintain them between inspections.

6 Air cleaner

The three nuts which secure the cleaner top cover to the cleaner canister should be removed and the air cleaner cartridge lifted out.

Shake the cartridge repeatedly to free the dust and then blow the dust away with a low pressure blast. (Some vacuum cleaners are fitted with means of using the expelled air from the cleaner as a low pressure blast).

If dusty conditions prevail the cleaner cartridge may need replacing at 3,000 mile intervals.

7 Spark plugs

The spark plugs should be removed and carbon deposits on the electrodes and procelain insulation cleaned off. This is best done at your local garage which will probably have a spark plug sanding machine which also checks the plug electrically. Finally before refitting the plugs into the engine, the electrode gap should be checked to 0.020 - 0.024 inches.

Every 6 months or 6,000 miles

1 Carry out those tasks listed for maintenance at weekly, monthly and three monthly intervals.

2 Run the engine until thoroughly warmed up, and then remove the protective panel underneath the front of the engine. Place a container of 7 Imp pints (4 US quarts) under the engine sump drain plug located in the middle of the rear edge of the sump. Remove the drain plug and its copper sealing washer, and allow the old oil to drain out for 10 minutes. Whilst the old oil is draining, the oil filter cartridge located on the right hand side of the engine, can be unscrewed and discarded. Smear the rubber seal on the new cartridge with a little engine oil ann then refit it to the filter head. Once the seal of the new filter contacts the filter head, tighten a further ¾ turn.

Clean the oil filter cap and refit the sump drain plug fitted with the copper washer.

Refill the engine with 6½ Imp pints (4 US quarts) of Castrol GTX and clean off any oil which may have been spilt over the engine or its components.

Restart the engine and run for a few minutes, then inspect the cartridge oil filter joint for leaks, and check the level of oil in the sump. Finally refit the protective panel underneath the front of the engine.

RM3. Coolant expansion plastic bottle, showing minimum level of coolant. Engine cold

RM4. Electropneumatic horn compressor lubricating plug

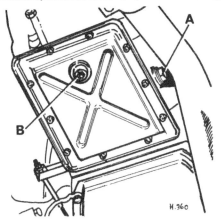

RM5. Gearbox filler/level plug A, drain plug B

RM6. Rear axle oil level/filler plug (A), and drain plug (B)

The interval between oil changes should be reduced in very hot or dusty conditions or during cool weather with much slow and stop/start driving.

On cars fitted with the 124AC00 series 1438 cc engine, the centrifugal filter on the oil bypass line need only be cleaned at major engine overhauls - 60,000 mile intervals.

3 **Valve tappet clearance:** Remove the air cleaner, and spark plug leads so that the cam and valve tappet covers may be removed. Unscrewing the two knobs which secure each cover. NO disassembly of the camshaft is required.

With a cold engine the gap between the cam lobe and the valve tappet, when the cam lobe rise is pointing directly away from the tappet, should be 0.018 inch for intake valves, and 0.020 inch for exhaust valves.

The tappet/cam gap should be measured with feeler gauges and the difference between the measured gap and the specified gap noted. If the tappet/cam gap requires adjustment proceed as directed in Chapter 1 of this manual.

4 **Air cleaner:** It will be necessary to renew the filter cartridge at 6,000 mile intervals and at lesser intervals if the environment is dusty. Remove and replace the cartridge as described in para 6 of the 3 month maintenance tasks list.

5 **Carburettor:** Check the tune of the carburettor and clean the jets and petrol strainer with an air blast as described in Chapter 3.

6 **Ignition distributor:** The distributor is situated on the right hand side of the engine at the front on 1438 cc, 124ACOO; 1592 cc and 1755 cc engines. On 1608 cc, 125BCOO series engines the distributor is driven by the exhaust cam shaft and is situated on the top left hand corner of the engine.

There is no vacuum advance mechanism on either pattern of distributor and maintenance involves pouring a few drops of engine oil through the hole in contact breaker mechanism base to lubricate the top distributor cam bearing and checking the contact breaker gap. The gap should be 0.017 to 0.019 inches when the cam follower is resting on top of one of the four rises on the distributor cam. For adjusting the contact breaker gap and cleaning the distributor mechanism refer to Chapter 4.

7 **Clutch pedal free travel:** This should be checked and adjusted to about 1 inch. It is adjusted by a special nut and a locknut on the clutch actuating lever end of the clutch cable. The cable end and lever are to be found on the right hand side of the clutch bellhousing. On right hand drive cars the cable is routed from the pedal assembly forwards around the front of the engine and then down the carburettor side of the engine to the clutch actuating lever projecting out of the bottom right hand side (as a person in front of the vehicle facing the engine) of the bellhousing.

Loosen the locknut on the cable, turn the special nut to achieve the desired free travel and then secure with the locknut.

8 **Braking system:** The thickness of the brake pad friction material must **not** be less than 0.08 in (2 mm). The pads, calipers and flexible hoses may be inspected and brushed clean once the wheels have been removed. Refer to Chapter 10 for the pad change procedure and the brake system fault diagnosis chart.

The handbrake system should be brushed clean and lubricated with engine oil, The handbrake should apply the rear brakes after 3 notches have been passed on the handle ratchet. If the handbrake handle travel is excessive refer to Chapter 10 for the adjustment procedure.

9 **Suspension and steering:** Jack up the front of the car until the wheels are off the ground and place chassis stands underneath the front suspension support structure. Grasp the top and bottom of the wheel and move to check for wear in the wheel bearing and sloppiness in any of the front suspension joints. Finally grasp the wheel by its forward and rearward edges and move to check for wear and slackness in the steering mechanism. Wear in the wheel bearing will also be noticed with this last check. Examine each suspension and steering joint for corrosion and deterioration of the dust boots or seals. Steering joints should be checked for looseness individually.

The front wheel bearings should be adjusted as described in Chapter 9, if necessary, and in any event the bearing cap should be removed and a little Castrol LM Grease added. Refer to

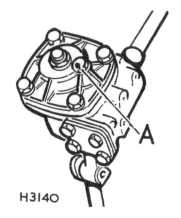

H3140

RM7. Steering box oil filler plug (A)

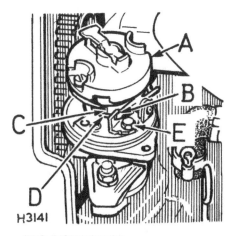

H3141

RM8. DISTRIBUTOR

A Rotor and centrifugal advance mechanism
B Contact breaker points
C Distributor cam
D Distributor shaft lubrication hole
E Contact breaker securing screws

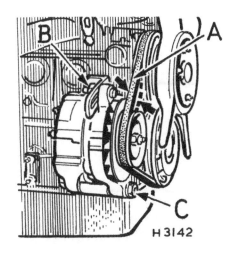

H3142

RM9. FAN BELT TENSION ADJUSTMENT

A Fan belt deflection with moderate hand pressure should be 3/8 inch
B Top alternator/belt tensioning bolt
C Alternator pivot bolt

Chapter 9 for replacement procedures for suspension and steering system components.

10 **Front wheel geometry**: Always pay particular attention to the manner in which the front tyres wear. Uneven wear is a definite indication of wheel misalignment and once noticed, the car should be taken to a FIAT agent who will have the specialised equipment to make an accurate job of wheel alignment. Chapter 9 includes illustrations of uneven tyre wear and cites their causes.

11 **Battery**: As well as checking the electrolyte levels, the terminals and clamp connections should be separated, cleaned and refitted after being given a liberal coat of vaseline.

12 **Cooling system**: Carefully examine the cooling and heater systems for signs of leaks. Make sure that all hose clips are tight and that none of the hoses have cracked or perished. Do not attempt to repair a leaking hose, always fit a new one. Generally inspect the exterior of the engine for signs of leaks or stains. The method of repair will depend on its location. This check is particularly important before filling the cooling system with anti-freeze as it has a greater searching action than water and is bound to find any weak spots.

13 **Fan belt adjustment**: The fan belt must be tight enough to drive the alternator and water pump without overloading their bearings. The method of tension adjustment is given in Chapter 2 and it is correct when it can be pressed 3/8 inch under moderate hand pressure at the lid point of the run from the alternator to the fan pulley.

14 **Windscreen wipers and washer**: Unscrew the washer jet retaining nut and clean the jet hole with an air blast. Once clean the jet piece may be refitted and aligned. The gauze in the suction end of the bottle pipe should be cleaned at this time too. Lubricate the wiper arms and spindles with a few drops of glycerine.

15 **General oiling**: Boot, bonnet and hood hinges and mechanisms all require drops of engine oil at 6 monthly intervals to minimize wear. Door hinges, locks, tiltable seat articulations and limiting arms need lubricating with engine oil also, but the striker blocks, latches, and seat rails need smearing lightly with grease. Avoid a too liberal lubrication of items around the door and inside the car because it may get transferred to passengers clothing.

16 **Seat belts**: Inspect the seat belts for damage to the webbing and check that the anchorages are secure.

17 **Bodywork**: Wash the bodywork and chromium fittings and clean out the interior of the car. Wax polish the bodywork including all the chromium and bright trim. Force wax polish into any joints in the body work to prevent rust formation.

Every 12,000 miles or 12 months

---

1 In addition to the weekly, monthly, 3 monthly and 6 monthly checks the following inspection tasks should be carried out.

2 **Carburettor**: Refer to Chapter 3 for full instructions on cleaning and tuning the carburettor(s).

3 **Crankcase emission control system**: The procedure for cleaning this system is given in Chapter 3.

4 **Exhaust system**: The condition of the elastic mountings of the pipe system should be checked, and each joint should be secure. Minor holes and corrosion may be repaired with many proprietary compounds, but in the majority of cases it is cheaper in the long run to renew the relevant section of pipe.

5 Thoroughly inspect and ensure the integrity of all joints of mechanical units to the body shell. Rusted suspension anchorages are a common cause of failure of the Ministry of Transport vehicle test in Britain. Remember the steering system and damper anchorages, failure of these would also be catastrophic. It is worth while having the underside of the car steam cleaned so that the condition of the undersealing and chassis can be determined more thoroughly. Always entrust chassis repairs and renovations to a reputable Bodywork Repair garage.

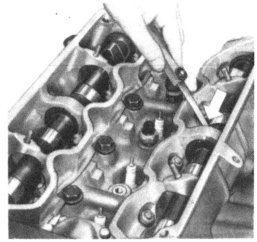
RM10. Measuring camshaft/tappet clearance

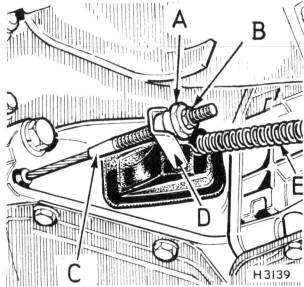
RM11. CLUTCH PEDAL FREE TRAVEL
ADJUSTMENT

A Special nut on clutch cable
B Special nut locking nut
C Clutch cable
D Clutch actuating lever
E Lever return spring

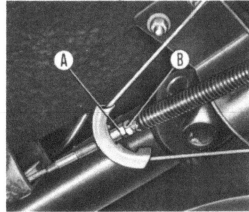
RM12. HANDBRAKE ADJUSTMENT

A Adjusting nut
B Locking nut

Every 18 months or 18,000 miles

1   In addition to those checks listed for weekly, monthly, three monthly and six monthly interval services the following tasks should be performed.

2   **Transmission oil change:** Drain the oil from both gearbox and rear axle after a short run when the oil should drain completely.

Refill the gearbox with approximately 3 Imp pints (3½ US pints) of Castrol GTX Oil.

Refill the rear axle with approximately 3¼ Imp pints - 3½ US pints Castrol EP90 Gear Oil.

Wipe the gearbox and back axle clean and ensure that both drain and level/filler plugs are replaced securely.

3   **Rear axle inspection:** Refer to Chapter 8 for the backlash check procedure if it is felt that the axle gears are running noisily.

4   **Front propeller shaft slip yoke:** The flexible joint end of the slip yoke requires lubricating with Castrol LM Grease. See Chapter 7 for details of the propeller shaft.

5   **Starter motor:** The starter motor should be removed from the car and cleaned. Refer to Chapter 11 for cleaning and inspection of the motor commutator and the brushes. Replace the brushes if necessary.

The drive pinion and slide, and shaft bushes should be lubricated with light oil as directed in Chapter 11.

Every 2 years or 24,000 miles

1   Carry out all those maintenance tasks listed for weekly, monthly, 3 monthly, 6 monthly and yearly servicing.

2   **Timing belt**

Every 24,000 miles or at the most 36,000 miles the timing belt should be replaced. Chapter 1 details the procedure for belt replacement with the engine in the car.

3   **Alternator**

Remove the alternator and inspect the brushes and clean the slip rings on the alternator as directed in Chapter 11.

4   **Brake system**

Completely drain the brake hydraulic fluid from the system. All the seals and flexible hoses throughout the braking circuits should be examined and preferably renewed.

The working surfaces of the master cylinder, wheel and caliper should be inspected for signs of wear or scoring and new parts fitted as necessary. Refill the hydraulic system with recommended brake fluid. See Chapter 10.

The brake servo unit should be inspected as directed in Chapter 10.

5   **Engine:** Test the cylinder compression and if necessary remove the cylinder head and decarbonise. Grind in the valves and fit new valve springs. See Chapter 1.

RM13. Checking the thickness of the brake pad friction material

RM14. The exhaust system elastic mountings

Lubrication chart

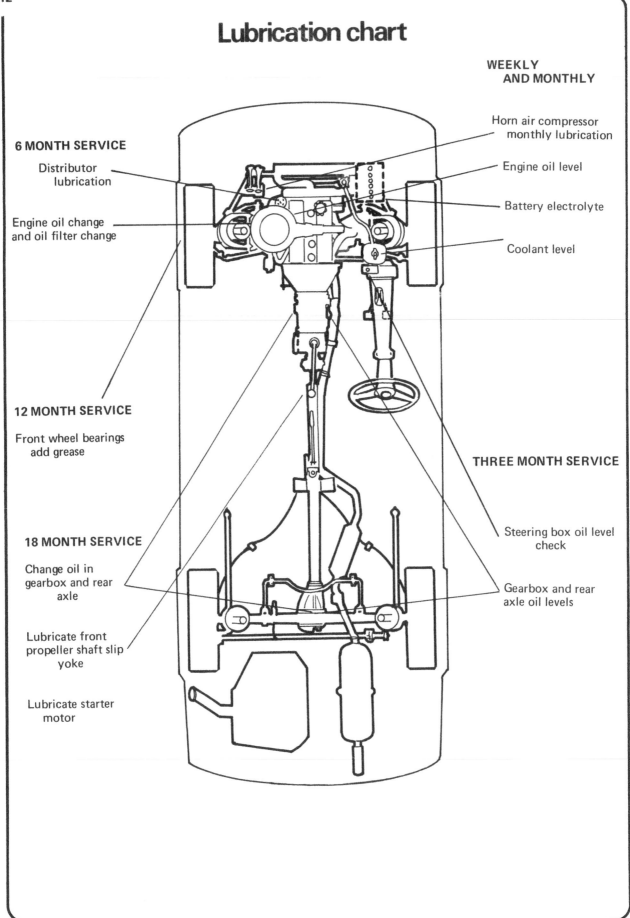

**WEEKLY
AND MONTHLY**

Horn air compressor
monthly lubrication

Engine oil level

Battery electrolyte

Coolant level

6 MONTH SERVICE

Distributor
lubrication

Engine oil change
and oil filter change

12 MONTH SERVICE

Front wheel bearings
add grease

THREE MONTH SERVICE

Steering box oil level
check

Gearbox and rear
axle oil levels

18 MONTH SERVICE

Change oil in
gearbox and rear
axle

Lubricate front
propeller shaft slip
yoke

Lubricate starter
motor

Recommended lubricants and fluids

COMPONENT	TYPE OF LUBRICANT OR FLUID	CORRECT CASTROL PRODUCTS
ENGINE	Multigrade. To API SE specification ...	Castrol GTX
GEARBOX	Multigrade (20/50)	Castrol GTX
REAR AXLE	High quality EP90 gear oil	Castrol Hypoy B
GREASE	Multi-purpose high melting point lithium based grease	Castrol LM Grease
DISTRIBUTOR LUBRICATION ...	See text for details of application ...	Vaseline petroleum jelly and Castrol LM Grease
WINDSCREEN WIPER SPINDLES ...	Glycerine	
COOLING SYSTEM	Anti-freeze solution complying with BS 3151 or 3152	Castrol Anti-freeze
BRAKE HYDRAULIC SYSTEM ...	Hydraulic brake fluid exceeding SAE specification J1703C	Castrol Girling Universal Brake and Clutch Fluid
ELECTROPNEUMATIC HORN COMPRESSOR	Paraffin-base oil containing EP additives	

Additionally Castrol 'Everyman' oil can be used to lubricate door, boot, bonnet, hood, hinges, locks and pivots etc.

Safety first!

Professional motor mechanics are trained in safe working procedures. However enthusiastic you may be about getting on with the job in hand, do take the time to ensure that your safety is not put at risk. A moment's lack of attention can result in an accident, as can failure to observe certain elementary precautions.

There will always be new ways of having accidents, and the following points do not pretend to be a comprehensive list of all dangers; they are intended rather to make you aware of the risks and to encourage a safety-conscious approach to all work you carry out on your vehicle.

Essential DOs and DON'Ts

DON'T rely on a single jack when working underneath the vehicle. Always use reliable additional means of support, such as axle stands, securely placed under a part of the vehicle that you know will not give way.

DON'T attempt to loosen or tighten high-torque nuts (e.g. wheel hub nuts) while the vehicle is on a jack; it may be pulled off.

DON'T start the engine without first ascertaining that the transmission is in neutral (or 'Park' where applicable) and the parking brake applied.

DON'T suddenly remove the filler cap from a hot cooling system – cover it with a cloth and release the pressure gradually first, or you may get scalded by escaping coolant.

DON'T attempt to drain oil until you are sure it has cooled sufficiently to avoid scalding you.

DON'T grasp any part of the engine, exhaust or catalytic converter without first ascertaining that it is sufficiently cool to avoid burning you.

DON'T allow brake fluid or antifreeze to contact vehicle paintwork.

DON'T syphon toxic liquids such as fuel, brake fluid or antifreeze by mouth, or allow them to remain on your skin.

DON'T inhale dust – it may be injurious to health (see *Asbestos* below).

DON'T allow any spilt oil or grease to remain on the floor – wipe it up straight away, before someone slips on it.

DON'T use ill-fitting spanners or other tools which may slip and cause injury.

DON'T attempt to lift a heavy component which may be beyond your capability – get assistance.

DON'T rush to finish a job, or take unverified short cuts.

DON'T allow children or animals in or around an unattended vehicle.

DO wear eye protection when using power tools such as drill, sander, bench grinder etc, and when working under the vehicle.

DO use a barrier cream on your hands prior to undertaking dirty jobs – it will protect your skin from infection as well as making the dirt easier to remove afterwards; but make sure your hands aren't left slippery. Note that long-term contact with used engine oil can be a health hazard.

DO keep loose clothing (cuffs, tie etc) and long hair well out of the way of moving mechanical parts.

DO remove rings, wristwatch etc, before working on the vehicle – especially the electrical system.

DO ensure that any lifting tackle used has a safe working load rating adequate for the job.

DO keep your work area tidy – it is only too easy to fall over articles left lying around.

DO get someone to check periodically that all is well, when working alone on the vehicle.

DO carry out work in a logical sequence and check that everything is correctly assembled and tightened afterwards.

DO remember that your vehicle's safety affects that of yourself and others. If in doubt on any point, get specialist advice.

IF, in spite of following these precautions, you are unfortunate enough to injure yourself, seek medical attention as soon as possible.

Asbestos

Certain friction, insulating, sealing, and other products – such as brake linings, brake bands, clutch linings, torque converters, gaskets, etc – contain asbestos. *Extreme care must be taken to avoid inhalation of dust from such products since it is hazardous to health.* If in doubt, assume that they *do* contain asbestos.

Fire

Remember at all times that petrol (gasoline) is highly flammable. Never smoke, or have any kind of naked flame around, when working on the vehicle. But the risk does not end there – a spark caused by an electrical short-circuit, by two metal surfaces contacting each other, by careless use of tools, or even by static electricity built up in your body under certain conditions, can ignite petrol vapour, which in a confined space is highly explosive.

Always disconnect the battery earth (ground) terminal before working on any part of the fuel or electrical system, and never risk spilling fuel on to a hot engine or exhaust.

It is recommended that a fire extinguisher of a type suitable for fuel and electrical fires is kept handy in the garage or workplace at all times. Never try to extinguish a fuel or electrical fire with water.

Note: *Any reference to a 'torch' appearing in this manual should always be taken to mean a hand-held battery-operated electric lamp or flashlight. It does NOT mean a welding/gas torch or blowlamp.*

Fumes

Certain fumes are highly toxic and can quickly cause unconsciousness and even death if inhaled to any extent. Petrol (gasoline) vapour comes into this category, as do the vapours from certain solvents such as trichloroethylene. Any draining or pouring of such volatile fluids should be done in a well ventilated area.

When using cleaning fluids and solvents, read the instructions carefully. Never use materials from unmarked containers – they may give off poisonous vapours.

Never run the engine of a motor vehicle in an enclosed space such as a garage. Exhaust fumes contain carbon monoxide which is extremely poisonous; if you need to run the engine, always do so in the open air or at least have the rear of the vehicle outside the workplace.

If you are fortunate enough to have the use of an inspection pit, never drain or pour petrol, and never run the engine, while the vehicle is standing over it; the fumes, being heavier than air, will concentrate in the pit with possibly lethal results.

The battery

Never cause a spark, or allow a naked light, near the vehicle's battery. It will normally be giving off a certain amount of hydrogen gas, which is highly explosive.

Always disconnect the battery earth (ground) terminal before working on the fuel or electrical systems.

If possible, loosen the filler plugs or cover when charging the battery from an external source. Do not charge at an excessive rate or the battery may burst.

Take care when topping up and when carrying the battery. The acid electrolyte, even when diluted, is very corrosive and should not be allowed to contact the eyes or skin.

If you ever need to prepare electrolyte yourself, always add the acid slowly to the water, and never the other way round. Protect against splashes by wearing rubber gloves and goggles.

When jump starting a car using a booster battery, for negative earth (ground) vehicles, connect the jump leads in the following sequence: First connect one jump lead between the positive (+) terminals of the two batteries. Then connect the other jump lead first to the negative (–) terminal of the booster battery, and then to a good earthing (ground) point on the vehicle to be started, at least 18 in (45 cm) from the battery if possible. Ensure that hands and jump leads are clear of any moving parts, and that the two vehicles do not touch. Disconnect the leads in the reverse order.

Mains electricity and electrical equipment

When using an electric power tool, inspection light etc, always ensure that the appliance is correctly connected to its plug and that, where necessary, it is properly earthed (grounded). Do not use such appliances in damp conditions and, again, beware of creating a spark or applying excessive heat in the vicinity of fuel or fuel vapour. Also ensure that the appliances meet the relevant national safety standards.

Ignition HT voltage

A severe electric shock can result from touching certain parts of the ignition system, such as the HT leads, when the engine is running or being cranked, particularly if components are damp or the insulation is defective. Where an electronic ignition system is fitted, the HT voltage is much higher and could prove fatal.

Chapter 1 Engine

Contents

Specifications

Manufacturers Type number:	124AC000	125BC000	132 series	
Swept volume	1438 cc	1608 cc	1592 cc	1756 cc
Bore	3.150 inch	3.150 inch	3.150 inch	3.308 inch
Stroke	2.815 inch	3.150 inch	3.110 inch	3.110 inch
Firing order	1 3 4 2			
Compression ratio	8.9 : 1	9.8 : 1	9.8 : 1	9.8 : 1
Engine rotation	Clockwise when viewed from front			
Max. horsepower (DIN)	90	110	104	114

Crankshaft

Main journal diameter	...	...	...	...	...	...	1.9990/1.9998	2.0860/2.0868
Regrind diameters	...	...	...	...	...	...	−0.010, −0.020, −0.030, −0.040 inch	
Crankpin journal diameter ...	...	...	...	...	...	1.7917/1.7925	'A' 1.8990/1.8994 'A' 1.9997/2.0001	
								'B' 1.8986/1.8990 'B' 1.9993/1.9997

Class 'A' identified by a red paint dot on crank.
    'B' identified by a blue paint dot on crank.

Crankpin regrind diameters	...	...	...	...	...	−0.010, −0.020, −0.030, −0.040	
Crankshaft end thrust taken by washers on either side of rear main journal							
Crankshaft end float	...	...	...	...	...	...	0.002 to 0.010 inches
Maximum journal taper and ovality	...	...	...	...	0.0002 inches	

Main bearings

Number	...	...	...	...	...	...	...	5
Diametrical clearance	...	...	...	...	...	...	0.002/0.0037 inch
Undersizes	...	...	...	...	...	...	...	−0.010, 0.020, 0.030, 0.040 inch

Big end bearings

Diametrical clearance	...	...	...	...	...	...	0.00102/0.00299 'A' 0.0018/0.0031
							'B' 0.0019/0.0032

Gudgeon pin:

Type	...	...	...	...	...	...	Clearance fit in piston

						1438 cc engines	1608, 1592, 1756 cc engines
Pin outer diameter -	Class 1	...	...	...	...	0.8649/0.8651	0.8658/0.8659
	Class 2	...	...	...	...	0.8651/0.8652	0.8659/0.8660
	Class 3	...	...	...	...	0.8652/0.8654	No Class 3

Class indicated on pin end and on underside of piston boss.

							1438 cc	1608 cc	1592 & 1756 cc
Fit in piston:	...	...	...	...	...	...	0.0003	0.0001	0.0004
							0.0006	0.0003	0.0006

Pistons ... ...

Piston diameter - (measured at same point as clearance with cylinder bore) ... ...

Steel belted design

							1438 cc	1608 cc	1592 cc	1756 cc
Class A	...	...	...	...	...	...	3.1457/	3.1467/	3.1464/	3.3050/
							3.1461	3.1470	3.1468	3.3055
Class C	...	...	...	...	...	...	3.1465/	3.1474/	3.1472/	3.3059/
							3.1468	3.1478	3.1476	3.3062
Class E	...	...	...	...	...	...	3.1472/	3.1482/	3.1479/	3.3066/
							3.1476	3.1486	3.1483	3.3070

Class indicated underneath piston skirt.

Oversize pistons	...	...	...	...	...	...	+0.0079, 0.0157, 0.0236 inch			
Piston/bore	...	...	...	...	...	...	**1438 cc**	**1608 cc**	**1592 cc**	**1756 cc**
							0.0035	0.0026	0.0027	0.0016
Clearance	...	...	...	...	...	...	0.0043	0.0033	0.0035	0.0024
Measured at	...	...	...	...	...	...	2.05" down from piston crown	0.905" up from skirt bottom	2.057" down from piston crown	1.181" up from skirt bottom

Piston rings (3 off)

							Piston ring/groove clearances	Width of grooves
Piston ring gap in bore:								
Top No. 1	...	0.0118/0.0177	...	...	...		0.0018/0.0030	0.0604/0.0612
No. 2	...	0.0078/0.0137	...	...	...		0.0011/0.0027	0.0798/0.0806
No. 3	...	0.0078/0.0137	...	...	...		0.0011/0.0027	0.1561/0.1569

Piston ring thickness	-	1st compression	...	...	...	0.0582/0.0587
	-	2nd oil control	...	...	...	0.0779/0.0787
	-	3rd oil control	...	...	...	0.1544/0.1549

Piston ring oversizes as per pistons.

Camshafts

Journal diameters:	-	Front	...	...	...	...	...	1.1789/1.1796
	-	Centre	...	...	...	...	...	1.8014/1.8020
	-	Rear	...	...	...	...	...	1.8172/1.8177
Bearing inside diameter	-	Front	...	...	...	...	1.1815/1.1824
	-	Centre	...	...	...	...	1.8032/1.8041

Rear	1.8189/1.8199
Endthrust	Taken on plate locating rear end of shaft
Drive	Toothed belt and wheel

Tappets

Type	Bucket with recessed top to accept tappet discs which can be changed to vary the tappet/camshaft gap
Tappet/camshaft gap - Inlet (cold)	0.017 inch
- Exhaust (cold)	0.019 inch
Thickness of tappet discs	Nominal 0.1575 ± 0.0004
Available	0.1299 inches thick to 0.1924 inches in 0.002 increments
Tappet bore in housing	1.4566/1.4576
Tappet bucket outside diameter	1.4557/1.4565

Valves

Seat angle - Inlet and Exhaust	45º ± 5'
Head diameter:	132 series
Inlet - 1.629 inches	1.6220/1.6378
Exhaust - 1.417 inches	1.4115/1.4350
Valve stem diameter - Inlet	0.3140/0.3146
- Exhaust	0.3137/0.3143
Stem to guide clearance - Inlet	0.0012/0.0025
- Exhaust	0.0015/0.0028
Valve lift	0.3824 inches

Valve guides

Valve guide seat bore in cylinder head	0.5886/0.5896
Outer diameter of valve guide	0.5905/0.5912
Oversize guide O.D.	+0.0079 inches
Valve guide bore	0.3158/0.3165 inches

Snap ring on guide outer surface sets the height of the guide above the cylinder head surface.

Valve springs ... ... | 2 springs per valve

Free height - Inner	1.646 inches
- Outer	2.122 inches
Spring stiffness - Inner	Length 1.22 inches - load 32.7 lb f
- Outer	Length 1.417 inches - load 75.5 lb f

Valve timing

Timing marks	Holes on camshaft wheels
Inlet valve - Opens	26º before TDC
- Closes	66º after BDC
Exhaust valve - Opens	66º before BDC
- Closes	26º after TDC

Auxiliary drive shaft

Bush bores in crankcase - Front	2.0126/2.0138
- Rear	1.6547/1.6559
Inside bores of bushes - Front	1.8930/1.8938
- Rear	1.5354/1.5362
Shaft journal diameter - Front	1.8903/1.8913
- Rear	1.5326/1.5336
Journal/bore clearance	0.0018/0.0036 inches

Lubrication

System	Wet sump - pressure fed
Pressure - running	57 to 85 p.s.i.
Oil pump	Meshing gear type
Sump oil capacity	6.6 Imp. pints; 7.9 U.S. pints
Oil filter	Full flow disposable cartridge type
By-pass valve opens:	
By-pass valve spring - Length	0.886 inch
- Load	10.16 lb f
Oil pump:	
Clearance gears/pump cover	0.0012/0.0045 in
Clearance gears/pump housing wall	0.0043/0.0071 in

Torque wrench settings:

	ft lbs	
Crankshaft end bolt (pulley and timing wheel)	88	
Main bearing cap bolts	59	
Big end bearing cap nuts	38	47 ft lbs (132 series)
Flywheel bolts	59	
Cylinder head bolts	56	
Camshaft housing/head stud nuts	21	
Camshaft wheel bolts	36	
Timing belt idler locking nuts	34	
Fan and water pump drive pulley nut	88	
Inlet and exhaust manifold nuts	18	
Sump attachment bolts	5.8	
(gasket should not be punched)		
Alternator upper bracket screw	38	
Alternator lower bracket nut	31	
Alternator upper screw nut (adjusting)	31	
Alternator lower screw nut	50	
Sparking plugs	30	
Water pump retaining bolts	17	
Oil pump securing bolts	14	
Cylinder head water outlet elbow bolts	8	
Clutch bellhousing to engine bolts	62	
Engine support bracket nut	18	
Engine support bracket to pad plate nut	21.7	
Pad to plate nut	5.76	
Pad to crossmember nut	21.7	
Gearbox to pad bolt	18	
Pad to crossmember nut	18	
Crossmember to chassis nut	10.84	
Oil pressure switch and temperature senders	20	

1 General description

All the engines fitted to the 124 Sport Coupe and 124 Sport Spider models are basically the same design. The design of the engines is similar to that of the Fiat 125 and provides a four cylinder high performance engine with considerable tuning scope.

The carburation system comprises normally of one or two twin choke downdraught Weber carburettors. The inlet manifold feeds four inlet ports on the cylinder head and it has fittings to provide for the vacuum operated servo assisted brakes. The inlet ports lead to valves angled at 45° and a hemisphertical combustion chamber in the cylinder head. The exhaust valves lead to four exhaust ports and a cast iron exhaust manifold. The two sets of valves are operated by separate camshafts situated on the cylinder head and the camshafts are driven by a toothed belt running on a pulley mounted on the front end of the crankshaft. Belt tension is maintained by an idler wheel. The same toothed belt also drives auxiliary shaft mounted on the right hand side of the cylinder block. This shaft motors the fuel pump, oil pump and distributor. On 1608 cc (125 BC000) engines however the distributor is situated on the rear left hand corner of the engine and is driven by the exhaust camshaft.

The statically and dynamically balanced steel crankshaft rotates in five replaceable main bearings which are individually fed with oil at pressure from the oil pump. Drillings in the crankshaft take oil from the main journals to the crankpin/big end bearings. Drillings in the cylinder block cum crankcase take oil from the crankshaft main bearings to the camshaft bearings.

The auxiliary drive shaft bearings are also lubricated with oil direct from the pump. On the early 1438 cc (124 AC000) engines the oil pressure relief valve is situated in a centrifugal oil filter mounted within the crankshaft end pulley. This pulley and its 'V' belt motors the water pump and alternator. The oil pressure relief valve on the other engines is incorporated in the oil pump cover in the sump.

Various electrical transducers are used to monitor water temperature, oil pressure and on cars sold in the United States, brake servo vacuum. The engine controls include a hand throttle as well as the usual choke and accelerator. The ignition system is statically timed and is advanced as required by a centrifugal advance mechanism in the distributor. There is no vacuum advance.

Starting the engine is accomplished by a pre-engaged design of starter motor. It is situated low on the rear right hand end of the engine and the pinion meshes with a ring gear shrunk onto the steel flywheel attached to the rear end of the crankshaft.

Power is taken from the engine from the face of the flywheel by a single dry plate clutch retained by a diaphragm spring. A four or five speed gearbox transmits the power to the propeller shaft and rear wheels. The whole engine and gearbox assembly is supported on 4 rubber blocks and the whole assembly may be removed if work needed on either major unit.

2 Major operations - engine in place

The following major operations can be carried out on the engine with it in place in the car:
1 Removal and replacement of cylinder head assembly.
2 Removal and replacement of timing belt.
3 Removal and replacement of camshafts.
4 Removal and replacement of water pump.

For the following major operations it may be necessary to remove the forward support soft mountings and hoist the engine from above to provide the necessary clearance beneath:
1 Removal and replacement of sump.
2 Removal and replacement of big end bearing shells.
3 Removal and replacement of oil pump.
4 Removal and replacement of pistons and con-rods.

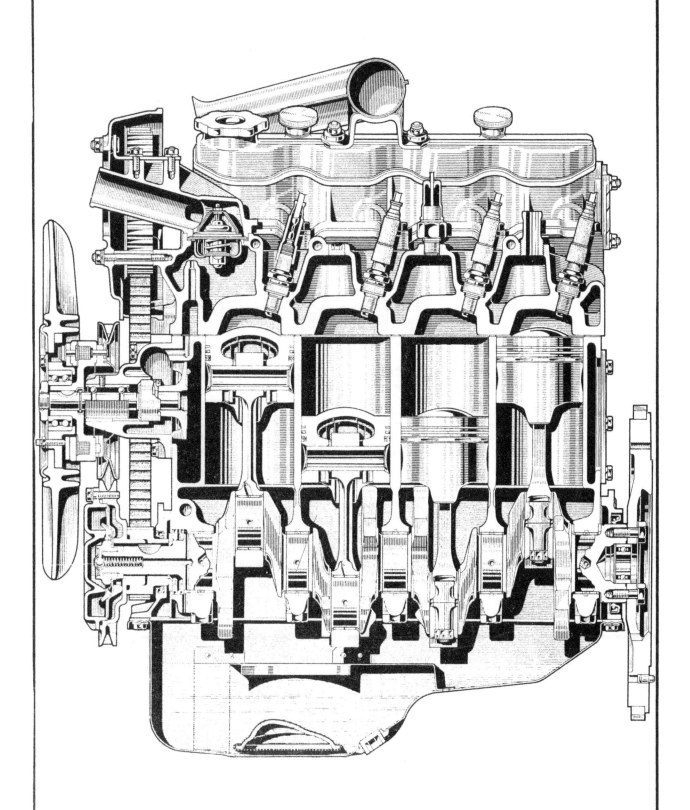

Fig.1.1. A longitudinal section through the 124ACOOO engine - the 125BCOOO and 132 series engines are virtually the same

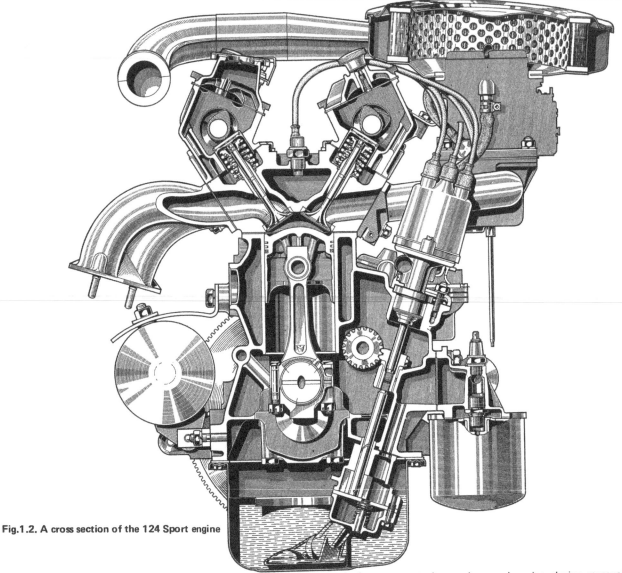

Fig.1.2. A cross section of the 124 Sport engine

3 Major operations - engine removed

The following operations can only be carried out with the engine out of the car and on a bench or the floor:
1   Removal and replacement of main bearings.
2   Removal and replacement of crankshaft.
3   Removal and replacement of auxiliary drive shaft.
4   Removal and replacement of flywheel.

4 Methods and equipment for engine removals,

1   The engine can be removed either attached to the gearbox or disconnected from it. Both methods are described but the photo sequence shows removal with gearbox attached because it was considered to be the easiest method and did not need special tools.
2   Essential equipment includes chassis stands to retain the car at a height sufficient for access to the exhaust and steering systems in particular. If the gearbox is being removed separately then the car will need to be raised to a height approximately 2 feet at the front to allow the gearbox to be slid out from underneath the car. The next piece of essential equipment is a hoist, the engine and gearbox together weigh approximately 200 lbs, therefore ensure that all the hoists equipment is sufficient for that load. A hydraulic jack or comparable lift scissor jack will be

required to support the gearbox and engine during removal operations. Finally if it is desired to remove the engine without the gearbox attached, then a special articulated spanner Fiat No. A55035 will be required to undo the engine/gearbox attachment bolts with the engine and gearbox in place in the car.

5 Engine removal with gearbox

1   The removal sequence begins with the detachment of the front grill, which is retained by four screws to small brackets in the radiator aperture.
2   With the grill clear, undo the two bolts from each bonnet hinge that secures them to the body shell. Lift the bonnet clear of the car after removing the bonnet springs from the underside of the bonnet (photo).
3   The engine compartement is now fully exposed and the task of removing ancillary engine components commences with the air filter. Undo the nuts that retain the top cover, lift the cleaning elements clear and proceed to remove the filter housing securing plates which are held by nuts to the top of the carburettors (photo).
4   Take care when lifting the air cleaner housing clear, there are several pipes attached to the underside including a breather pipe from the crankcase. (photo).
5   The carburettor trumpets need only be removed if it is necessary to work on the instruments themselves. They are

better left in place to prevent damage to pipes emerging from the top of the carburettor.

6 Next remove the distributor cap, together with the HT leads to the spark plugs and from the coil. Disconnect the LT leads to the distributor. (photo).

7 The carburettor controls should be removed next beginning with the choke cable and accelerator cables (photo).

8 Proceed then to remove the carburettor fuel feed pipe, the breather pipes and the servo vacuum pipe which connects to the inlet manifold (photos).

9 Remove both battery connections, identifying as necessary. Continue by removing the electrical connections to the oil pressure and water temperature sender units mounted on the oil filter fitting and cylinder head respectively. Disconnect the electrical leads from alternator and starter motor and identify the leads if desired to ensure replacement onto the correct terminals later.

10 The various pipe connections to the engine must now be removed, beginning with the water pipes. Remove the radiator filler caps and unscrew the radiator drain tap at the bottom right hand side of the radiator. If anti-freeze solution is in the water fix a small pipe on the tap and collect the coolant. Do not allow the anti-freeze solution to splash on the bodyshell - it corrodes paintwork. When draining the radiator ensure that the car heater/air conditioner controls are set at 'hot', this will enable the heater radiator to drain by syphon action with the main radiator. Finally there is a plug on the right hand side of the cylinder block that allows the cylinder water jacket to be completely drained. (photos).

11 On models fitted with an electric radiator fan, disconnect the electrical leads and once the cooling system has been drained, the upper and lower hoses may be removed from the engine and radiator.

On early models the thermostat is mounted in a housing positioned midway along the bottom hose. Finally the pipe from the expansion tank to the radiator needs to be detached from the radiator (photo).

12 The radiator is now clear to be removed, it is secured by two nuts, one on each side. Take care not to lose the spacers, washers and mounting rubbers, when the radiator is lifted clear. (photo).

13 Now the car should be positioned over a pit, or raised onto chassis stands to gain access to the front underside of the car. The body shell is particularly strong for chassis stands at the points where the front suspension frame is attached.

14 From underneath the car, drain the sump of oil and remove the dust shield from around the forward area of the sump.

15 Next slacken the clamp on the exhaust pipe beneath the engine gearbox assembly and drive the rearward section of the pipe back off the pipe from the manifold. (photo).

16 The front section of exhaust pipe should then be removed from the manifold by unscrewing the retaining nuts. (photo).

17 The centre steering track rod needs to be removed from the relay lever and steering box levers. The ball joint nuts should be removed and a joint extractor used to force the ball pin out of the mating lever.

18 Turning your attention to the clutch and gearbox, the clutch operating cable should be removed from the clutch actuating lever and cable housing location. Then the speedometer drive cable should be disconnected (photos).

19 The last major disconnection from underneath the car is the doughnut coupling on the end of the gearbox. Remove nuts and bolts from the shaft coupling yoke so that the doughnut remains on the gearbox shaft. (photo).

20 The last disconnections include the fuel feed pipe to the fuel pump - plug it with a suitable metal or wooden rod. The top half of the gear lever should be removed. This is accomplished by slackening the lever leather boot and moving it down to expose the lever joint. A sharp tug upwards should release the lever from the spring connection in the lower portion of lever. (photo).

21 Begin removing the engine and gearbox assembly by supporting the rear of the gearbox on a jack and unscrewing the two nuts which secure the rear gearbox support crossmember to the bodyshell (photo).

22 Next position a tough hoist sling around the clutch bell housing and front of the engine. The photo sequence shows a chain attached to the top of the engine on a strip of metal between cam cover nuts. The lifting arrangement should allow the engine and gearbox to tilt considerably without becoming unsafe.

23 Take the engine weight with the hoist, and finally disconnect the engine support brackets from the front rubber mounts. Finally check that all connections - electrical, cable and pipe are free from the engine and safely stowed so that they will not be damaged during engine removal. (photo).

24 The engine and gearbox should now be free to be lifted from the car. The assembly should be tilted so that the gearbox can pass beneath the rear engine compartment bulkhead whilst the engine rises above the forward section of the body shell. (photo).

25 Before the engine can be raised finally clear the gear lever extension fitting has to be removed from the top of the gearbox. (photo).

26 Once the engine and gearbox unit is out of the car the gearbox and engine can be separated. Undoing the four major nuts and bolts that secure the crankcase and clutch bellhousing together, remove the flywheel protecting plate from the bottom of the bellhousing and finally remove the starter motor. The engine and gearbox can now be moved apart. Take care to support the gearbox as the two are separated; the clutch friction plate can be damaged if the gearbox and engine become misaligned (photo).

5.2 Bonnet hinge detachment

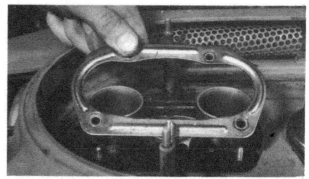

5.3 Removing air cleaner box retaining plate

5.4 Removing air cleaner box

5.6 Remove distributor cap and HT leads

5.7 Remove carburettor control cables

5.8 Disconnect carburettor fuel feed pipe

5.8a Disconnect vacuum servo pipe from inlet manifold

5.10 Radiator drain tap, unscrew to drain cooling system

5.10a Remove water hoses from engine and radiator

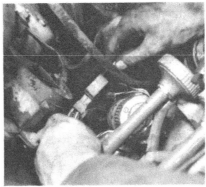

5.11 Disconnect leads to electric radiator fan if fitted

5.12 Lift away radiator

5.15 Moving rearward section of exhaust pipe away

5.16 Forward section of exhaust pipe being removed

5.18 Disconnect clutch actuating cable

5.18 Disconnect speedometer drive

5.19 Uncouple doughnut drive coupling

5.20 Pull off gear lever extension

5.21 Remove rear gearbox support cross member

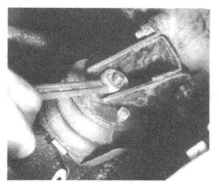
5.23 Disconnect engine support brackets

5.24 Engine and gearbox free

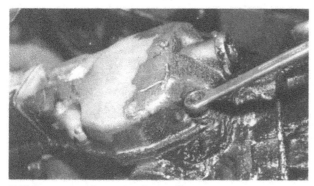
5.25 Remove gearbox extension so that assembly will come free

5.26 Engine and gearbox separated

6 Engine removal - without gearbox

1   This method of engine removal is fairly straightforward (provided that a long reach socket spanner or Fiat Tool No. A55035 is available). Its advantage is that the hoisting of the engine does not involve precarious angles of support. The disadvantage is that it can be tedious and possibly damaging to the clutch to reunite the engine to the gearbox with the gearbox in place in the car.

2   The procedure for separate engine removal follows that outlined in paragraphs 1 to 17 of Section 5. Then in addition:-

3   Remove the last connections of the car to the engine which include fuel pump feed pipe; block with a metal or wooden rod.

4   Next support the gearbox with a jack and then position a tough sling around the engine. The sling may pass around the exhaust and inlet manifolds close to the cylinder head. Take the weight of the engine with the hoist.

5   From underneath the car, with a long reach socket spanner or tool No. A55035, undo and remove the four major nuts and bolts that secure the crankcase to the clutch bellhousing.

6   Remove the cover plate on the bottom of the bellhousing and then the starter motor should be removed. Finally check that all connections, electrical, cable and pipe are free and safely stowed away.

7   Lastly the nuts which retain the engine support brackets to the support rubbers should be unscrewed and removed so that the engine may be raised a little and moved forward to disengage from the gearbox and clutch bellhousing.

8   Once the engine is free from the gearbox unit it may be raised from the engine compartment and clear of the car.

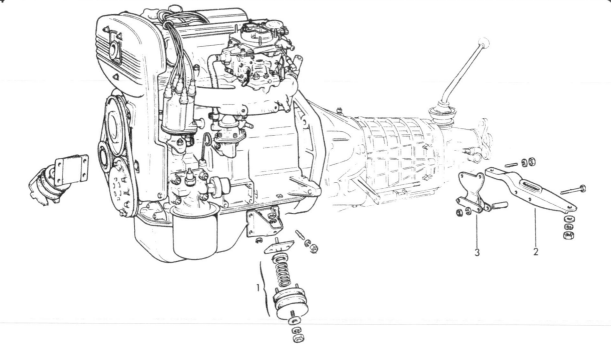

FIG.1.3. ENGINE AND GEARBOX SUPPORT COMPONENTS

1   *Forward support soft mounting*   2   *Rear support crossmember*   3   *Rear support soft mounting*

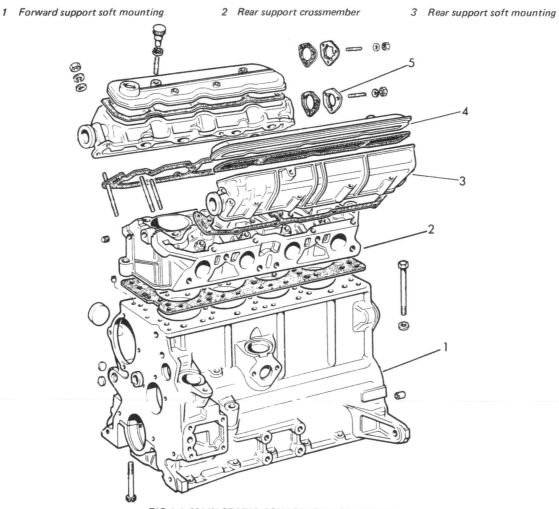

FIG.1.4. MAIN STATIC COMPONENTS OF ENGINE

1   *Crankcase/cylinder block*   3   *Camshaft housing*   4   *Camshaft cover*   5   *Camshaft locating plate*
2   *Cylinder head*

7 Engine dismantling

1 It is best to mount the engine on a dismantling stand, but if one is not available stand the engine on a strong bench, to be at a comfortable working height. It can be dismantled on the floor but it is not easy.

2 During the dismantling process greatest care should be taken to keep the exposed parts free from dirt. As an aid to achieving this, thoroughly clean down the outside of the engine, removing all traces of oil and congealed dirt.

3 Use paraffin or Gunk; The latter compound will make the job much easier for, after the solvent has been applied and allowed to stand for a time, a vigorous jet of water will wash off the solvent with all the grease and dirt. If the dirt is thick and deeply embedded, work the solvent into it with a wire brush.

4 Finally wipe down the exterior of the engine with a rag and only then, when it is quite clean, should the dismantling process begin. As the engine is stripped, clean each part in a bath of paraffin or Gunk.

5 Never emmerse parts with oilways (for example the crankshaft), in paraffin, but to clean wipe down carefully with a petrol dampened cloth. Oilways can be cleaned out with nylon pipe cleaners. If an airline is available, all parts can be blown dry and the oilways blown through as an added precaution.

6 Re-use of the old engine gaskets is false economy and will lead to oil and water leaks, if nothing worse. Always use new gaskets throughout.

7 Do not throw the old gasket away for it sometimes happens that an immediate replacement cannot be found and the old gasket is then very useful as a template. Hang up old gaskets as they are removed.

8 To strip the engine it is best to work from the top down. The underside of the crankcase when supported on wood blocks acts as a firm base. When the stage is reached, where the crankshaft and connecting rods have to be removed, the engine can be turned on its side, and all other work carried out in this position.

9 Whenever possible, replace nuts, bolts and washers finger tight from wherever they were removed. This helps avoid loss and muddle later. If they cannot be replaced, lay them out in such a fashion that it is clear from whence they came.

8 Removing the ancillary engine components

Before basic engine dismantling begins it is necessary to strip it of ancillary components as follows:-
1 Alternator
2 Distributor
3 Thermostat
4 Oil filter cartridge
5 Exhaust manifold
6 Inlet manifold and carburettor(s)
7 Crankcase breather
8 Fuel pump.

It is possible to remove any of these components with the engine in place in the car, if it is merely the individual items that require attention.

Presuming the engine is to be out of the car and on a bench and that the items mentioned are still on the engine follow the procedure described below:-

1 Slacken off the alternator retaining bolts and nuts, and remove the fan belt. Remove the alternator together with its retaining nuts and bolts.

2 To remove the distributor undo the nuts which press the distributor clamp on the mounting flange. Once the clamp plate is removed the distributor may be lifted free. If it is not anticipated to disturb the distributor drive from the auxiliary drive shaft, then mark the distributor flange and cylinder block so that it may be immediately refitted to the position from which it was taken.

3 One of the design differences between the earlier 124AC000, 1438 cc and 125BC000, 1608 cc engines and the latest 1592 and 1756, 132 series is the location of the thermostat.

4 On 124AC000 and 125BC000 engines the thermostat would have been removed preparatory to removing the engine, on the latest 132 series engines it is located beneath an elbow in the centre of the forward end of the cylinder head.

5 To remove the thermostat on the 132 series engines, and the cylinder head water outlet elbow on 124 and 125 series engines, the timing belt cover should first be removed. It is retained by 2 nuts and 2 bolts to brackets within the cover.

6 The cylinder head elbow then is removed after unscrewing the four retaining bolts, and the thermostat extracted if a 132 series engine is being dismantled.

7 The next item to be removed is the cartridge oil filter. The filter canister and filter is secured to the cylinder block by studs and four nuts. Once the nuts have been unscrewed the filter assembly can be removed.

8 Having already removed the timing belt cover to release the cylinder head outlet elbow, the timing belt itself can be removed. Slacken the idle wheel position lock nuts to slacken the belt itself and remove it from the engine.

9 The exhaust manifold may be removed now after undoing the six nuts which secure it to the cylinder head.

10 The inlet manifold is similarly retained on the cylinder head by six nuts and can be removed complete with the carburettor assembly. The photo sequence shows that the inlet manifold and carburettor may be left on the cylinder head until it has been removed from the cylinder block.

11 The crankcase breather vapour/oil separator is fitted to the cylinder block immediately above the fuel pump, and it is removed after undoing the single central bolt and slipping the forward breather pipe off the union on the crankcase.

12 The fuel pump is removed after undoing the two bolts which secured it to the side of the crankcase; keep the spacer and bolts with the fuel pump.

13 The engine is now stripped of ancillary components and ready for the major dismantling tasks to begin.

9 Timing belt removal, refitting and adjustment

The timing belt will need replacement at 3 yearly or 36,000 mile intervals at the most and therefore will certainly be one of the more frequent major tasks to be completed on this car. Fiat Tool No. A60319 is required to set cam timing.

1 Begin by draining the cooling system as described in Chapter 2.

2 Remove all the water hoses to the engine and radiator from the engine and radiator.

3 Slacken the alternator mounting bolts and remove the fan belt.

4 Remove the electrical connections from the battery - an important safety measure - identify them as necessary and stow them safely beside the battery.

5 It will make the later tasks much easier if the radiator is removed. It rests on rubber mounts and is secured by two nuts. On 1600 and 1800 models the electrical leads to the fan motor will also need disconnecting before the radiator is free to be lifted clear. Take care not to lose the spacers and rubber mountings as the radiator is removed from the car.

6 There is now ample space for work to commence on the timing mechanism itself.

7 The cover is retained by two nuts and two bolts and their removal will free the cover.

8 The timing belt is now exposed; the engine should now be turned so that the timing marks on the camshaft wheels are aligned with the pointers on the cover support fitting. The timing mark on the auxiliary shaft wheel should be about 34^0 to the right hand side of the vertical. Leave the car in gear and apply the handbrake to prevent the engine turning and fit Tool No. A60319 to lock the camshaft and auxiliary shaft wheels in position whilst the timing belt is removed and a new one fitted. The belt tension is maintained by the idler wheel below the two cam wheels. Slackening the two idler wheel locking nuts will slacken the belt and allow it to be removed. It may also be necessary to remove the crankshaft pulley.

9 Refitting is the reversal of the procedure to remove the belt, except that two essential checks must be made before and immediately after fitting.

10 The static timing of the crankshaft, camshaft and auxiliary drive shafts must be checked before the belt is refitted. The belt must be tensioned immediately after fitting.

11 The camshafts, auxiliary shaft and crankshaft are correctly positioned relative to one another when the crankshaft has the No 1 and No 4 pistons at TDC, the timing hole in the auxiliary shaft drive wheel is approximately 34 degrees clockwise from the vertical, and lastly the timing marks on the camshaft wheels are aligned with the pointers on the timing cover top support fitting. FIAT tool number 60319 will hold the camshaft and auxiliary shaft wheels in position whilst the timing belt is fitted and tightened.

To ensure that the crankshaft is in the correct position to accept the new timing belt, replace the timing belt cover temporarily. The relative position of the crankshaft timing notch on the fan belt pulley and the static timing marks on the base of the belt cover can then be checked. The crankshaft timing mark on the pulley should be aligned with the longest timing mark on the cover (the farthest clockwise mark) for the number 1 and 4 pistons to be at TDC.

If the instructions given in paragraph 8 of this section have been followed, then it should not be necessary to readjust the position of the crankshaft before the belt is fitted.

Should it be that the timing belt is being refitted during an engine overhaul, an easier indication of the No 1 and No 4 piston TDC position, is the white spot on the periphery of the flywheel: it should be at TDC.

12 Adjusting the belt tension should be completed before the cover is replaced if fitting a new belt.

13 If the task is mere adjustment with the belt remaining in place then the tasks outlined in parapgraphs 1 to 8 should be carried out and then the following procedure followed.

14 On 124AC series and 125 series engines before engine number 42577, the belt tension was applied by attaching a spring balance to the idler wheel mounting plate in the hole provided and exerting a load of 60 lbs in a direction just above the horizontal. The nuts which lock the mounting plate and idler wheels position are then tightened and the load removed.

15 The procedure on later series engines is similar except that the tensioning spring is a built-in feature of the engine. A mouse-trap type spring has been mounted on a pin adjacent to the idler wheel plate and the spring acts through the same hole as was previously used for the spring balance.

16 Using the built-in spring loosen the two idler wheel locking nuts and allow the spring to tension the belt. Tighten the lock nuts to the specified torque to maintain the tension.

17 Having fitted the new belt, checked the camshaft timing and tensioned the belt with the crankshaft, camshafts and auxiliary positions maintained; the special wheel locking tool may be removed.

18 Check the belt tension two or three times again, turning the crankshaft half turn to three quarters of a turn in its direction of rotation between each check.

19 The essential checks having been completed, the timing cover, radiator, water hoses and fan belt can be refitted and the cooling system refilled.

See the specification at the beginning of this chapter for the tightening torques of the various nuts and bolts. Chapter 2 details the procedure for refilling the cooling system.

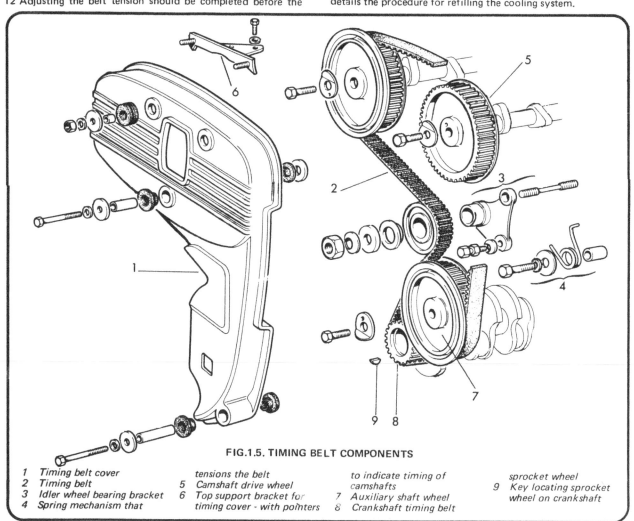

FIG.1.5. TIMING BELT COMPONENTS

1 Timing belt cover	tensions the belt	to indicate timing of	sprocket wheel
2 Timing belt	5 Camshaft drive wheel	camshafts	9 Key locating sprocket
3 Idler wheel bearing bracket	6 Top support bracket for	7 Auxiliary shaft wheel	wheel on crankshaft
4 Spring mechanism that	timing cover - with pointers	8 Crankshaft timing belt	

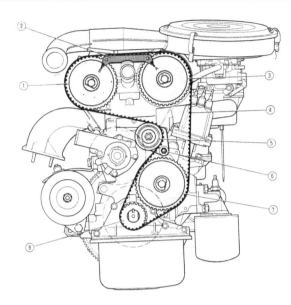

FIG.1.6. TIMING GEAR AND AUXILIARY DRIVE

1   Exhaust camshaft wheel   5   Tensioning idler pulley
2   Pointer fixed   6   Idling clamping nut
3   Inlet camshaft wheel   7   Auxiliary drive wheel
4   Toothed belt   8   Crankshaft wheel/sprocket

10 Cylinder head removal - engine in car

1   Drain the cooling system as described in Chapter 2.
2   For safety reasons disconnect the battery connections and tuck the leads safely away.
3   Remove all the water hoses connecting to the radiator and engine from both engine and radiator.
4   Slacken the alternator position retaining nut and bolt, push the alternator towards the engine and remove the fan belt.
5   It will make the later tasks easier if at this stage the radiator is removed. It is secured by two nuts and locates in mounting rubbers. Take care not to lose the spacers and rubbers when the radiator is lifted clear. On later series cars an electric radiator fan is mounted on the radiator; its electrical power leads should be disconnected prior to removal of the radiator.
6   There should be ample space now for the timing belt cover retaining nuts to be removed and the cover lifted off the engine.
7   The nuts which lock the timing belt idler wheel in position should be slackened to allow the belt itself to be slackened and removed.
8   Remove all HT leads to the spark plugs and identify as necessary. Remove the electrical connections to the water temperature sender units in the top of the cylinder head.
9   The air cleaner should now be removed as described in Section 5.
10 The choke and accelerator cables should be disconnected from the carburettor assembly and the fuel feed pipe from the pump removed and plugged with a wood or metal rod.
11 Pipe connections to the inlet manifold - including the brake servo connection be removed.
12 On the exhaust side the nuts securing the twin exhaust pipe flange to the manifold should be removed, and the pipes eased downwards clear of the manifold. It may become necessary to disconnect the first pipe support shackle underneath the car to allow the twin pipe to be moved sufficiently far away from the exhaust manifold.
13 The cylinder head should now be cleared of all connections to the engine and car and ready for the ten cylinder head bolts to be slackened and removed.
14 The cylinder head bolts should be slackened in the reverse order to which they are tightened refer to Fig. 1.7.

15 The cylinder head assembly may now be lifted clear. If it appears stuck to the cylinder block, do not try to free it by prising it off with a screwdriver or cold chisel, but tap the cylinder head firmly with a plastic or wooden headed hammer.
    The mild shocks should break the bond between the gasket, the head and cylinder block, allowing the cylinder head to be lifted clear.

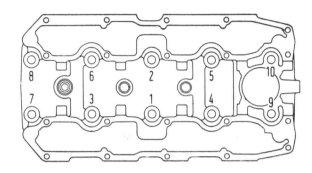

Fig.1.7. Cylinder head hold-down bolt tightening sequence

11 Cylinder head removal - engine out of car

Remove the ancillary components as described in Section 8 and then proceed as directed in paragraphs 13,14,15 in the preceding Section 10.

12 Dismantling the cylinder head

1   The cylinder head assembly is made of aluminium alloy and it comprises three sub assemblies: the basic head assembly, and the two overhead camshaft assemblies.
    The basic head assembly incorporates the combustion chambers, inlet and exhaust ducts and the valve assemblies. The two camshaft assemblies comprise a camshaft, the housing, the tappets and housing cover.
    Once the cylinder head has been removed from the engine, place it on a clean bench and commence removal of the exhaust manifold, inlet manifold and carburettor assembly. Refer to Chapter 3 'Fuel system' for instructions on maintainance of the carburettors.
2   Both manifolds are held to the cylinder head on six studs and retained by six nuts.
3   Once the manifolds have been removed, the camshaft housing covers are next removed with a large hexagon Allen key.
4   If the cylinder head water outlet elbow has not yet been removed, it should be now. It is secured by four long bolts.
5   The camshaft housings are located on the cylinder head by ten long pins retained by nuts. The tappet barrels should remain in their bores in the camshaft housings.

13 Valve removal

1   The valves can be removed from the cylinder head by the following method: With a valve spring compressor, compress each spring in turn until the two halves of the collets can be removed. Release the compressor and remove the valve spring cap, springs (inner and outer), and lower spring seating. The inlet valve guides have oil seals fitted which should be removed together with the inlet and exhaust valves themselves.
2   If when the valve spring compressor is screwed down, the valve spring cap refuses to free and expose the split collets, do not continue to screw down on the compressor as there is a likelihood of damaging it.
3   Gently tap the top of the tool directly over the cap with a light hammer. This will free the cap. To avoid the compressor jumping off the spring cap when it is tapped, hold the compressor firmly in position with one hand.
4   It is essential that the valves keep to their respective places in

the head, unless they are so badly worn that they need to be renewed. If they are going to be kept and used again, place them in a sheet of card having two rows of four holes numbered 1 to 4 exhaust and 1 to 4 inlet corresponding with the positions the valves occupied in the cylinder head. Keep the valve springs, caps in their correct order too.

14 Valve guide removal

1   Only remove the guides if after inspection they have been found to have worn excessively. The correct bore size is in the specifications.

2   The valve guides are an interference fit in the cylinder head block. If it is wished to remove the valve guides they can be removed from the cylinder head in the following manner.

3   Place the cylinder head upside down on a bench on two stout blocks of wood ensuring that the studs in the cylinder head to not contact the bench.

4   With a suitable diameter drift, (maximum diameter 14 mm, minimum 12 mm) carefully drive the guides from the cylinder head.

5   It should be noted that the inlet and exhaust valve guides are not interchangeable.

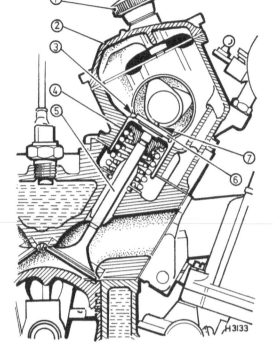

FIG.1.8. CROSS SECTION OF THE INLET SIDE OF THE CYLINDER HEAD

1	Camshaft cover retaining bolt	4	Tappet
2	Camshaft cover	5	Inlet valve
3	Notch in tappet for extract-ing capping plate	6	Tappet disc plate
		7	Cam/tappet gap

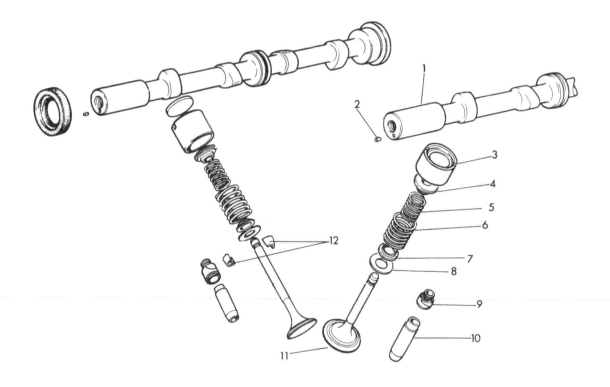

FIG1.9. VALVE OPERATING MECHANISM COMPONENTS

1	Camshafts	4	Spring cover/cap	7	Inner spring seat	10	Valve guide
2	Camshaft/wheel locating dowel	5	Inner spring	8	Outer spring seat	11	Valve
3	Tappet	6	Outer spring	9	Oil seal	12	Collets

15 Camshaft, camshaft wheel, tappet removal

1 Since the camshafts are withdrawn from the rear end of the housings, it will be necessary to begin by removing the camshaft wheel at the forward end.

2 The wheel is retained on the shaft by a single bolt which is locked by the tab on the washer. Tap back the tab, grasp the wheel gently in a soft jawed vice and undo the bolt.

3 Remove the bolt and washer which keeps the wheel/shaft alignment dowel in place. Ease the wheel off the shaft and take care not to lose the alignment dowel.

4 Next at the rear end of the housing, the camshaft retaining plate is held on by three nuts. Once removed the camshaft can be carefully withdrawn from the housing.

5 Be careful not to damage the camshaft bearing bores in the housing when removing the shaft. The housing is of aluminium alloy whereas the shaft is steel.

6 The tappets can be simply pushed out of there bores and as with the valves it is wise to store the tappets and shims so that they may be refitted into the bores from which they were removed.

16 Cylinder block dismantling - engine out of car

1 Having removed all the ancillary items as described in Section 8 and removed the cylinder head as described in Section 10 the cylinder block is best positioned upside down on a strong bench.

2 Refer to Chapter 5 and remove the clutch from the flywheel. Mark the position of the clutch assembly on the flywheel so that it may be refitted later into the same position from which it was removed.

3 The sump is removed after undoing all the bolts which secure the pressed steel sump to the crankcase. Clean the joint surfaces of the old gasket.

4 The flywheel is removed next. It is secured to the crankshaft by six bolts.

5 The rear crankshaft oil seal is mounted on a carrier fitting which is held onto the crankcase by six bolts. Undo the bolts and lift off the carrier and seal.

6 Working on the front of the engine begin by removing the crankshaft pulley. Tap back the tag on the locking washer and undo the crankshaft bolt. Then ease the pulley off the end of the crankshaft.

7 On 124 AC000 (1438 cc) engines the pulley incorporates a centrifugal oil filter. The filter must be dismantled before the pulley/filter is removed. Six bolts retain the centrifugal filter cover in place and once the cover is removed the impellor can be extracted to expose the main crankbolt and its locking washer.

8 The timing wheel can be removed immediately after the 'V' belt pulley has been removed, the main crankshaft end bolt retains both onto the shaft.

9 The water pump pulley should be detached from the pump shaft after undoing the three retaining bolts.

10 The pump itself can then be removed from the cylinder block after undoing the four bolts that retain it.

11 The auxiliary shaft wheel should be removed next by tapping back the retaining bolt locking tag so that the bolt can be undone.

12 The locking washer and bolt hold the auxiliary wheel aligning dowel in place and when the wheel is eased off, take care not to lose the dowel.

13 The forward crankshaft oil seal is mounted in an aluminium alloy housing and a housing carries the auxiliary shaft forward oil seal. With both crankshaft pulley and timing wheel removed, and the auxiliary shaft wheel removed the two oil seal housings may be unbolted from the cylinder block and lifted clear of the engine.

14 The cylinder block is now sufficiently stripped for the removal of the basic moving parts to be commenced.

17 Big-end bearings

1 The big end bearings are one of those items which may require attention more frequently than any other moving part on the engine. They are the usual shell bearings, running on a hardened steel crankpin and the shell surface is very soft; it is not uncommon to be advised to renew the shells every 25,000 miles or so. This interval is probably too short if the engine oil has been good and changed regularly, a more reasonable life would be 40 to 50,000 miles when it is also advisable to decoke the cylinder head and thoroughly inspect the valve mechanisms.

2 The big ends may be dismantled without removing the engine from the car, it is only necessary to drain the engine oil, remove the dust shield and unbolt the sump from the crankcase for full access to be gained to the big end bearings.

3 Having arrived at the stage when the dismantling of the big end bearings is the next step - either with the engine in the car and with the tasks in paragraph 2 complete - or with the engine out of the car and the dismantling as far as described in Section 16, complete, proceed as follows.

4 Check that the connecting rod and big end bearing cap assemblies are correctly marked. Normally the number 1 to 4 are stamped on the adjacent sides of the big end caps and connecting rods indicating which cap fits on which rod and which way round the cap fits on the rod. The number 1 should be stamped on the con-rod and cap operating No 1 piston at the front of the engine, 2 for the second piston/cylinder and so on.

5 If for some reason no numbers or lines can be found, then scratch mating marks on across the rod cap joint with a sharp screwdriver. One line for rod and cap on No 1 cylinder, two lines for No 2 and so on.

6 Undo and remove the big end cap retaining bolts and keep them in their respective order for correct refitting.

7 Lift off the big end caps; if they are difficult to remove they may be gently tapped with a soft face hammer.

8 To remove the shell bearings press the bearing opposite the groove in both the connecting rod and bearing cap and the shells will slide out easily.

9 If the bearings are being attended to with the engine in the car, make sure that the pistons do not fall too far down the cylinder bores, the piston could reach a position when the lower piston rings pass out from the bottom of the cylinder bores. It is difficult and tedious using a 'jubilee' clip inside the crankcase to compress the piston rings so that the piston can be eased back into the cylinder.

18 Big end shell bearing renewal - engine in car

1 Having removed the old bearing shells as described in the previous section, the crankpins on the crankshaft should be inspected.

2 The specification at the beginning of this chapter sets down the crankpin diameters for the engines fitted to the '124 Sport as well as the taper and ovality tolerances. The taper tolerance is the maximum difference in crankpin diameter measured at each end of the bearing and the ovality tolerance is the maximum difference in diameters measured in any plane at right angles to the centre line axis of the crankpin.

3 Always use a good quality micrometer when measuring the crankpin and remember to check it by measuring a standard length gauge.

4 Having measured the crankpins, if they do not match up to the specification then the engine must be removed, the crankshaft removed and reground at a automobile workshop. See later sections describing the renovation and reassembly of the engine.

5 Fit the appropriate size of new shell to the con-rod and cap. Ensure that the slots in the shells match the slots in the rod and cap and also ensure that the holes in the upper shell coincide with the holes in the rod, which provide for splash lubrication of the cylinder bores.

6 Take great care when handling the new shells, the bearing

surface is soft and can easily be damaged.

7 Before offering the connecting rods back onto the crank-shaft, coat the crankpins liberally with engine oil.

8 With the con-rod and shell bearing in place refit the appropriate caps and shell bearing. Ensure that the identification numbers stamped on the con-rods and caps tally and are all facing the same side of the crankcase.

9 Replace the big end bearing bolts and tighten in pairs to the torque specified.

10 Clean the joint faces of the sump and crankcase of the old gasket. Smear the faces of the new gasket with heavy grease and stick it to the crankcase joint face. Fit the sump and tighten the sump bolts.

11 It is worth changing the oil filter cartridge when the engine bearings have been renewed.

12 Refit the dust shield, check the tightness of the sump drain plug and finally fill the engine with Castrol GTX or equivalent oil.

19 Removal and separation of pistons - piston rings and con-rods

1 The procedure for piston and con-rod removal is the same whether the engine is in or out of the car.

2 In either case the cylinder head needs to have been removed (see Section 10,11) and the big end bearings dismantled (see Section 17).

3 The engine having been stripped down as indicated, giving complete access to the basic moving parts of the engine - proceed as follows.

4 The pistons and con-rods are pushed out of the top of the cylinder block together. Be careful not to allow the rough edges of the connecting rod score the fine bore of the cylinder.

5 The pistons will not pass the crankshaft downwards out of the crankcase.

6 Once the piston/connecting rod has been removed from the engine the piston rings may be slid off the piston.

7 Slide the piston rings carefully over the surface of the piston taking care not to scratch the aluminium alloy from which the piston is made. It is all too easy to break the cast iron piston ring if they are pulled off roughly, so this operation must be done with extreme care. It is helpful to make use of an old feeler gauge.

8 Lift one end of the piston ring to be removed out of its groove and insert the end of a feeler gauge under it.

9 Move the feeler gauge slowly around the piston and apply slight upward pressure to the ring as it comes out of the groove so that it rests on the land above. It can then be eased off the piston and the feeler gauge used to prevent it slipping into other grooves as necessary.

10 The piston and connecting rod are joined by a gudgeon pin. The pin is an interference fit in the connecting rod and a clearance fit in the piston.

11 FIAT provide a special mandrel A60308 for the task of removing the gudgeon pin and a fixture A95605 to support the piston and con-rod whilst the pin held in the special mandrel is being forced out by a hydraulic press.

12 The method to remove the gudgeon pin with the FIAT tools is as follows:-

13 Remove the knurled nut from the mandrel, leave the spacer on and insert the centre shaft through the centre of the gudgeon pin. Refit the knurled nut.

14 Heat the piston and connecting rod assembly in an oil bath to around 200°C and then swiftly transfer them to the support fixture so that the pin may be drifted out of the piston - rod assembly.

15 Providing one has the equipment a wooden support cradle could be made for the piston and rod assembly, and a convention drift used to drive out the gudgeon pin from the connecting rod and piston once they have been heated.

16 It will be appreciated that the removal of the gudgeon pin requires quite a few bits of specialised equipment and therefore in all probability it will be more practical if the gudgeon pin really must be removed, to take the piston/rod assembly to your local garage or workshop and use their facilities.

20 Removal of oil pump unit

1 The oil pump unit comprises the main housing, the pump shaft and pumping gears, the cover and oil collector funnel.

2 Two long bolts which project right through the pump unit secure it to the inside of the crankcase. The pump shaft locates into the centre of a spiral gear wheel which is driven by the auxiliary drive shaft. Once the two bolts have been removed, lift the oil pump unit out of the engine.

3 The oil pump may be removed with the engine in the car, after the sump has been drained and unbolted from the crankcase.

4 It is necessary to remove the oil pump in order to gain the space required for removal of the crankshaft.

21 Crankshaft and main bearing removal

1 The crankshaft may be removed after the engine has been stripped of the following items:-
1 Clutch and flywheel.
2 Oil pump and sump.
3 Big end bearings.
4 Timing belt, fan belt and the respective crankshaft pulleys.
5 Auxiliary drive shaft pulley wheel.
6 Front and rear crankshaft oil seal carrier fitting.

2 It is not essential to remove the cylinder head in order to complete the task of crankshaft removal, but it makes the cylinder block easier to handle and the opportunity of inspecting the cylinder head assembly should not be missed.

3 Before beginning to actually remove the crankshaft, it is wise to check the end float on the crankshaft.

4 Use feeler gauges to measure the gap between the rear most main bearing journal wall and the thrust washers on the shoulder of the rear main bearing housing.

5 Move the crankshaft forwards as far as it will go with two tyre levers to obtain a maximum reading. The end float should not exceed 0.0024 to 0.0102 inches. If this is exceeded new oversize thrust washers should be fitted when the crankshaft is refitted. See the 'specification' at the beginning of this chapter for thrust washer and main bearing data.

6 Ensure that the main bearing caps are marked in a manner that will enable them to be refitted later into the exact positions from which they were removed. Main bearing caps are NOT interchangeable and should not even be refitted the opposite way round.

7 Undo the bolts which retain the five main bearing caps by one turn only, to begin with, and once all have been loosened proceed to unscrew and remove them, together with the washer if fitted.

8 The main bearing caps are then lifted away, together with the bearing shells inside them.

9 The crankshaft can be lifted out once the main bearing caps are clear - and collect the semi-circular thrust washers on the crankcase shoulder of the rear main bearing.

22 Auxiliary drive shaft removal and oil pump/distributor drive

1 The auxiliary drive shaft can be removed once the drive wheel and oil seal carrier fitting have been removed. (Section 8 and 9). In the extremely unlikely event that this shaft may need individual attention it is possible to remove it with the engine remaining in the car.

2 As the auxiliary shaft oil seal carrier fitting is lifted away the shaft retaining plate is exposed. This plate is retained by two bolts onto the cylinder block.

3 Remove the two bolts and the shaft retaining plate. Be careful not to scratch or score the bearing bores in the cylinder block as the auxiliary drive shaft is drawn out.

4 Once the auxiliary drive shaft is out, the pinion which drives

the oil pump and distributor may be removed. Push the pinion toward the distributor aperture out of the bush in which it runs. The pinion is held in place when the engine is running by the thrust from its helical gear arrangement.

23 Lubrication system

Basically the same lubrication system has been designed into each of the four engines that have been fitted to the FIAT 124 Sport series of cars. The only type of engine which has anything unusual is the 124 AC000, 1438 cc engine. This was provided with a centrifugal oil filter; the oil flows through this filter when a pressure actuated valve in the filter outlet is opened. The excess oil would then flow from the pump through the centrifugal oil filter and back into the sump. The oil supply to the crankshaft and other major rotating components passed through a cartridge full flow filter as on the later three types of engines.

The pressure relief valve on the 1608, 1592, 1755 cc engines

is located in the oil pump cover and it provides for the excess oil to be discharged directly into the sump.

On all engines the oil is pumped through the full flow filter into a major gallery of oil drillings which direct it to each of the five main crankshaft bearings, the auxiliary drive shaft bearings and the camshaft bearings adjacent to the timing wheels.

Drilled galleries in the cylinder block tap oil from the main bearings on the crankshaft and feed it to the camshaft bearings augmenting the supply direct from the pump to the forward camshaft bearings.

Drillings in the crankshaft tap oil from the main journals to supply it to the big end bearings, and holes in the connecting rod allow oil to squirt from the big ends onto the cylinder walls.

Holes drilled into the camshaft bearing housings allow oil to spray out from the bearings on to the cams and tappets.

Finally the oil pressure transducer and low pressure indicator switch are screwed into the oil filter fitting and monitor the pressure of oil after it has flowed through the filter.

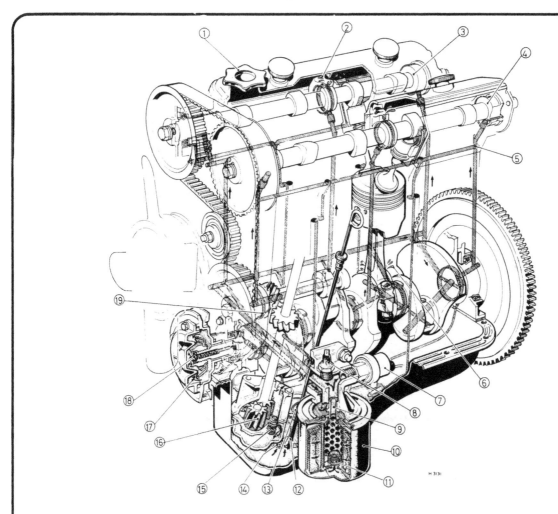

FIG.1.10. LUBRICATION SYSTEM – 124AC000 ENGINE SHOWN

Centrifugal oil filter by-pass system not included on 125BC000 and 132 series engines, the system otherwise remains unchanged

1 Oil filler cap	6 Oil squirts to interior of cylinders
2 Oil squirt to camshaft lobes and tappets	7 Oil gauge sending unit
3 Exhaust valve drive camshaft	8 Low oil pressure indicator sending unit
4 Intake valve drive camshaft	9 Oil delivery duct from filter to engine working parts
5 Oil return ducts from tappets	10 Full-flow cartridge oil filter

11 By-pass valve for oil filter in case of obstruction	18 Pressure regulating valve in sump centrifugal filter
12 Oil dipstick	19 Oil way for spray lubrication of oil pump and ignition distributor gears
13 Oil drain plug from sump	
14 Oil pump suction filter	
15 Oil pressure relief valve	
16 Oil pump gears	
17 By-pass centrifugal oil filter	

24 Centrifugal oil filter, 124 AC000 engines

The oil filter should be dismantled and cleaned at least every 30,000 miles (50,000 Km) especially in cold operating conditions or under heavy duty service.

1 Undo and remove the six securing bolts and spring washers and lift away the cover.

2 Wash the hub/pulley in paraffin and carefully wipe dry. Take care not to allow paraffin to pass down the centre of the crankshaft.

3 Inspect the rubber 'O' ring and if it shows signs of perishing or hardening it should be renewed. Whenever possible a new 'O' ring should be obtained to prevent the possibility of subsequent oil leaks.

4 Inspect the baffle ring, cover and hub/pulley for signs of damage, cracks or distortion and, if evident a new part should be obtained.

5 Reassembly is the reverse sequence to removal. Check the dynamo and water pump drive belts tension and adjust the tension until there is a deflection of 0.39 to 0.59 inch (10 to 15 mm) under a load of 22 lb (10 Kg).

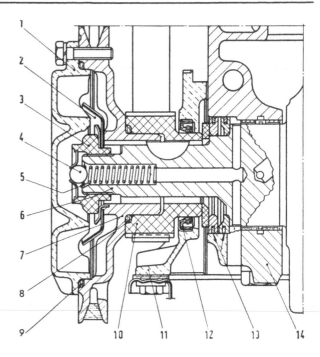

FIG.1.11. SECTION THROUGH THE CENTRIFUGAL OIL FILTER ON THE 124ACOOO ENGINES

1 Cover of centrifugal filter	8/9 'O' rings
2 Impellor	10 Driving wheel (sprocket)
3 Nut	for timing belt
4 Ball	11 Cover
5 Spring	12 Gasket
6 Crankshaft	13 Oil seal rings
7 Pulley hub	14 Front main bearing cap

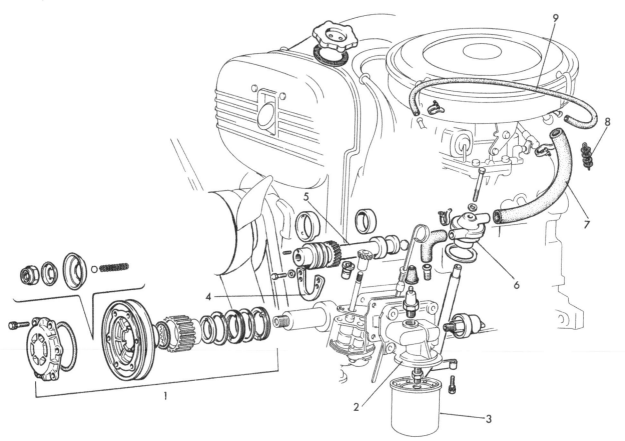

FIG.1.12. SOME ENGINE COMPONENTS

1 Centrifugal oil filter and by-pass valve	3 Cartridge oil filter	5 Auxiliary shaft	7 Vapour hose to air cleaner
2 Cartridge oil filter fitting	4 The auxiliary shaft locating plate	6 Crankcase vapour/oil separator	8 Wire brush flame arrester
			9 Vacuum hose

25 Dismantling the oil pump

1 The oil pump assembly is held together by three short studs and the two long bolts which secure it to the inside of the crankcase. Having already removed the two long bolts all that remains to dismantle the pump is the three studs.
2 The pump comprises the main body, two gears - housed in the main body, a cover plate incorporating the pressure relief valve and finally the suction horn with a metal filter screen across the large aperture.
3 Once the retaining bolts have been removed all of these oil pump parts may be separated for cleaning and inspection.

26 Oil pump pressure relief valve and centrifugal filter regulating valve

1 The relief valve is of caged ball spring design. Examination necessitates inspection of the ball and orifice for signs of wear and checking the free length of the actuating spring - 1.58 inches.
2 On 124 AC000 engines a relief/regulating valve is mounted in the centre of the centrifugal oil filter fitted to that engine. The valve opens at similar pressures to the relief valve in the pump. It allows the excess oil to flow through the centrifugal filter and be cleaned before discharging it into the sump.
3 The main crnakshaft nut incorporates the orifice plate against which the ball is pressed by the valve spring. Inspect the ball and orifice surfaces for wear and indentations - renew components if necessary. Check the free length of the valve spring.
4 The oil pressure relief valves should maintain the pressure to between 50 and 71 lbf/sq in.

27 Oil filter cartridge unit

1 When work is of a routine maintenance nature and the filter is due for renewal, a useful tool to loosen an unobliging filter cartridge is a small chain or strap wrench. This type of wrench will grip the old filter firmly and enable the most obstinate of cartridges to be freed. Do not use such a wrench to tighten the new filter into position.
2 If the engine has been completely dismantled the opportunity of cleaning the oil filter support fitting interior and exterior should not be missed. Both oil pressure and low oil pressure indicator transducers are mounted on the support fitting and these should be replaced or checked if possible at your local electrical workshop if there is any doubt as to their operation.

28 Engine examination and renovation

1 With the engine stripped and all parts thoroughly cleaned, every component should be examined for wear. The items listed in the sections following should receive particular attention and where necessary renewed or renovated.
2 So many measurements on engine components require accuracies down to tenths of a thousanth of an inch. It is advisable therefore to either check your micrometer against a standard gauge occasionally to ensure that the instrument zero is set correctly, or use the micrometer as a comparative instrument. This last method however necessitates that a comprehensive set of slip and bore gauges is available.

29 Crankshaft - examination and renovation

1 Examine the crankpin and main journal surfaces for signs of scoring or scratches and check the ovality and taper of the crankpins and main journals. If the bearing surface dimensions do not fall within the tolerance ranges given in the specification at the beginning of this chapter, the crankpins and/or main journals will have to be regoround.

Big end and crankpin wear is accompanied by distinct metallic knocking particularly noticable when the engine is pulling from low revs.
Main bearing and main journal wear is accompanied by severe engine vibration - rumble - getting progressively worse as engine revs increase.
If the crankshaft is reground the workshop should supply the necessary undersize bearing shells.

30 Big end and main bearing shells - examination and renovation

1 Big end bearing failure is accomplished by a noisy knocking from the crankcase and a slight drop in oil pressure. Main bearing failure is accompanied by vibration which can be quite severe as the engine speed rises and falls, and a drop in oil pressure.
2 Bearings which have not broken up, but are badly worn will give rise to low oil pressure and some vibration. Inspect the big ends, main bearings and thrust washers for signs of general wear, scoring, pitting and scratches. The bearings should be matt grey in colour. With lead-indium bearings, should a trace of copper colour be noticed the bearings are badly worn as the lead bearing material has worn away to expose the indium underlay. Renew the bearings if they are in this condition or if there is any sign of scoring or pitting.
3 The undersizes available are designed to correspond with the regrind sizes, i.e. 0.02 inch (0.508 mm) bearings are correct for a crankshaft reground - 0.02 inch (0.508 mm) undersize. The bearings are in fact, slightly more than the stated undersize as running clearances have been allowed for during their manufacture.
4 Very long engine life can be achieved by changing big end bearings at intervals of 30,000 miles (48,000 Km) and main bearings at intervals of 50,000 miles (80,000 Km) irrespective of bearing wear. Normally, crankshaft wear is infinitesimal and regular changes of bearings will ensure mileages in excess of 100,000 miles (160,000 Km) before crankshaft regrinding becomes necessary. Crankshafts normally have to be reground because of scoring due to bearing failure.

31 Cylinder bores - examination and renovation

1 The cylinder bores must be examined for taper, ovality, scoring and scratches. Start by carefully examining the top of the cylinder bores. If they are at all worn a very slight ridge will be found on the thrust side. This marks the top of the piston travel. The owner will have a good indication of the bore wear prior to dismantling the engine, or removing the cylinder head. Excessive oil consumption accompanied by blue smoke from the exhaust is a sure sign of worn cylinder bores and piston rings.
2 Measure the bore diameter just under the ridge with a micrometer and compare it with the diameter at the bottom of the bore, which is not subject to wear. If the difference between the two measurements is more than .006 inch (0.15 mm) then it will be necessary to fit special piston rings or to have the cylinders rebored and fit oversize pistons and rings. If no micrometer is available remove the rings from a piston and place the piston in each bore in turn about three quarters of an inch below the top of the bore. If an 0.010 inch (0.25 mm) feeler gauge can be slid between the piston and the cylinder wall on the thrust side of the bore then remedial action must be taken. Oversize pistons are available in the following sizes:-

plus 0.0079 inch (0.2 mm)
plus 0.0157 inch (0.4 mm)
plus 0.0236 inch (0.6 mm)

3 These are accurately machined to just below these measurements so as to provide correct running clearances in bores bored out to the exact oversize dimensions.
4 If the bores are slightly worn but not so badly worn as to justify reboring them special oil control rings can be fitted to the existing pistons which will restore compression and stop the

engine burning oil. Several different types are available and the manufacturer's instructions concerning their fitting must be followed closely.

32 Piston and piston rings - examination and renovation

1 If the old pistons are to be refitted carefully remove the piston rings and thoroughly clean them. Take particular care to clean out the piston ring grooves. At the same time do not scratch the aluminium. If new rings are to be fitted to the old pistons, then the top ring should be stepped to clear the ridge left above the previous top ring. If a normal but oversize new ring is fitted, it will hit the ridge and break, because the new ring will not have worn in the same way as the old, which will have worn in unison with the ridge.

2 Before fitting the rings on the pistons each should be inserted approximately 3 inches down the cylinder bore and the gap measured with a feeler gauge as shown in Fig. 1.13. This should be as detailed in the specification at the beginning of this chapter. It is essential that the gap is measured at the bottom of the ring travel. If it is measured at the top of a worn bore and gives a perfect fit, it could easily seize at the bottom. If the ring gap is too small rub down the ends of the ring with a fine file, until the gap, when fitted, is correct. To keep the rings square in the bore for measurement, line each up in turn with an old piston in the bore upside down, and use the piston to push the ring down about three inches. Remove the piston and measure the piston ring gap.

3 When fitting new pistons and rings to a rebored engine the ring gap can be measured at the top of the bore as the bore will now not taper. It is not necessary to measure the side clearance in the piston ring groove with rings fitted, as the groove dimensions are accurately machined during manufacture. When fitting new oil pistons it may be necessary to have the groove widened by machining to accept the new wider rings. In this instance the manufacturer's representative will make this quite clear and will supply the address to which the pistons must be sent for machining.

4 When new pistons are fitted, take great care to fit the exact size best suited to the particular bore of your engine. FIAT go one stage further than merely specifying one size of piston for all standard bores. Because of very slight differences in cylinder machining during production it is necessary to select just the right piston for the bore. A range of different sizes are available either from the piston manufacturers or from the dealer for the particular model of car being repaired.

5 Examination of the cylinder block face will show, adjacent to each bore, a small diamond shaped box with a number stamped in the metal. Careful examination of the piston crown will show a matching diamond and number. These are the standard piston sizes and will be the same for all four bores. If standard pistons are to be refitted or standard low compression pistons change to standard high compression pistons, then it is essential that only pistons with the same number in the diamond are used. With larger pistons, the amount of oversize is stamped in an ellipse in the piston crown.

6 On engines with tapered second and third compression rings, the top narrow side of the ring is marked with a 'T', always fit this side uppermost and carefully examine all rings for this mark before fitting.

33 Camshaft and camshaft bearing - examination and renovation

1 Carefully examine camshaft bearings for wear. If the bearings are obviously worn or pitted or the metal underlay just showing through, then they must be renewed. This is an operation for your local Fiat agent or automobile engineering works, as it demands the use of specialised equipment. The bearings are removed using a special drift after which the new bearings are pressed in, care being taken that the oil holes in the bearings line up with those in the block. With another special tool the bearings are then reamed in position.

2 The camshaft itself should show no signs of wear, but, if very

slight scoring marks on the cams are noticed, the score marks can be removed by very gentle rubbing down with a very fine emery cloth or an oil stone. The greatest care should be taken to keep the cam profiles smooth.

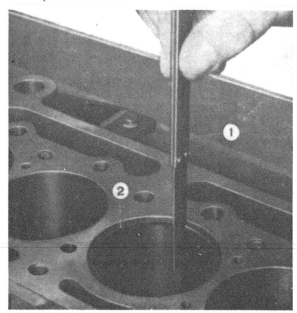

Fig. 1.13. Measuring piston ring gap
1 Feeler gauge   2 Piston ring

34 Valves and seats - examination and renovation

1 Examine the heads of the valves for pitting and burning especially the heads of the exhaust valves. The valve seatings should be examined at the same time. If the pitting on the valves and seats is very slight the marks can be removed by grinding the seats and valves together with coarse, and then fine, valve grinding paste. Where bad pitting has occurred to the valve seats it will be necessary to recut them to fit new valves. If the valves seats are so worn that they cannot be recut, then it will be necessary to fit new valve seat inserts. These latter two jobs should be entrusted to the local Fiat agent or autombile engineering works. In practice it is very seldom that the seats are so badly worn that they require renewal. Normally, it is the valve that is too badly worn for replacement, and the owner can easily purchase a new set of valves and match them to the seats by valve grinding.

2 Valve grinding is carried out as follows:-
    Place the cylinder head upside down on a bench, with a block of wood at each end to give clearance for the valve stems. Alternatively, place the head at 45° to a wall with the combustion chambers facing away from the wall.

3 Smear a trace of coarse carborundum paste on the seat face and apply a suction grinder tool to the valve heads as shown in the photo. With a semi-rotary action, grind the valve head to its seat, lifting the valve occasionally to redistribute the grinding paste. When a dull matt even surface finish is produced on both the valve seat and the valve, then wipe off the paste and repeat the process with fine carborundum paste, lifting and turning the valve to redistribute the paste as before. A light spring placed under the valve head will greatly ease this operation. When a smooth unbroken ring of light grey matt finish is produced, on both valve and the valve seat faces, the grinding operation is complete.

4 Scrape away all carbon from the valve head and the valve stem. Carefully clean away every trace of grinding compound, taking great care to leave none in the ports or in the valve guides. Clean the valves and valve seats with a paraffin soaked rag, then with a clean rag, and finally, if an airline is available, blow the valve, valve guides and valve ports clean.

35 Timing wheels and belt - examination

1   The belt has a rubber surface and is reinforced for strength, and as a consequence it is most likely that the wheels will need anything more than a clean in an oil grease solvent and wiping dry.

2   The idler wheel which maintains the tension in the timing belt runs on a prepacked ball bearing race. Whenever the timing belt is being changed or attended to, it is advisable to check the excessive play of the wheel and bearing. The wheel bearing assembly should be renewed if discernable play is felt.

3   The belt however does not wear and fatigues, and because failure of this belt would be catastrophic for the engine, it must be renewed at regular intervals even though on the surface it might appear serviceable.

4   The timing belt must be renewed at intervals NOT exceeding 37,000 miles - FIAT justifiably recommend that it should be replaced every 25,000 miles.

5   It is important to remember that when handling a new belt - when fitting - to avoid bending it to a sharp angle or the fibres which reinforce the belt will be seriously weakened.

36 Tappets - examination and renovation

Examine the bearing surface of the tappets which lie on the camshaft. Any indentation in this surface or any cracks indicate serious wear, and the tappets should be renewed. Thoroughly clean them out removing all traces of sludge. It is most unlikely that the sides of the tappets will be worn, but, if they are a very loose fit in their bores and can be readily rocked, they should be discarded and new tappets fitted. It is very unusual to find worn tappets and any wear present is likely to occur only at very high mileages.

37 Flywheel and starter ring - examination

1   If the teeth on the flywheel peripheral gears are badly worn, or if some are missing, it will be necessary to renew the whole flywheel. The ring gear is not supplied as an individual spare.

38 Oil pump - dismantling, examination and renovation

1   With the oil pump away from the engine, undo and remove the three bolts and spring washers securing the suction horn to the pump housing.

2   Thoroughly clean all the component parts in petrol and then check the gear end float and tooth clearances in the following manner.

3   With the two gears in position in the pump, place the straight edge of a steel rule across the joint face of the housing, measure the gap between the bottom of the straight edge and the top of the gears with a feeler gauge. The clearance should not exceed 0.006 inch (0.15 mm).

4   Check the backlash between the two gears and this should not exceed 0.006 inch (0.15 mm).

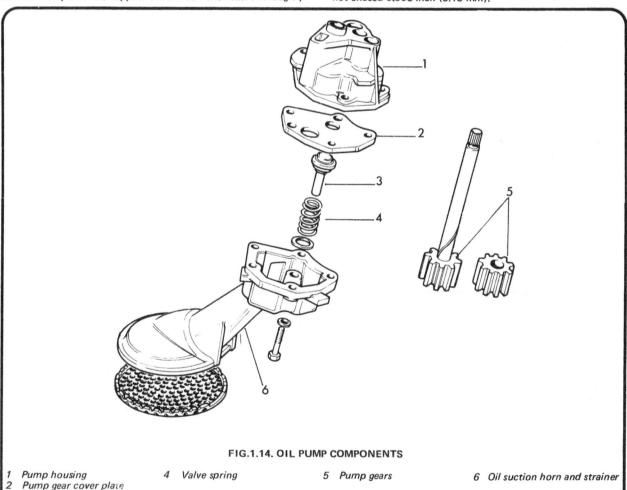

FIG.1.14. OIL PUMP COMPONENTS

1 Pump housing	4 Valve spring	5 Pump gears	6 Oil suction horn and strainer
2 Pump gear cover plate			
3 By-pass valve			

39 Cylinder head - decarbonisation

1 This operation can be carried out with the engine either in or out of the car. With the cylinder head off, carefully remove with a wire brush and blunt scraper all traces of carbon deposits from the combustion spaces and the ports. The valve stems and valve guides should also be freed from any carbon deposits. Wash the combustion spaces and posts down with petrol and scrape the cylinder head surface of any foreign matter with the side of a steel rule or a similar article. Take care not to scratch the surface.

2 Clean the pistons and top of the cylinder bores. If the pistons are still in the cylinder bores then it is essential that great care is taken to ensure that no carbon gets into the cylinder bores as this could scratch the cylinder walls or cause damage to the piston and rings. To ensure that this does not happen first turn the crankshaft so that two of the pistons are at the top of the bores. Place clean non-fluffy rag into the other two bores or seal them off with paper and masking tape. The water-ways and push rod holes should also be covered with a small piece of masking tape to prevent particles of carbon entering the cooling system and damaging the water pump, or entering the lubrication system and damaging the water pump or bearing surface.

3 There are two schools of thought as to how much carbon ought to be removed from the piston crown. One is that a ring of carbon should be left around the edge of the piston and on the cylinder bore wall as an aid to keep oil consumption low. Although this is probably true for early engines with worn bores, on later engines the tendency is to remove all traces of carbon during decarbonisation.

4 If all traces of carbon are to be removed, press a little grease into the gap between the cylinder walls and the two pistons which are to be worked on. With a blunt scraper carefully scrape away the carbon from the piston crown, taking care not to scratch the aluminium. Also scrape away the carbon from the surrounding lip of the cylinder wall. When all carbon has been removed, scrape away the grease which will now be contaminated with carbon particles, taking care not to press any into the bores. To assist prevention of carbon build up the piston crown can be polished with a metal polish such as 'Brasso'. Remove the rags or masking tape from the other two cylinders and turn the crankshaft so that the two pistons which were at the bottom are now at the top. Place non-fluffy rag into the other two bores or seal them off with paper and masking tape. Do not forget the waterways and oil ways as well. Proceed as previously described.

5 If a ring of carbon is going to be left round the piston then this can be helped by inserting an old piston ring into the top of the bore to rest on the piston and ensure that carbon is not accidentally removed. Check that there are no particles of carbon in the cylinder bores. Decarbonising is now complete.

40 Valve guides - examination and renovation

1 Examine the valve guides internally for wear. If the valves are a very loose fit in the guides and there is the slightest suspicion of lateral rocking, then new guides will have to be fitted. If the valve guides have been removed compare them internally by visual inspection with a new guide as well as testing them for rocking with the valves. It will be seen that the exhaust valve guides are different from the inlet valve guides as they are threaded throughout their length for lubrication.

When fitting new guides the drift used must have spigot to locate the new guide in the guide bore. Carefully drive new guides in from the top until the snap rings locate on the face of the cylinder head.

41 Auxiliary drive shaft and oil pump/distributor drive examination

1 The auxiliary drive shaft runs into bushes which are interference fits in bores in the cylinder block. There are not any undersize bushes supplied and therefore if the bush internal bores are found to be worn the cylinder block should be taken to your local Fiat agent who will have the necessary equipment to fit new bushes into the cylinder block.

2 The diameters of the shaft journals should also be checked for wear, and if found the only course of action is to renew the shaft.

3 The fit clearance between the bushes and shaft journals should be between 0.002 and 0.0035 of an inch.

4 Together with the inspection of the shaft bearings the gears for the distributor drive and oil pump drive, and the cam lobe for actuating the fuel pump should also be examined for wear.

   As stated earlier if wear is found the only recourse is to replace the shaft.

5 Finally the oil pump and distributor drive pinion bearing journal and gear teeth should be examined for wear. The pinion remains in the bush - which is interference fit in the cylinder block - by the end thrust from the helical gear arrangement. It is necessary therefore to ensure that the thrust faces on the pinion and bush are not scored or worn. Like the drive shaft if wear is found, renewal of the relevant component is the only action possible.

42 Engine reassembly - general

1 To ensure maximum life with minimum trouble from a rebuilt engine, not only must every part be correctly assembled, but everything must be spotlessly clean, all the oil-ways must be clear, locking washers and spring washers must always be fitted where indicated and all bearings and other working surfaces must be thoroughly lubricated during assembly. Before assembly begins renew any bolts or studs whose threads are in any way damaged; whenever possible use new spring washers.

2 Apart from your normal tools, a supply of non-fluffy rag, an oil can filled with engine oil (an empty washing-up fluid plastic bottle thoroughly clean and washed out will invariably do just as well), a supply of new spring washers, a set of new gaskets and a torque wrench should be collected together.

43 Crankshaft - replacement

Ensure that the crankcase is thoroughly clean and that all oil ways are clear. A thin twist drill is useful for cleaning them out. If possible blow them out with compressed air. Treat the crankshaft in the same fashion and then inject engine oil into the crankshaft oil-ways.

Commence work on rebuilding the engine by replacing the crankshaft and main bearings.

1 Never replace old main bearing shells (a false economy to do so). Fit the upper halves of the main bearing shells to their location in the crankcase, after wiping the location clean. (photo).

2 Note that on the back of each bearing is a tab which engages in locating grooves in either the crankcase or the main bearing cap housings.

3 If new bearings are being fitted, carefully wipe away all traces of protective grease with which they are coated.

4 With the five upper bearing shells securely in place, wipe the lower bearing cap housings and fit the three lower shell bearings to their caps ensuring that the right shell goes into the right cap if the old bearings are being refitted.

5 Wipe the recesses either side of the rear main bearing which locates the upper halves of the thrust washers.

6 Smear some grease onto the plain sides of the upper halves of the thrust washers and carefully place them in their recesses. (photo).

7 Generously lubricate the crankshaft journals and the upper and lower main bearing shells and carefully lower the crankshaft into position. Make sure that it is the right way round.

8 Lubricate the crankshaft journals, injecting oil into the oil ways to ensure adequate lubrication upon the initial start of the engine.

9 Fit the main bearing caps into position ensuring that they locate properly. The mating surfaces must be spotlessly clean or the caps will not seat correctly (photos).

10 When replacing the rear main bearing cap ensure that the thrust washers, generously lubricated, are fitted with their oil grooves facing outwards and the locating tab of each washer in the slot in the bearing cap.

11 Replace the main bearing cap bolts and screw them up finger tight.

12 Test the crankshaft for freedom of rotation. Should it be very stiff to turn, or possess high spots, a most careful inspection must be made, preferably by a skilled mechanic with a micrometer to trace the cause of the trouble. It is very seldom that any trouble of this nature will be experienced when fitting the crankshaft.

13 Tighten the main bearing bolt to a torque wrench setting of 59.3 ft lbs (8.2 kgm) and recheck the crankshaft for freedom of rotation. (photo).

14 Using feeler gauges check the crankshaft end float which should not exceed 0.0024 to 0.012 inch (0.06 to 0.26 mm). Oversize thrust washers 0.005 inch (0.127 mm) thick are available.

15 Using new seals and gaskets, fit the front and rear crankshaft oil seals and carriers. (photos).

16 On 124 AC000 engines fitted with a centrifugal oil filter, two special ring seals with a separating spring need to be positioned on the land just forward of the front main bearing journal, **before** the crankshaft is lowered into the engine.

The front crankshaft washer which fits between the special twin seals and the main crankshaft oil seal needs to be in place **before** the main oil seal and carrier is fitted.

17 Finally fit the timing wheel and crankshaft pulley (photo) to the front of the crankshaft and tighten centre bolt to 88 ft lbs.

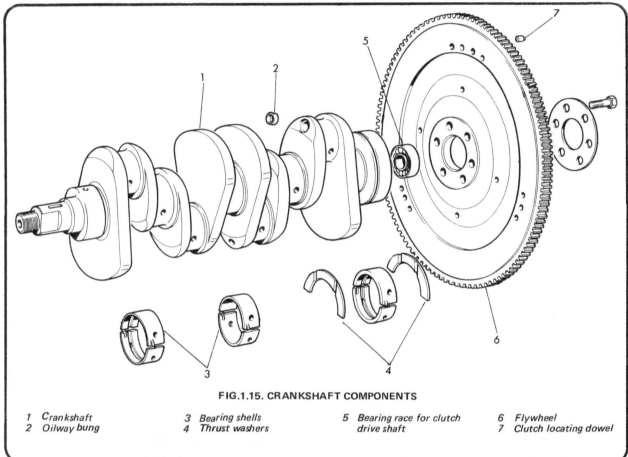

FIG.1.15. CRANKSHAFT COMPONENTS

1 Crankshaft	3 Bearing shells	5 Bearing race for clutch
2 Oilway bung	4 Thrust washers	drive shaft
		6 Flywheel
		7 Clutch locating dowel

43.1 Insert new main bearing shells

43.6 Position the thrust washers beside the rear main bearing

43.9 Fit the main bearing caps

43.9 Check main bearing cap identification

43.13 Tighten main bearing cap bolts

43.15 Fit the front crankshaft oil seal and carrier

43.15a Fit the rear crankshaft oil seal and carrier

43.17 Fit 'O' ring on crankshaft pulley boss

43.17a Fit crankshaft pulley and timing wheel

44 Piston and connecting rod - assembly

If the same pistons are being used, then they must be mated to the same connecting rod with the same gudgeon pin. If new pistons are being fitted it does not matter with which connecting rod they are used.

Because the gudgeon pin is a tight fit in the little end bearing it will be necessary to use a special tool to draw the gudgeon pin into position. Alternatively it is possible to drift the gudgeon pin into position provided it is not too tight a fit.

1   Using the tool, first remove the knurled nut and slide off the sleeve. Fit the gudgeon pin over the tool as shown in the photo.
2   Slide the spacer onto the tool and secure with the knurled nut.
3   Secure the connecting rod in the vice as shown in the photo and place the piston the correct way round on the connecting rod. This may be seen in Fig. 1.16.
4   Carefully drift the gudgeon pin into the little end until it is central in the connecting rod.
5   This photo shows the gudgeon pin in its central position.
6   The second method of fitting the gudgeon pin is to heat the piston in an oil bath and then place it on its side on soft wood blocks. Fit the connecting rod into the piston making sure it is the correct way round as shown in Fig. 1.17.
7   Using a soft metal drift, carefully drive the gudgeon pin in until it is central in the connecting rod.
8   Whichever method if used make sure that the piston is free to move axially on the connecting rod. A little stiffness is permissible when new parts have been fitted.

FIG.1.16. CONNECTING ROD/PISTON ASSEMBLY

1   Number indicating gudgeon pin diameter class
2   Letter indicating piston diameter and cylinder bore class
3   Numbers stamped on rod to show to which cylinder the rod belongs

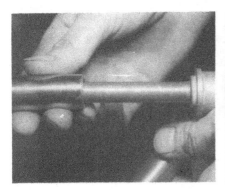

44.1 Slide the gudgeon pin on special tool

44.3 Make sure rod and piston are facing correct way

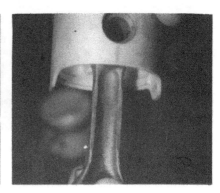

44.5 Gudgeon pin in centre of piston

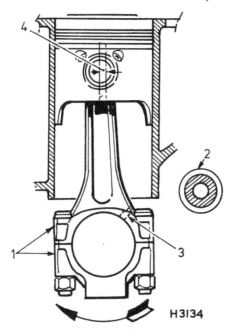

FIG.1.17. INSTALLATION OF PISTON — CONNECTING ROD ASSEMBLY

1 Location of connecting rod matching number with cylinder
2 Auxiliary shaft
3 Lubrication hole
4 Piston pin offset

45 Piston rings - replacement

1 Check that the piston ring grooves and oil ways are thoroughly clean and unblocked. Piston rings must always be fitted over the head of the piston and never from the bottom.
2 The easiest method to use when fitting rings is to wrap a 0.020 inch feeler gauge round the top of the piston and place the rings one at a time starting from the bottom oil control ring, over the feeler gauge.
3 The feeler gauge, complete with ring can then be slid down the piston over the other piston ring grooves until the groove is reached. The piston ring is then slid gently off the feeler gauge into the groove. Set all ring gaps 120° to each other.
4 An alternative method to fit the rings is by holding them slightly open with the thumb and both index fingers. This method requires a steady hand and great care as it is easy to open the ring too much and break it.
5 The two top rings are suitably marked to ensure that they are not fitted the wrong way round. The left ring should be to the top of the piston.

46 Piston- replacement

1 Lay the piston and connecting rod assembly in the correct order ready for refitting into their respective bores. There are numbers stamped on the big end bearing bosses as shown in the photo.
2 With a wad of clean non-fluffy rag wipe the cylinder bores clean.
3 Position the piston rings so that their gaps are 120° apart and then lubricate the rings.
4 Fit the piston ring compressor to the top of the piston, making sure it is tight enough to compress the piston ring.
5 Using a piece of fine wire double check that the little jet hole in the connecting rod is clean.
6 The pistons, complete with connecting rods, are fitted to their bores from above.
7 As each piston is inserted into its bore ensure that it is the correct piston - connecting rod assembly for the particular bore and that the connecting rod is the right way round, also that the front of the piston is towards the front of the bore, ie towards the front of the engine. Lubricate the piston with clean correct grade oil.
8 The piston will slide into the bore only as far as the bottom of the piston ring compressor. Gently tap the top of the piston with a wooden or plastic hammer whilst the connecting rod is guided into approximate position on the crankshaft. (photo).
9 Repeat the previous sequence for all four pistons and connecting rod assemblies.

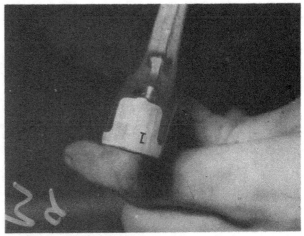

46.1 Ensure that numbers match

47 Connecting rod to crankshaft - reassembly

1   Wipe clean the connecting rod half of the big end bearing and the underside of the shell bearing. Fit the shell bearing in position with its locating tongue engaged with the corresponding groove in the connecting rod.
2   If the old bearings are nearly new and are being refitted then ensure they are replaced in their correct locations in the correct rods.
3   Generously lubricate the crank pin journals with engine oil, and turn the crankshaft so that the crank pin is in the most advantageous position for the connecting rod to be drawn into it.
4   Wipe clean the connecting rod bearing cap and back of the shell bearing and fit the shell bearing in position ensuring that the locating tongue at the back of the bearing engages the locating groove in the connecting rod cap.
5   Generously lubricate the shell bearing and offer up the connecting rod bearing cap to the connecting rod. (photo).
6   Fit the connecting rod bolts and tighen in a progressive manner to a final torque wrench setting of 38 ft lbs (5.2 kgm).

48 Refitting the auxiliary drive shaft and oil pump

1   First of all insert the oil pump and distributor drive pinion into the bush in the crankcase. It is advisable to liberally coat the bearing surfaces of the pinion gear with engine oil before fitting.
2   Carefully insert the auxiliary drive shaft into the cylinder block taking care not to scratch the bearing bores in the bushes with the gears and cam lobes machine onto the shaft.
3   With the auxiliary shaft in place, it is retained by the thrust plate which is secured by two bolts onto the cylinder block. (photo).
4   The shaft oil seal should be renewed at major overhaul of the

engine, then press out the old seal from the carrier fitting and fit a new one. Smear the carrier gasket with grease, and offer the seal carrier assembly onto the engine (pnoto).
5   Secure the shaft oil seal carrier with the five bolts and tighten them progressively and evenly. Then fit the auxiliary shaft wheel lightly tightening the centre bolt at this time (photo).
6   The oil pump assembly comprising the pump, intake funnel and filter, and drive shaft is offered up into the crankcase. Be careful not to push the drive pinion out of its bush when trying to engage the pinions internal spline with the pump shaft spline. (photo).
7   Once the pump is in place it is secured by two long bolts which pass through the pump body into the crankcase. Tighten these bolts progressively to the specified torque.

49 Sump refitting

   Before refitting the sump ensure that the following items have been assembled into and onto the crankcase, and that all the securing bolts have been tightened to the specified torques.
a) Crankshaft - end float.
b) Main bearing caps.
c) Big end caps, con-rods and piston assembly.
d) Crankshaft freedom to turn.
e) Rear crankshaft oil seal and carrier.
f) Front crankshaft oil seal and carrier.
g) Auxiliary drive shaft.
h) Distributor and oil pump drive pinion.
i) Oil pump assembly.
j) Oil return pipe.
1   Clean mating sump and crankcase surfaces of old gasket, and place new gasket - smeared with grease - onto the crankcase face.
2   Offer the sump onto the crankcase taking care not to disturb the new gasket resting on the joint face. (photo).
3   Secure the sump with the bolts and tighten progressively.

46.8 Insert pistons and con rods after compressing piston rings

47.5 Having positioned the big end bearing shells, fit the bearing caps

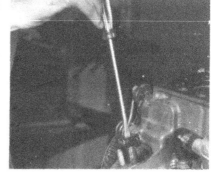
48.1 Insert oil pump and distributor drive pinion

48.3 Insert the auxiliary drive shaft and position locating plate

48.4 Fit auxiliary shaft oil seal and carrier

48.5 Fit auxiliary shaft drive wheel

48.6 Fit oil pump

49.2 Fit sump on new gasket

50.4 Fit the flywheel and torque bolts

50 Flywheel and clutch refitting

1 Stand the engine the right way up on its sump.
2 Turn the crankshaft so that No 1 and No 4 pistons are at top dead centre and clean the mating surfaces of the flywheel and crankshaft.
3 Offer up the flywheel onto the crankshaft mounting flange so that the small white dot near the periphery of the wheel is vertically above the centre of the flywheel.
4 Secure the flywheel with the six bolts which are progressively tightened to a torque of 58 ft lbs (photo).
The crankshaft may be prevented from turning by grasping the front end of the shaft with a spanner on the flat machine for the timing wheel location.
5 If the cold pressure plate assembly is being refitted then its correct position on the flywheel is given by the mark made when it was removed.
6 Refit the clutch disc and pressure plate assembly and lightly secure the position with the six bolts and spring washers.
7 If the first motion shaft is available use this to line up the clutch discs with the crankshaft spigot bearing. A suitable piece of wood dowel may be used as an alternative.
8 Firmly tighten the clutch assembly bolts in a diagonal fashion locking the crankshaft as described in paragraph 4.

51 Water pump, fuel pump and crankcase breather refitting

Water pump
1 Make sure the mating surfaces of the water pump and cylinder block are free of old gasket or jointing compound.
2 Smear a little grease onto the water pump joint face and place on a new gasket.
3 Fit the water pump to the cylinder block securing it with five bolts and spring washers. Tighten these bolts progressively to the specified torque (photo).

Fuel pump
4 Assemble the fuel pump, and spacer fitting new gaskets.
5 Fit the fuel pump onto the engine ensuring that as the pump is offered up to the cylinder block the pump actuating arm is positioned over the auxiliary shaft cam load. The pump will be damaged if the lever arm is allowed to pass beneath the cam(photo).
6 Secure the pump with the two bolts and spring washers and tighten to the specified torque.

Crankcase breather
7 Fit the rubber breather tubes to the breather housing and clean the mating faces of the housing and cylinder block.
8 Fit a new gasket onto the breather housing and position the assembled breather system onto the cylinder block immediately above the fuel pump (photo).
9 Tighten the single bolt which secures the breather housing to the block and fit the short angled breather tube to the union beside the distributor position. Tighten the pipe clips.

51.3 Fit water pump and new gasket

51.5 Fit fuel pump

51.8 Fasten crankcase breather assembly

52 Valve and spring reassembly

To refit the valves and the valve springs to the cylinder head proceed as follows:-

1 Rest the cylinder head on its side, or if the manifold studs are fitted with the gasket surface downwards.
2 Fit the valves into the same guide from which it was removed. On inlet valves slip the oil seal over the protruding stem onto the top of the guide (photos).
3 Place the lower spring seat in position. (photo).
4 Fit both inner and outer valve springs. (photo).
5 Fit the springs cap onto the spring and position the spring compression tool onto the valve assembly. (photo).
6 Compress the springs sufficiently for the cotters to be slipped into place in the cotter groove machine into the top of the valve stem. (photo).
7 Remove the valve spring compressor and repeat this procedure until all eight valves have been assembled into the cylinder head.

53 Camshafts, tappets and camshaft wheel - refitting

1 Wipe the camshaft bearing journals and cam lobes and liberally lubricate with engine oil.
2 Wipe the inside bearing bores of each camshaft housing and lubricate with engine oil.
3 Carefully insert the camshafts into the housings so that the surfaces of the bearings in the housing are not scratched by cam lobes as they pass through.
4 Fit the camshaft end plate and tighten the three securing bolts to the torque specified.
5 Refit the respective camshaft drive wheel onto the shaft projecting through the front of the housing. The wheel is located

by a dowel pin and secured by a single bolt. A tag washer is fitted beneath the bolt and this serves not only to lock the bolt when it is tightened later, but also to retain the dowel pin in its bore.
6 Generously lubricate the tappets internally and externally and insert them into the bores from which they were removed.

54 Cylinder head - reassembly

The cylinder head assembly comprises:-
1 Basic head block with valves fitted.
2 The two camshaft assemblies.
3 Camshaft housing covers.
4 Water outlet elbow and thermostat.
5 Spark plugs.
6 Water temperature transducers.
7 Exhaust manifold.
8 Inlet maifold and carburettor assembly.

Before commencing reassembly, ensure that the correct new gaskets have been obtained. Compare their shape with the old gaskets and the joint faces to which they are to be fitted. Fiat have made numerous small changes to the cylinder head assembly and therefore it is essential that you state the engine type and number when obtaining new gaskets.
1 Clean the mating faces of the head block and camshaft housing. Smear the head joint face with grease and position the new gasket.
2 Place the camshaft assembly onto the head, over the studs which locate it in the correct position.
3 Fit the spring washers and nuts and tighten progressively to the torque specified - 21 ft lbs (photo).
4 Do not fit the camshaft housing covers until the tappet cam gap has been adjusted see Section 55.
5 The water elbow and cylinder head joint faces should be

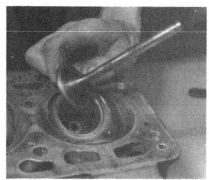
52.2 Valve ready for insertion into guide

52.2a Valve fitted and stem oil seal in position

52.3 Valve spring seats being positioned

52.4 Both valve springs being fitted

52.5 Valve spring cap being fitted

52.6 Collets in position with springs compressed

wiped next and a new gasket positioned on the head joint face (photo).

6 On the latest engines (1592 and 1755 cc) the thermostat is located beneath the elbow, and therefore on these engines the thermostat should be placed into the cylinder head outlet orifice before the elbow is fitted.

7 Secure the elbow with the four bolts with spring washers and tighten to the specified torque.

8 The spark plugs should be screwed into the cylinder head next and tightened to the specified torque. 30 ft lbs.

9 The water temperature transducers can usefully be fitted to the cylinder head at this point.

10 The exhaust manifold is fitted to the cylinder head next. As many times before, ensure that the mating surfaces are clean and free from fragments of old gaskets (photo).

11 Fit a new gasket - dry - over the studs and then offer up the exhaust manifold. Secure with the nuts which tighten to 18 ft lbs.

12 The inlet manifold may or may not be fitted complete with the carburettor assembly, but in any event use new gaskets and ensure that the mating faces are free from fragments of old gaskets. Air leaks on the inlet manifold system can cause a lot of bother! (photo).

13 The nuts which secure the inlet manifold to the cylinder head tighten to 18 ft lbs.

55 Tappet/camshaft gap setting - engine in or out of car

1 Remove the camshaft housing covers.

2 Rotate the camshaft; (by rotating the crankshaft - if this task is performed with the engine in the car) until the cam which moves the tappet to be checked is perpendicular to the tappet face plate and the valve is closed.

3 Measure the existing tappet gap and record the measurement so that the required thickness of the cap plate can be calculated.

4 If the gap is not 0.017" inlet valve or 0.019" exhaust valve -

when the engine is cold, proceed as follows.

5 Rotate the camshaft until the valve is fully open then insert FIAT Tool A60318 or alternatively a bent screwdriver to hold the tappet down (photo).

6 Rotate the camshaft again so that the cap plate may be extracted from the top of the tappet with a thin small screwdriver inserted into two grooves in the side of the tappet barrel.

7 Measure the thickness of the tappet cap plate with a micrometer. Add or subtract as appropriate the difference between the measured gap (paragraph 3) and the desired gap (paragraph 4) from the thickness of the cap plate to derive the thickness which should produce the desired gap.

8 The specification lists the range of thicknesses of tappet cap plates available from FIAT agents. There are 30 different thicknesses and the thickness is marked on one face of the plate. This face must be on the tappet barrel side when fitted, and it is well to check that the plate is the thickness marked.

9 Insert the new cap plate into the recess in the top of the tappet barrel. Turn the camshaft again so that the spacing tool or bent screwdriver (paragraphs 5) can be extracted. Turn the camshaft again to the position indicated in paragraph 2 and check that the desired tapper/cam gap has been achieved.

10 Repeat this procedure on each of the eight valve tappets. Refit the camshaft housing covers.

56 Cylinder head - refitting to cylinder block

1 It is essential to position the camshafts in the cylinder head and the crankshaft in the engine block in their correct relative positions before the head is lowered onto the engine. Once the head has been fitted it is equally important that the camshafts are not turned until they have been coupled by the timing belt to the crankshaft. There will be severe - possibly damaging mechanical interference of pistons and valves if these precautions are not taken.

54.3 Nuts retaining camshaft housing to cylinder head being tightened

54.5 Tappet adjustment completed, the water elbow fitting is assembled onto the 'head'

54.10 Exhaust manifold being fitted

54.12 Inlet manifold and carburettor assembly fitted

55.5 The tappet cap plate being removed, with the tappet barrel being held down with a bent screwdriver, and the cam pointing directly away from the tappet

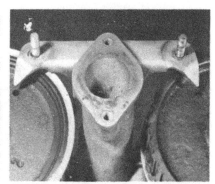

56.2 Timing holes in camshaft wheels in correct position for the cylinder head to be fitted to the engine block

2 Turn the camshafts so that the reference marks, ie the small indentations in the faces of the camshaft wheels, are adjacent to the pointers on the water elbow fitting. The cylinder head is now in the correct condition for fitment to the engine. (photo). Note that the letters cast in the faces of the wheels merely identify them, ie A = inlet, S = exhaust.

3 Turn the engine so that the No 1 and 4 pistons are at top dead centre. The engine is now in the correct condition to accept the cylinder head.

4 Place a new cylinder head gasket on the engine block, having smeared it with a little grease first. The gasket is marked as to which way up it should be.

5 Lower the cylinder head carefully onto the engine. Take great care not to knock the open valves against the engine block face, they may bend.

6 A couple of dummy studs screwed into the cylinder block will aid the task of lowering the cylinder head safely onto the block.

7 Once the head is in place the securing bolts may be screwed in and tightened to a torque of 55 ft lbs (Fig. 1.7).

8 The timing belt should be fitted and tensioned next.

57 Timing belt, wheels and tensioner reassembly

1 Ensure that the tensioner idler wheel is bolted loosely in position.

2 Fit the auxiliary shaft drive wheel onto the shaft and lightly secure with the end bolt.

3 Refer to Section 9 of this chapter to fit and tension the timing belt.

4 Once the belt is fitted the end bolts on the camshaft wheels, auxiliary shaft wheels and crankshaft wheels may be tightened to their respective specified torque.

5 Refit the timing belt cover (photo).

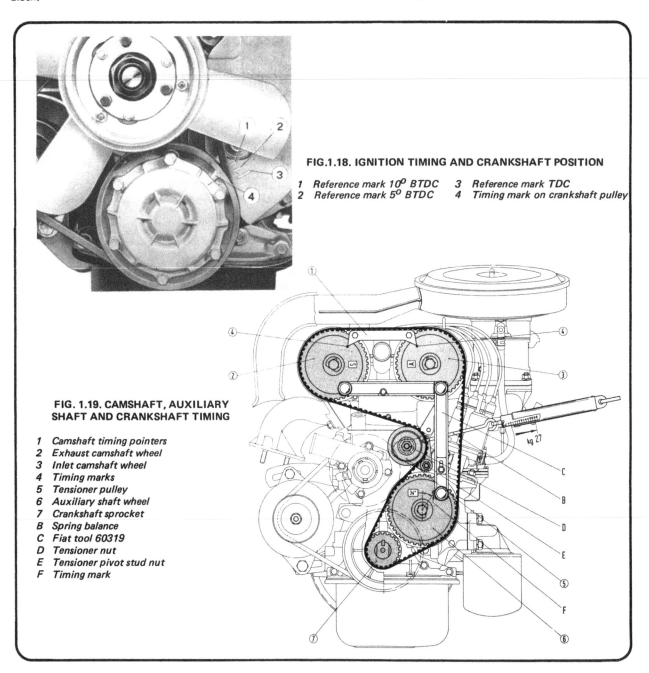

FIG.1.18. IGNITION TIMING AND CRANKSHAFT POSITION

1   Reference mark 10° BTDC   3   Reference mark TDC
2   Reference mark 5° BTDC   4   Timing mark on crankshaft pulley

FIG. 1.19. CAMSHAFT, AUXILIARY SHAFT AND CRANKSHAFT TIMING

1   Camshaft timing pointers
2   Exhaust camshaft wheel
3   Inlet camshaft wheel
4   Timing marks
5   Tensioner pulley
6   Auxiliary shaft wheel
7   Crankshaft sprocket
B   Spring balance
C   Fiat tool 60319
D   Tensioner nut
E   Tensioner pivot stud nut
F   Timing mark

58 Distributor drive - reassembly

1   The crankshaft should be turned to bring the No 1 piston to Top Dead Centre on the firing stroke. This can be identified by inspecting the position of the cam lobes operating the number 1 cylinder valves. The lobes should be pointing very nearly upwards and both valves closed. The white spot on the flywheel should be at its highest point - vertically above the centre of the flywheel.

2   The distributor should be taken in hand and the rotor arm turned towards the position occupied by the No 1 spark plug terminal.

3   Then turn the crankshaft back - anti-clockwise to the static advance required - 10⁰. This is indicated by lines on the timing belt cover. The first line anti-clockwise is 10⁰ advance.

4   With the engine rotating components in the appropriate position for the No 1 cylinder spark, the distributor is inserted into the cylinder block into engagement with the drive pinion.

5   On/125 AC000 engines the distributor is inserted into the exhaust camshaft housing into engagement with the camshaft gear wheel (photo).

6   The distributor is then lightly clamped into position, prior to a accurate adjustment of the static timing as described in Chapter 4.

59 Final assembly

1   Refit the carburettor assembly if not already completed.
2   Refit the water pump pulley.
3   Refit the alternator and fan belt (photo).
4   Refit the centrifugal oil filter on 124 AC000 engines.
5   Refit the cartridge oil filter fitting and element.

60 Engine replacement

Although the engine or engine and gearbox can be replaced by one man and a suitable hoist, it is easier if two are present. Generally replacement is the reverse sequence to removal. In addition however:-

1   Ensure all the loose leads, cables etc., are tucked out of the way. It is easy to trap one and cause much additional work after the engine is replaced.

2   Refit the following:-

a) Gearbox to engine - if split.
b) Mounting nuts bolts and washers.
c) Propeller shaft coupling.
d) Reconnect clutch actuating cable.

e) Reconnect the speedometer cable.
f) Refit gear lever extension.
g) Refit gear lever surround if disturbed.
h) Both oil pressure transducer leads.
i) Water temperature transducer leads.
j) Reconnect leads to alternator, and distributor.
k) Starter motor and electrical leads.
l) Air cleaner and breather pipes.
m) Carburettor controls.
n) Pipe connection from the inlet manifold to vacuum servo brake unit.
o) Exhaust down pipe.
p) Earth leads.
q) Radiator and water hoses.
r) Car heater hoses.
s) Battery.
t) Fuel lines to carburettor and fuel pump.
u) Bonnet and grill.

61 Engine - initial start up after overhaul or major repair

1   Make sure that the battery is fully charged and that all lubricants, coolant and fuel are replenished.

2   If the fuel system has been dismantled it will require several revolutions of the engine on the starter motor to pump the petrol up to the carburettor. An initial prime of about 1/3 of a cup full of petrol poured down the air intake of the carburettor will help the engine to fire quickly, thus relieving the load on the battery. Do not overdo this, however, as flooding may result.

3   As soon as the engine fires and runs keep it going at a fast tick over only, (no faster) and bring it up to the normal working temperature.

4   As the engine warms up there will be odd smells and some smoke from parts getting hot and burning off oil deposits. The signs to look for are leaks of water or oil which will be obvious if serious. Check also the exhaust pipe and manifold connections, as these do not always 'find' their exact gas tight position until the warmth and vibration have acted on them, and it is almost certain that they will need tightening further. This should be done, of course, with the engine stopped.

5   When normal running temperature has been reached adjust the engine idling speed as described in Chapter 3.

6   Stop the engine and wait a few minutes to see if any lubricant or coolant is dripping out when the engine is stationary.

7   Road test the car to check that the timing is correct and that the engine is giving the necessary smoothness and power. Do not race the engine - if new bearings and/or pistons have been fitted it should be treated as a new engine and run in at a reduced speed for the first 300 miles (500 km).

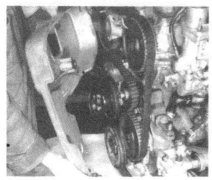
57.5 Fit timing belt cover

58.5 Fasten distributor blank 125BC000 engines only

59.2 Fit alternator and water pump pulley

62 FAULT DIAGNOSIS

Symptom	Reason/s	Remedy
Engine will not turn over when starter switch is operated	Flat battery Bad battery connections Bad connections at solenoid switch and/or starter motor	Check that battery is fully charged and that all connections are clean and tight.
	Starter motor jammed	Rock car back and forth with a gear engaged. If ineffective remove starter (not automatic).
	Defective solenoid Starter motor defective	Remove and check solenoid. Remove starter and overhaul.
Engine turns over normally but fails to fire and run	No spark at plugs	Check ignition system according to procedures given in Chapter 4.
	No fuel reaching engine	Check fuel system according to procedures given in Chapter 3.
	Too much fuel reaching the engine (flooding)	Check fuel system if necessary as described in Chapter 3.
Engine starts but runs unevenly and misfires	Ignition and/or fuel system faults	Check the ignition and fuel systems as though the engine had failed to start.
	Incorrect valve clearances Burnt out valves	Check and reset clearances. Remove cylinder head and examine and overhaul as necessary.
Lack of power	Ignition and/or fuel system faults	Check the ignition and fuel systems for correct ignition timing and carburettor settings.
	Incorrect valve clearances Burnt out valves	Check and reset the clearances. Remove cylinder head and examine and overhaul as necessary.
	Worn out piston or cylinder bores	Remove cylinder head and examine pistons and cylinder bores. Overhaul as necessary.
Excessive oil consumption	Oil leaks from crankshaft oil seal, rocker cover gasket, drain plug gasket, sump plug washer	Identify source of leak and repair as appropriate.
	Worn piston rings or cylinder bores resulting in oil being burnt by engine Smoky exhaust is an indication	Fit new rings or rebore cylinders and fit new pistons, depending on degree of wear.
	Worn valve guides and/or defective valve stem seals	Remove cylinder head and recondition valve guides and valves and seals as necessary.
Excessive mechanical noise from engine	Wrong valve to rocker clearances Worn crankshaft bearings Worn cylinders (piston slap)	Adjust valve clearances. Inspect and overhaul where necessary.
Unusual vibration	Misfiring on one or more cylinders Loose mounting bolts	Check ignition system. Check tightness of bolts and condition of flexible mountings.

NOTE: When investigating starting and uneven running faults do not be tempted into snap diagnosis. Start from the beginning of the check procedure and follow it through. It will take less time in the long run. Poor performance from an engine in terms of power and economy is not normally diagnosed quickly. In any event the ignition and fuel systems must be checked first before assuming any further investigation needs to be made.

Chapter 2 Cooling system

Contents

Specifications

Type of system	Pressurized with centrifugal circulation pump, electric cooling fan, wax thermostat, a sealed system
Capacity	1.6 Imp. galls. including heater and expansion chamber
Expansion chamber	Plastic translucent bottle

Thermostat

Opening temperature	87° C, 189° F
Travel	0.3 inches approx.

Radiator ... ...   Vertical tube grilled layout

    Electromagnetic fan drive gap ... ...   0.010 to 0.014 inches

Water pump

    Heater ... ...   See Chapter 12

Torque wrench settings: **ft lbs**

Fan and water pump drive pulley nuts	88
Water pump retaining bolts	17
Cylinder head water outlet elbow bolts	8

1 General description

The engine cooling water is circulated by a thermo-syphon, water pump assisted system, and the coolant is pressurised. This is primarily to prevent premature boiling in adverse conditions and to allow the engine to operate at its most efficient running temperature; this being just under the boiling point of water. The overflow pipe from the radiator is connected to an expansion chamber which makes topping up virtually unnecessary. The coolant expands when hot, and instead of being forced down an overflow pipe and lost, it flows into the expansion Chamber. As the engine cools the coolant contracts and because of the pressure differential flows back into the radiator.

The cap on the expansion chamber is set to a pressure of 15 lb ft (1.05 kg m) which increases the boiling point of the coolant to 230°F. If the water temperature exceeds this figure and the water boils, the pressure in the system forces the internal valve of the cap off its seat thus exposing the expansion tank overflow pipe, down which the steam from the boiling water escapes and so relieves the pressure. It is therefore important to check that the expansion chamber cap is in good condition and that the spring behind the sealing washers has not weakened. Check that the rubber seal has not perished and its seating in the neck is clean, to ensure a good seal. A special tool which enables a cap

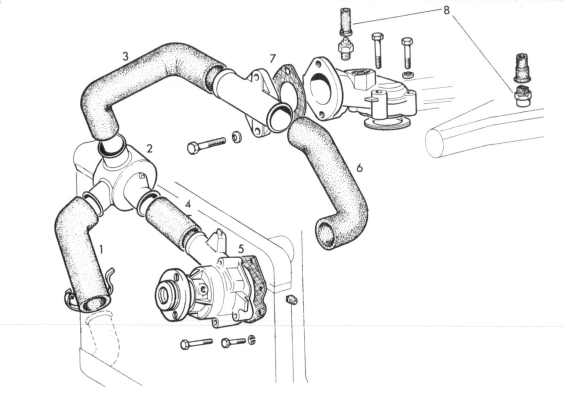

FIG.2.1. THE COOLING SYSTEM HOSES AND COMPONENTS

1	Lower radiator hose to the thermostat	3	Cylinder head by-pass hose
2	Two-way thermostat assy.	4	Thermostat to water pump hose
		5	Water pump
		6	The cylinder head water outlet to radiator hose
		7	Cylinder head water outlet elbow components
		8	Water temperature sender units

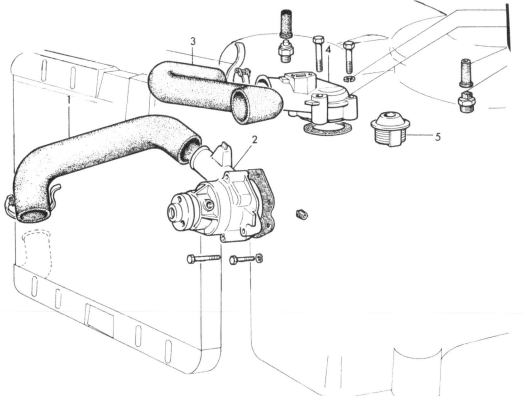

FIG.2.2. EXPLODED VIEW OF LATER COOLING SYSTEM

1	Lower radiator hose	3	Top radiator hose	4	Cylinder head water outlet elbow	5	Thermostat
2	Water pump						

to be pressure tested is available at some garages.

The cooling system comprises the radiator, top and bottom hoses, heater hoses, the impeller water pump (mounted on the front of the engine it carries the fan blades and is driven by the fan belt) and, the thermostat.

The system functions as follows on 124 AC000 1438 cc and 125 BC000 1608 cc engines: the water flow round the system is controlled by a special double valve type thermostat, located in the junction of the lower radiator, bi-pass and water pump hoses. This valve governs the flow of water from the cylinder head bi-pass hose and the lower radiator hose into the water pump. If the engine is cool the pump forces the water through the cylinder head and into the bi-pass hose to the thermostat. As the engine temperature reaches its proper running level, the thermostat bellows move to close the bi-pass hose and open the lower radiator hose. The hot water from the cylinder head will then flow down through the radiator up through the lower radiator hose and the thermostat and upwards to the pump.

The cooling system is slightly different in the latest 132 series engines 1598 cc and 1755 cc. The thermostat is a single valve type and is positioned in the more usual place of beneath the cylinder head water outlet elbow. The mode of operation is quite conventional, once the temperature of the water in the cylinder head has reached the proper running level the thermostat valve opens to allow it to flow into the radiator and back to the pump via the lower radiator hose.

The method of supplying hot water to the car interior heating system is common to all engines; a union in the top of the cylinder head enables hot water to flow to the heater and it is returned to a cooling system at the water pump inlet.

2 Cooling system - draining

With the car on level ground drain the system as follows:
1 With the cooling system cold unscrew and remove the radiator filler plug. DO NOT REMOVE THE PLUG WHILST THE ENGINE IS HOT.
2 If anti-freeze is being used in the cooling system it should be collected in a bowl located under the bottom hose. Do not carry out the instructions in paragraph 5.
3 Turn the car heater control to 'HOT', to ensure that the heater system will drain and fill without air locks forming.
4 Release the expanse tank pressure cap.
5 Undo and remove the cylinder block drain plug. This is located to the rear of the left hand side of the cylinder block.
6 Slacken the hose clip and carefully ease the bottom hose at its connection on the radiator.
7 When the coolant has finished running out of the cylinder block drain hole, probe the orifice with a short piece of wire to dislodge any particles of rust or sediment which may be causing a blockage preventing complete draining.

3 Cooling system - flushing

1 Generally even with proper use, the cooling system will gradually lose its efficiency as the radiator becomes choked with rust scale, deposits from the water and other sediment. To clean the system out, remove the radiator filler plug, cylinder block plug and bottom hose and leave a hose running in the radiator filler plug hole for fifteen minutes.
2 Reconnect the bottom hose, refit the cylinder block plug and refill the cooling system as described in Section 4, adding a propreitary cleaning compound. Run the engine for fifteen minutes. All sediment and sludge should now have been loosened and may be removed by then draining the system and refilling again.
3 In very bad cases the radiator should be reverse flushed. This can be done with the radiator in position the cylinder block plug left in position and a running hose placed in the bottom hose union orifice allow water out of the filler plug hole.
4 The hose is then removed and placed in the filler plug hole and the radiator washed out in the usual manner.

4 Cooling system - filling

1 Fit the cylinder block drain plug and if the bottom hose has been removed it should be reconnected.
2 Fill the system slowly to ensure that no air locks develop. Check that the valve to the heater unit is open, otherwise an air lock may form in the heater. The best type of water to use in the cooling system is rain water.
3 Fill up the radiator to the level of the filler plug and top up the level of coolant in the expansion tank to the level indicated.
4 Run the engine at a fast idle until the thermostat opens (coolant level drops sharply), then top up the radiator again.
5 Add coolant to the expansion tank to bring it up to the required level.
6 Stop the engine, then loosen the heater hose clip on the water delivery line (see Fig. 12.4, Chapter 12).
7 Run the engine again, checking that coolant is present at the loosened connection. Stop the engine and tighten the hose clip.
8 Loosen the heater hose clip on the water return line (see Fig. 12.4).
9 Briefly run the engine again, allowing air to bleed from the loosened connection. When water flows from the connection without any air being present, tighten the hose clip. Take care that the hot water does not cause scalding; protect the hands, if necessary, with an absorbent cloth. (At this point, the temperature of the delivery and return lines should be approximately the same, and will probably be too hot to hold comfortably.) Stop the engine.
10 Allow the engine to cool, then top up the radiator and expansion tank to the required levels. Refit the radiator filler plug and expansion tank cap.
11 Run the engine again checking for leaks from any hose connections which have been disturbed.

5 Radiator - removal and refitting

1 Drain the cooling system.
2 Slacken the clip securing the expansion tank hose to the radiator. Carefully ease the hose from the union pipe on the radiator.
3 Slacken the clips securing the radiator top and bottom hoses to the radiator pipes and carefully ease the two from them.
4 Undo and remove the two nuts and bolts which secure the radiator to the radiator aperture in the body shell. (photo).
5 On models where electric fans are fitted, the electric leads to the motor should be disconnected.
6 The radiator may now be lifted up from its lower mountings and away from the front of the car.
7 Take care not to loose the spacers, washers and mounting rubbers as the radiator is removed.
8 Refitting the radiator is the reverse seuqence to removal. Refill the cooling system as described in Section 4. Carefully check to ensure that all hose joints are water tight.

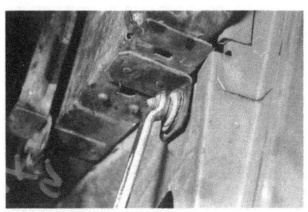

5.4 Removing radiator securing nuts

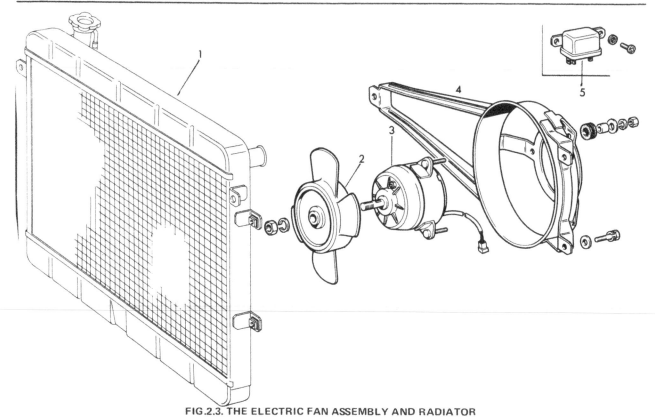

FIG.2.3. THE ELECTRIC FAN ASSEMBLY AND RADIATOR

1   Radiator   3   Electric motor   4   Motor and fan support frame   5   Motor circuit relay
2   Nylon fan

6   Radiator - inspection and cleaning

1   With the radiator out of the car, any leaks can be soldered up
or repaired with a compound such as Cataloy. Clean out the
inside of the radiator by flushing as described in Section 3.
2   When the radiator is out of the car, it is advanatageous to
turn it upside down for reverse flushing. Clean the exterior of
the radiator by hosing doen the radiator matrix with a strong jet
of water to clean away road dirt, dead flies etc.
3   Inspect the radiator hose for cracks, internal and external
perishing and damage caused by overtightening of the hose clips.
Replace the hoses as necessary. Examine the radiator hose clips
and renew them if they are rusted or distorted. The drain plugs
and washers should be renewed if leaking.

7   Thermostat - removal, testing and replacement

1   To remove the thermostat drain the cooling system as
described in Section 2 of this chapter. On later models with the
132 series engine it will only be necessary to partially drain the
cooling system to permit the thermostat below the cylinder head
water outlet elbow to be removed.
2   With sufficient water drained from the cooling system on the
1438 cc and 1608 cc engines; the clips securing the water hoses
to the thermostat housing are loosened and the hoses eased off
the housing.
3   The procedure for removing the thermostat on 132 series
engines is different. Once the cooling system has been partially
drained, the clip securing the top hose to the cylinder water
outlet elbow is slackened and the hose eased off the elbow.
4   The bolts securing the spark plug HT leads harness bracket
are removed and the bracket lifted aside. Next the bolts which
secure the water outlet elbow to the cylinder head are removed.
The elbow fitting can then be carefully lifted clear of the timing
belt cover and cylinder head, exposing the thermostat in the

cylinder head which can then be simply lifted out. (photo)
5   Test the thermostat for correct functioning by suspending it
together with a thermometer on a string in a container of cold
water. Heat the water and note the temperature at which the
thermostat begins to open. The opening temperature should be
$87^oC \pm 2^oC$ and the minimum travel when the water has reached
100^oC should be 7.5 mm, 0.295 inches. The maximum travel at
100^oC should be 11.0 mm, 0.433 inches.
    Discard the thermostat if it opens too early, or does nor close
when the water cools down. If the thermostat is stuck open
when cold it will be apparent when it is first exposed.
    Refitting the thermostat is the reverse procedure to removal.
Always ensure that the cylinder head and water elbow fitting
faces are clean and flat. If the thermostat elbow is badly
corroded and eaten away, fit a new elbow. A new gasket must
always be used. Thermostats should last for two to three years at
least between renewal.

7.4 The cylinder head water outlet elbow

8 Water pump - removal and refitting

1 For safety reasons disconnect the battery.
2 Refer to Section 5 and remove the radiator.
3 Slacken the top fixing bolt on the alternator so that it can be moved toward the engine and the fan belt loosened. The fan belt may now be removed.
4 On models fitted with electro-magnetic fan drive, the pulley and fan hub may be removed from the water pump shaft after the centre nut is undone. Do not loose the key which locates the fan hub assembly on the pump shaft.
5 On models fitted with an electically driven fan only the pulley needs to be removed to gain space to undo the bolts retaining the water pump on the cylinder block. The pulley is secured to the hub fitted by three bolts.
6 If you are fortunate, you may have a spanner that can reach the water pump body bolts with the pump pulley still attached; and in that case one should only remove the pulley assembly if absolutely necessary.
7 The two bolts securing the car heater return pipe to the pump inlet pipe should be undone and the heater pipe moved aside.
8 Finally the pump bolts themselves are removed and the pump lifted from the cylinder block. (photo)
9 Refitting is the reverse sequence to removal, but the following additional points should be noted:
10 Clean the mating faces of the pump and cylinder block free of all fragments of old gasket; to ensure a good water tight joint.
11 Always use a new gasket.
12 Adjust fan belt tension as described in Section 11.

9 Water pump - dismantling and overhaul

1 If the water pump starts to leak, shows excessive movement of the spindle, or is noisy during operation, the pump may be dismantled and overhauled. Make sure before you begin that the suspected faulty parts are readily available as spares - it is general policy nowaday not to hold pump parts but to supply the whole

8.8 The water pump away from the cylinder block

unit ready to fit on the engine.
2 Assuming that you have procured the spare parts necessary to renovate the pump, it may be dismantled as follows:
3 Hold the pump body lightly in a vice and with either a conventional puller or the special FIAT tool A40026 withdraw the pump impellor from the shaft.
4 Next the set screw which locates the bearing in the pump body is removed, to allow the shaft together with the pulley hub and bearings to be gently driven out.
5 The water seal may be tapped out of the pump body with an ordinary drift.
6 Again using a hub puller the pulley hub may be removed from the shaft if necessary.
7 Clean all the components and inspect them thoroughly for signs of wear or damage. Replace parts as necessary.
8 To assemble the pump first carefully insert the water seal into the pump body. Do not use excessive force, and be careful not to damage the carbon seal block and face. The seal block is fragile and if scratched or damaged will allow water to escape past the mating face of the pump impeller into the bearing assembly in the pump.
9 Refit the pulley hub onto the pump shaft which is complete with the bearing assembly; the shaft is now ready to be inserted into the pump body. When offering up the shaft assembly to the body ensure that the hole in the bearing outer race is aligned to accept the set screw which locates the shaft assembly in position in the pump. Carefully drive the shaft assembly into the pump and screw in the set screw.
10 Finally press the impellor onto the shaft until the free running clearance between the impellor vanes and the pump body is 0.040 in (1 mm).
11 The pump is now ready to be fitted to the engine see Section 8 of this Chapter.

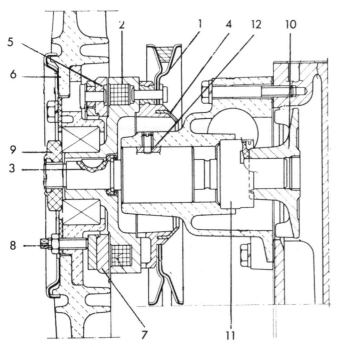

FIG.2.4. THE SECTION OF ELECTRO-MAGNETIC FAN DRIVE SYSTEM AND WATER PUMP

1 Water pump drive pulley
2 Pump pulley hub and electro-magnetic coil
3 Pump shaft
4 Contact ring
5 Electro-magnetic response ring, which drives fan when electricity if flowing through the coil in the pulley hub
6 Fan hub
7 Electro-magnetic fan drive gap
8 Fan drive ring position adjusting screws
9 Fan drive assembly securing nut
10 Water pump impellor
11 Water pump seal
12 Water pump bearing and locating screw

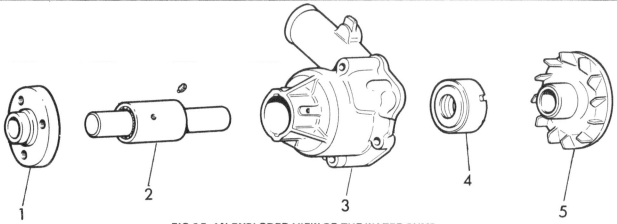

FIG.2.5. AN EXPLODED VIEW OF THE WATER PUMP

1 Pump drive fitting, the   2 Pump shaft and bearing   3 Pump housing   5 Centrifugal impellor
   hub of the pulley   assembly   4 Pump seal

10 Fan belt - removal and replacement

If the fan belt is worn or has over stretched it should be renewed. The most usual reason for replacement is that the belt has broken. It is therefore recommended that a spare belt is always carried in the car. Replacement is a reversal of the removal procedure, but if replacement is due to breakage:

1 Loosen the alternator pivot and slotted link bolts and move the alternator towards the engine.
2 Carefully fit the belt over the crankshaft, water pump and alternator pulleys.
3 Adjust the belt as described in Section 11 and tighten the alternator mounting bolts, NOTE: after fitting a new belt it will require adjustment 250 miles (400 km) later.

11 Fan belt- adjustment

It is important to keep the fan belt correctly adjusted and should be checked every 6,000 miles (9,600 km) or 6 months. If the belt is loose it will slip, wear rapidly and cause the alternator and water pump to malfunction. If the belt is too tight the alternator and water pump bearings will wear rapidly and cause premature failure.

The fan belt tension is correct when there is 0.5 in (13 mm) of lateral movement at the mid point position between the alternator pulley and the camshaft pulley.

To adjust the fan belt, slacken the securing bolts and move the alternator in or out until the correct tension is obtained. It is easier if the alternator bolts are only slackened a little so it requires some effort to move the unit. In this way the tension of the belt can be arrived at more quickly than by making frequent adjustments. If difficulty is experienced in moving the unit away from the engine a tyre lever placed behind the unit and resting against the block gives a good control so that it can be held in position whilst the securing bolts are tightened. Be careful of the alternator cover - it is fragile.

12 Expansion tank

The coolant expansion tank is mounted on the left hand side of the rear bulkhead of the engine compartment, and does not require any maintenance. It is important that the expansion tank filler cap is not removed whilst the engine is hot.

Should it be found necessary to remove the expansion tank disconnect the radiator to expansion tank hose from the radiator union.

Remove the bracket screws and carefully lift away the tank and its hose.

Refitting is the reverse sequence to removal. Add either water or anti-freeze solution until it is up to the level mark.

13 Temperature transducers, switches and gauges

1 There are several electrical systems associated with the cooling system.
a) Thermal switch operating the electro-magentic clutch to the fan, or the electric fan motor on some models.
b) Thermal sender unit giving a continuous signal to the temperature gauge.
c) Thermal sender unit that indicates high temperature and operates the temperature warning light.
2 The sender units for the gauge and warning light are screwed into the top of the cylinder head - between the spark plugs. The forward sender unit operates the warning light.
3 The thermal switch that operates the fan drive is screwed into the bottom of the right hand side of the radiator. (photo)
4 If unsatisfactory gauge readings are being obtained or the temperature warning system is suspected, the respective thermal sender units may be tested by removing the cable connection on the appropriate unit and placing the metal cable end on a good earthing point, for example a cleaned point on the bare metal cylinder head.
5 Switch on the ignition. If the temperature gauge sender unit is being tested and the gauge needle quickly moves to the 'hot' sector of the dial, a new temperature sender should be fitted. Alternatively should the temperature gauge not respond, check the continuity of the wire to the gauge, and finally if necessary check the gauge itself by substituting another.
6 The test procedure is similar for the warning system. Having directly bi-passed the sender unit and the light glows immediately the ignition is switched on, the sender unit should be renewed. If the warning light fails to glow then the continuity of the wire and bulb should be checked.
7 The thermal device which operates the electro-magnetic fan clutch, or on later models the electric fan motor, is simply a switch. To test the operation of this switch remove the two wires connecting to it and bring the metal ends together. It will be necessary to tape them together if a electro-magnetic clutch fan drive is fitted to the car.

Switch on the ignition and start the engine. If the fan operates when the engine is running (or when the ignition is switched on in the case of an electric motor driven fan) then the thermal switch should be renewed.

Should the fan not operate when the thermal switch has been shorted and the engine is running then check the continuity of the cable connections to the fan drive and finally check the electro-magnetic clutch or the motor as described in Sections 14 and 15 respectively of this Chapter.
8 Before removing any of the thermal devices from the engine or radiator drain the cooling system as described in Section 2.
9 Once new devices have been fitted the cooling system may be refilled as described in Section 4 of this Chapter.

13.3 The wire connections to the fan operating thermal switch at the base of the radiator

14 Electro-magnetic clutch driven fan

1 This device is mounted on the water pump pulley hub and provides for the fan to be motored only when current is supplied from the thermal switch in the bottom of the radiator. It is not an electric motor, but an electrically generated magnetic field which links the pump pulley to the fan so that the fan is rotated by the pulley. When there is not any current flowing through the clutch windings in the pulley hub, no magnetic field is formed to couple fan to the pulley. The fan will then merely feather around in the draught from the radiator.

2 The checks necessary on a suspect electro-magnetic clutch do not necessitate its removal from the engine.

Check

a) Examine the condition of the brush which is on the pump and contact ring on the water pump side of the pulley. Replace the brush if worn to a stub and clean the contact ring with fine emery and a dry cloth.

b) There should be a gap of 0.010 in (0.25 mm) between the pulley hub and electric winding body and the response face of the fan hub. See paragraph 3 for adjustment procedure.

c) Finally the electric windings of the clutch may be checked for a short or an open circuit. Should zero or infinite resistance be measured across the contact ring and pump body the clutch unit should be renewed.

3 The air gap between the clutch electrical winding body and the response plate on the fan hub is adjusted as follows.

a) There are three square headed adjusting bolts which found on the front face of the fan hub. The bolts have locking nuts fitted.

b) Unscrew the lock nuts and turn the adjusting bolts to move the response plate on the pump side of the fan hub so that the air gap is between 0.010 and 0.013 in. Insert the feeler gauges opposite the bolts that have been adjusted.

c) Once the adjustment has been made the adjusting bolts are locked by tightening the locknuts.
locked by tightening the locknuts.

4 In the event that the clutch is found faulty it may be removed from the pump as follows.

5 Remove the radiator as described in Section 5 of this chapter and then proceed to undo the central nut on the pump shaft that retains the fan clutch assembly on that shaft. Lift aside the brush in contact with the clutch.

6 The fan hub and clutch assembly may then be pulled off the shaft. Do not lose the key which locates the fan on the shaft.

7 The refitting of the fan and clutch assembly is the reversal of the removal procedure.

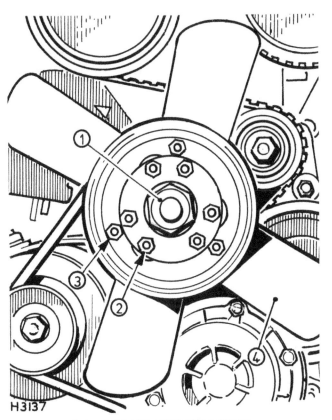

H3137

FIG.2.6. FRONT VIEW OF ELECTRO-MAGNETIC FAN DRIVE

1   Fan hub assembly securing nut
2   Electro-magnetic drive gap adjusting screws
3   Screws fixing fan to hub
4   Fan

15 Electric motor fan

1 The motor and fan is mounted on a frame on the rear face of the radiator.

2 To remove the fan and motor begin by removing the radiator as described in Section 5 of this chapter.

3 If the motor does not respond to being supplied directly from the battery the jump leads, it must be renewed.

4 The fan motor support frame is detached from the radiator after undoing the three nuts and bolts securing it to the radiator.

5 The fan motor is then finally removed from the frame, after unscrewing the three nuts which locate the motor.

6 Refitting the fan motor assembly follows the reversal of the removal procedure.

16 Anti-freeze

1 It is wise to have anti-freeze in the coolant all the year round. Castrol Anti-freeze or other glycol base anti-freeze which will not damage the aluminium cylinder head contain additives which ensure that the water ways through the engine and radiator remain free from corrosion.

2 Mixtures with anti-freeze such as that sold by Castrol may be left in cooling system for up to two years, but the strength of the mixture should be maintained.

3 Listed below are the amounts of anti-freeze which should be added to ensure adequate protection down to the temperature given.

33% solution	−19°C
50% solution	−36°C

17 Cooling system - FAULT DIAGNOSIS

Symptom	Reason/s	Remedy
Loss of coolant	Leak in system	Examine all hoses, hose connections, drain taps and the radiator and heater for signs of leakage when the engine is cold, then when hot and under pressure. Tighten clips, renew hoses and repair radiator as necessary.
	Defective radiator pressure seal	Examine cap for defective seal and renew if necessary.
	Overheating causing too rapid evaporation due to excessive pressure in system	Check reasons for overheating.
	Blown cylinder head gasket causing excess pressure in cooling system forcing coolant out	Remove cylinder head for examination.
	Cracked block or head due to freezing	Strip engine and examine. Repair as required.
Overheating	Insufficient coolant in system	Top up.
	Water pump not turning properly due to slack fan belt	Tighten fan belt.
	Kinked or collapsed water hoses causing restriction to circulation of coolant	Renew hose as required.
	Faulty thermostat (not opening properly)	Fit new thermostat.
	Engine out of tune	Check ignition setting and carburettor adjustments.
	Blocked radiator either internally or externally	Flush out cooling system and clean out cooling fins.
	Cylinder head gaskets blown forcing coolant out of system	Remove head and renew gasket.
	New engine not run-in	Adjust engine speed until run-in.
Engine running too cool	Missing or faulty thermostat	Fit new thermostat.

Chapter 3 Fuel system and carburation

Contents

Specifications

Fuel pump

    Make and type   ...   .   B.C.D.

Carburettors:

On 1438 cc 124 ACOOO engines (from engine No. 43458)

Dimensions in inches

	Weber 34 DFH		Weber 34 DHS/1	
	Primary barrel	Secondary barrel	Primary barrel	Secondary barrel
Venturi diameter	0.944	1.023	0.944	1.023
Main jet	0.049	0.047	0.049	0.047
Main air jet	0.070	0.059	0.070	0.067
Idling jet	0.017	0.023	0.017	0.023
Idling air jet	0.047	0.027	0.047	0.027
Float level from cover face to float	0.236		0.236	
Float travel	0.335		0.335	
Pump jet	0.015		0.016	
Extra fuel device:				
Mixture jet	0.043	0.075	0.043	0.075
Air jet	0.051	0.049	0.051	0.049
Fuel jet	0.943	0.075	0.043	0.075

On 1608 cc 125 BCOOO engines

	Weber 401DF	Solex C40 P116
	Each barrel	Each barrel
Venturi diameter	1.260	1.260
Auxiliary venturi	0.177	
Main jet	0.047	0.055
Idling jet	0.021	0.021
Main air jet	0.083	0.067

Idling air jet	0.045
Pump jet	0.015
Float level - cover to float	0.394

Engine 132 series

	1592 cc		1756 cc	
Weber	Prim. Sec. 32DHSA or Solex C34EIES4			
Venturi diameter	0.945	1.023	0.945	1.063
Main jet	0.049	0.061	0.049	0.059
Main air jet	0.071	0.071	0.059	0.055
Idling jet	0.019	0.027	0.018	0.031
Idling air jet	0.041	0.027	0.055	0.043
Float level	0.275		A95134	
Emulsion tube	F61	F61	0.138	0.138
Fuel tank capacity	10 Imperial gallons			

1 General description

The fuel system and carburation on this series of FIAT comprises a fuel tank positioned in the rear of the vehicle, a fuel pump on the right side of the cylinder block, driven by the auxiliary drive shaft, and finally the carburettor(s). The system also has several embellishments including an oil and petrol vapour recirculation system, air cleaner and electrical petrol quantity measuring system.

The fuel system components are conventional, and the WEBER (or Solex) carburettor is the usual fixed choke design with many additional features to ensure optimum fuel air mixture at all engine speeds and loads. Section 8 of this chapter describes in detail the way in which the WEBER carburettor works.

2 Air filter elements - removal and renewal

Although the configuration of the air filter is different with various models of the FIAT 124 Sports Spider the procedure for extracting and replacing the cleaner element is basically common.

1 Three or four nuts depending on the pattern of cleaner housing secure the top cover in position. Once these nuts have been removed, together with their washers the top cover can be lifted off.

2 The cleaner element is now exposed and can simply be lifted out of the housing. Wipe over the interior and fit the new element which should seat snugly over the locating ridges in the bottom of the housing.

3 When refitting cover ensure that the sealing ring is intact and in place.

4 On some patterns of cleaner the intake pipe may be positioned for 'Summer' or 'Winter' conditions by turning the pipe so that the open end faces forward to collect cool air for 'Summer' driving or downward so that it is immediately above the exhaust manifold which will warm the air for winter conditions.

3 Carburettor - description and principles

1 All the carburettors fitted to the FIAT 124 Sport Spider are of 'fixed choke' design and are manufactured by WEBER or in some instances SOLEX. The term 'fixed choke' relates to the throat (venturi) into which the main petrol jet sprays the fuel, and because on these carburettors this venturi is a fixed size, several other petrol jets are incorporated in the carburettor to enrich the fuel/air mixture passing into the engine as and when necessary. All fuel jets take the form of inserts screwed into the carburettor body.

2 The carburettors function as follows: Petrol is pumped into the float chamber and is regulated by the needle valve actuated by the float. As air is sucked into the engine a slight vacuum - proportional to air flow - is created in the venturi of the carburettor; this vacuum draws a corresponding amount of fuel from the float chamber through the emulsion tube where it mixes with the small amount of air coming from the air correction jet. The fuel air emulsion passes on through the main jet and into the main air stream in the venturi of the carburettor.

3 The flow of air through the carburettor is controlled by the butterfly valve operated by the accelerator linkage. The engine develops power in proportion to the amount of air and fuel drawn into the engine, and the economy is dependant on the relative proportions of fuel and air taken into the engine. On these fixed choke carburettors the relative proportions of the fuel and air are controlled by the sizes of the main jet, carburettor throat - 'venturi', and the other minor enrichment jets. All these sizes are fixed and decided by the engine designers. The only adjustment or trim to the fuel/air mixture available is provided by the accelerator's butterfly valve stop and the 'idling mixture control screw'. Section 5 sets down the carburettor adjustment procedure.

4 As mentioned earlier the fixed jet carburettors require additional fuel jets to enrich the air/fuel mixture when necessary. When power is required for acceleration the engine needs an enriched mixture of fuel and air. An acceleration jet is

provided in the throat of the carburettor and it is supplied with fuel under pressure from the accelerator pump. The pump is actuated by levers connected to the accelerator linkage. There is also compensating jet which introduces additional fuel into the throat of the carburettor to create the slightly rich mixture necessary to sustain the engine at high speeds.

5   In addition to the accelerator jet and compensating jets, there is the idling jet. The flow of fuel through this jet is adjustable. When the engine is idling and the accelerator butterfly valve almost closed, the slight vacuum in the inlet manifold sucks air and fuel via the pilot jet through the idling jet orifice. The three minor jets therefore provide the mixture enrichment necessary when accelerating, maintaining high speed, and engine idling.

6   Arrangements for cold starting consist of a choke flap above the main jet and venturi. To start the engine when cold, appreciably more fuel is required to overcome the losses due to condensation of the fuel vapour in the inlet maifold ducts. By closing the choke valve an increased vacuum is created in the carburettor barrel and this sucks a greater amount of fuel out of

the main jet system. The choke flap is mounted eccentrically on the support spindle so that when the engine demands more air the flap will partially open itself against a control spring to admit the air into the carburettor and engine. The SOLEX carburettors have a valve which admits fuel directly into the barrel for cold starting.

7   A feature of the carburettors of the 124 AC000 engine and the 132 series engine in the latest '1600 and '1800 cars is that they are what is normally described as 'Compound Carburettors'. In these carburettors the accelerator linkage does not open the butterfly valves in the two throats simultaneously. The principle employed here is that only one barrel is used when air requirements are low and the second barrel is opened only when extra power and air is required. This arrangement gives a smooth running engine at low speed whilst maintaining efficiency at high engine speeds. The FIAT Sport 1600, 1608 cc 125 BC000 engine is fitted with two twin choke WEBER down-draught carburettors - this arrangement provides that each carburettor barrel supplies a single cylinder.

Fig.3.1. The WEBER 40 IDF 10/11 carburettor installation on the 1608 cc 125BCOOO engine. (Solex C40 P116 very similar).

1 Carburettor fuel feed line
2 Twin trumpets to twin barrels in carburettor
3 Idling speed control screw
4 Screw which adjusts the relative position of the accelerator valve springs and therefore the balance of the two carburettors
5 Air correction adjusting screw
6 Idling mixture control screw
7 Vacuum monitoring
   hole plug
8 Fuel pump
9 Accelerator linkage to carburettors
10 Accelerator pump housing

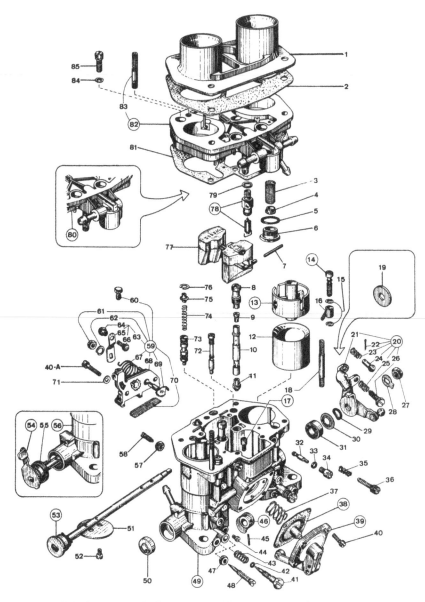

FIG. 3.1A. EXPLODED VIEW OF WEBER 40IDF 10/11 CARBURETTOR

1 Air horn
2 Gasket for 1
3 Filter
4 Plug for 3
5 Gasket for 4
6 Filter inspection plug
7 Float fulcrum trim
8 Emulsion jet carrier
9 Air restrictor jet
10 Emulsion tube
11 Main jet
12 Venturi
13 Secondary venturi
14 Pump inlet valve
15 Gasket for 16
16 Pump jet
17 Pump inlet and outlet valve
18 Screw
19 Washer
20 Butterfly control complete with:
21 Lever

22 Split trim
23 Spring for trim
24 Pin
25 Spring for screw
26 Screw
27 Fixing nut
28 Locking washer
29 Corrugated washer
30 Spring washer
31 Ball bearing
32 Idling jet
33 Gasket for idling jet carrier
34 Idling jet carrier
35 Spring for adjusting screw
36 Adjusting screw
37 Diaphragm return spring
38 Diaphragm
39 Pump cover
40 Fixing screw
40A Starter cover fixing screw

41 Idling adjusting screw
42 Gasket for above
43 Spring for above
44 Progression hole inspection plug
45 Cam locking pin
46 Pump operating cam
47 Nut for air adjusting screw
48 Air adjusting screw
49 Carburettor body
50 Sealing plug
51 Butterfly valve
52 Fixing screw
53 Main spindle
54 Main spindle
55 Valve return spring
56 Carburettor body
57 Nut for diffuser fixing screw
58 Diffuser fixing screw
59 Starter control complete with:
60 Outer cable fixing screw
61 Lever fixing nut
62 Spring washer
63 Lever complete with:

64 Cable fixing nut
65 Lever
66 Cable fixing screw
67 Lever return spring
68 Starter cover
69 Starter control spindle
70 Filter
71 Washer for fixing screw
72 Starter jet
73 Starter valve
74 Spring for above
75 Starter spring guide and stop
76 Spring ring and stop for spring
77 Float
78 Needle valve
79 Gasket for above
80 Carburettor cover
81 Gasket for above
82 Carburettor cover complete with:
83 Stud
84 Washer for 85
85 Cover fixing screw

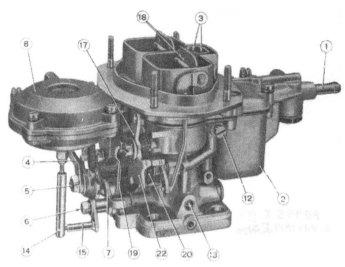

**FIG.3.2. THE 34 DFH, 34 DFH/4 WEBER CARBURETTOR
AS SEEN FROM THE CHOKE CONTROL SIDE**

1 Fuel line connection
2 Float chamber
3 Air corrector jet
4 Differential throttle
   control lever
5 Primary throttle
   spindle
6 Secondary throttle
   spindle
7 Secondary throttle
   differential control
   lever
8 Diaphragm device

12 Idling jet
13 Progression holes
14 Diaphragm device rod
15 Secondary throttle
    spindle actuating lever
17 Choke control lever
18 Choke valves
19 Primary throttle
    starting link
20 Choke valve return
    spring
22 Lobe of lever (17)

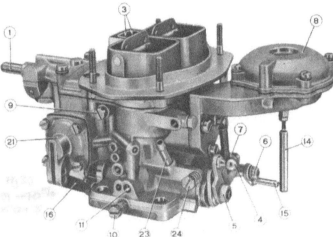

**FIG.3.3. THE 34 DFH, 34 DFH/4 CARBURETTOR AS SEEN
FROM THE ACCELERATOR PUMP SIDE**

1 Fuel line
3 Air corrector jets
4 Differential throttle
   opening lever
5 Primary throttle
   spindle
6 Secondary throttle
   spindle
7 Secondary throttle
   differential opening
   control lever
8 Diaphragm device
9 Idling jet

10 Idling jet adjusting
    screw
11 Progression holes
14 Diaphragm device rod
15 Secondary throttle
    control lever
16 Accelerator pump
    lever
21 Accelerator pump
23 Crankcase gas suction
    connection
24 Slow-running adjustment
    screw

4 Carburettor - removal and replacement

1 The carburettor may be removed easily with the engine in the
car for inspection and cleaning. Under some circumstances it
may be wise to remove the carburettor(s) from the inlet
manifold before removing the engine from the car.
2 To begin it will be necessary to remove the air cleaner.
First undo the nuts securing the cleaner cover. Once the cover is
off, the element is removed to expose the nuts and studs that
hold the cleaner box to the carburettors. Undo the nuts and lift
the box off the carburettor flanges, remember to slip the
breather hoses off the cleaner.
3 The carburettor(s) is now completely uncovered and it will
be easy to work accelerator and choke linkages. (photo).
4 Both accelerator and choke valves are operated by cables
from the driving position. Undo the clamps retaining the cables
and their sleeves on the linkages. (photo).
5 Pull off the petrol feed pipe from the float chamber union,
and the accelerator linkage return springs.
6 Finally the nuts which hold the carburettor(s) to the inlet
manifold can be undone, after which the assemblies may be
lifted clear. Retrieve the gasket, and plug the inlet manifold with
clean non-fluffy rag. Alternatively the carburettors may be
removed together with the inlet manifold. (photo).
7 Replacement of the carburettor is the exact reversal of

removal. You must be sure that the correct number of gaskets
have been replaced and that the control cables have been fitted
correctly to give the full range of movement the accelerator and
choke linkages require.

4.3 Accelerator linkage connections

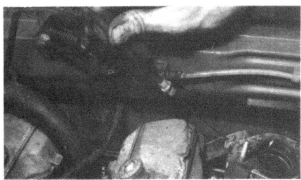

4.4 Choke cable removal

4.6 Carburettor and manifold removal

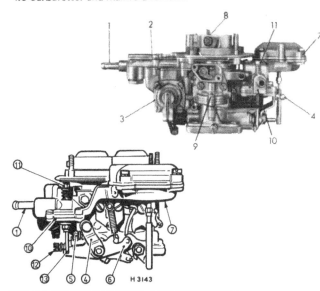

FIG.3.4. THE WEBER DHSA 1 CARBURETTOR

Fuel inlet union
Float chamber
Accelerator pump
Primary barrel accelerator valve lever
Slow idling speed adjusting screw
Secondary barrel accelerator valve clip
Primary venturi vacuum actuated device controlling secondary
barrel accelerator valve
Choke flap
Vacuum device over-riding manual choke valve control
Inlet manifold vacuum actuated diaphragm device controlling
fast idle selection
Adjustment screw for fast idle speed
Slow idling mixture control screw
Fast idle stop lever

5 Carburettor - setting and adjustment

1 **The WEBER 401DF 10/11 dual carburettor fitted to the 1608 cc 125 BC000 engine:—**
  It should be realised that these WEBER carburettors are extremely fine instruments; it is very likely that they will need attention often enough to justify the procurement of the special apparatus needed for their setting and adjustment. Therefore unless you have access to 'Sychrotest Equipment' or equivalent the task of setting and adjusting the WEBER carburettors should be entrusted to your local FIAT agent. The procedure set down here assumes that you have the specialised equipment mentioned. It is important to ensure that all other engine adjustments; tappet gaps, static ignition timing, spark plug gaps, contact breaker gap, are correct before commencing.
2 Carry out the carburettor setting and adjustments with the air cleaner off.
3 Temporarily remove the accelerator link from lever on the butterfly valve spindle. Move the lever to ensure that the spindles rotate easily and readily return to the idle stop.
4 The 'air correction screw' lock nuts should be loosened and the screws screwed in so that the air correction passage is closed off. The locking nuts may be tightened to secure the screws.
5 Turning your attention to the 'idling mixture screws' screw all four fully in, then back out by 4½ and 4¾ turns as indicated in Fig. 3.1.
6 Next undo the screw setting the idling speed of the engine and the relative position of the accelerator valve spindles in the two carburettors.
7 Press down the main lever which is actuated by the accelerator cable so that the spring of the reaction rod is fully compressed.
8 Then turn the screw which sets the relative position of the front and rear carburettor accelerator valve spindles so that it is in contact with the tag on the lever attached to the rear carburettor valve spindle.
9 Screw the idling speed setscrew into contact with the main actuating lever and then a further full turn.
10 Connect the Synchrotest equipment to the carburettor assembly. There are special tappings in the base of the carburettor to monitor the vacuum in the throat of each barrel.
11 Start the engine and check the vacuum readings for each carburettor barrel. Note that on this engine an individual barrel supplies each individual cylinder - there is no interconnection.
12 Then considering just one carburettor at a time note the vacuum readings for the two barrels in that carburettor. Undo the 'air correction screw' for that barrel which shows a lower vacuum reading until the vacuum in that barrel is brought to the same level as its twin. Secure the correction screw lock nut.
13 Having equalised the vacuum in the barrel pairs, equalise the two pairs by turning the screw which governs the relative position of the accelerator valve spindles in the two carburettors.
14 With the engine thoroughly warmed up, bring the idling speed up to 900 rpm with the idling speed adjusting screw.
15 Finally check the position of the idling mixture control screws by screwing them together in and out a little, to obtain the highest, smoothest idle running of the engine. This speed should be approximately 900 rpm, but once the smooth running has been obtained the idling speed adjusting screw should be turning to bring the speed to 900 rpm.
16 Finally before reconnecting the accelerator cable and refitting the air cleaner, check that the vacuum readings in the four carburettor barrels are equal.

The compound twin barrel carburettors on the 1438 cc 124 AC000 engine, 1592 and 1755 cc 132 series engine. Weber 34DHS: Weber 34DFH: Weber DHSA:
17 These carburettors have what is termed as a primary barrel and a secondary barrel. The carburettors operate in the following manner. The acelerator cable is coupled to the primary barrel accelerator butterfly valve. The valve spindle has a relay lever

fitted which operates the secondary barrel valve in one of two ways. Either the relay lever mechanically opens the secondary barrel valve when the primary has opened a pre-requisite amount, or it operates a trip device which allows a primary venturi vacuum actuated device to operate the secondary barrel valve.

18 In all cases there is little running adjustment for the carburettor save only the idling mixture control screw and idle speed control screw.

The method of adjusting the idling mixture is as follows:

19 Bring the idling speed to approximately 900 rpm with the idle speed adjusting screw. The screw the mixture control screw in or out to achieve the highest speed, smoothest running of the engine. The idling speed can then be retrimmed to the 900 rpm with the speed adjusting screw.

20 A final note on these compound carburettors is that the difference between the DHS, DFH and DHSA types is that the DHSA has an inlet manifold vacuum actuated choke override system. In this system when the vacuum in the inlet manifold reaches a certain level, and the choke has been closed for cold running, the vacuum actuated device opens the choke valve itself to admit the air demanded into the carburettor and engine.

21 It should also be noted that on the DHSA type the idling mixture control screw does not have the effect as on the more usual carburettor configurations because of the inclusion of exhaust emission devices.

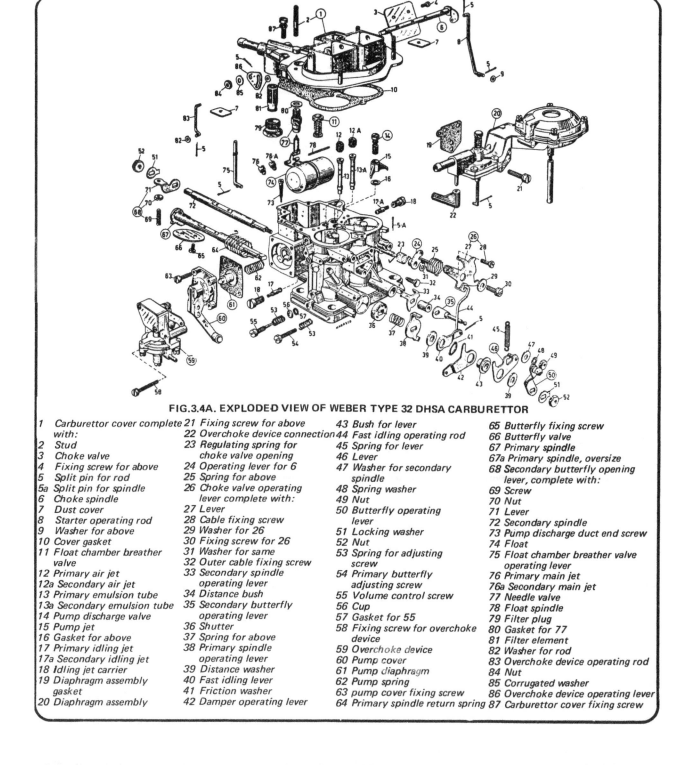

FIG.3.4A. EXPLODED VIEW OF WEBER TYPE 32 DHSA CARBURETTOR

1 Carburettor cover complete with:
2 Stud
3 Choke valve
4 Fixing screw for above
5 Split pin for rod
5a Split pin for spindle
6 Choke spindle
7 Dust cover
8 Starter operating rod
9 Washer for above
10 Cover gasket
11 Float chamber breather valve
12 Primary air jet
12a Secondary air jet
13 Primary emulsion tube
13a Secondary emulsion tube
14 Pump discharge valve
15 Pump jet
16 Gasket for above
17 Primary idling jet
17a Secondary idling jet
18 Idling jet carrier
19 Diaphragm assembly gasket
20 Diaphragm assembly

21 Fixing screw for above
22 Overchoke device connection
23 Regulating spring for choke valve opening
24 Operating lever for 6
25 Spring for above
26 Choke valve operating lever complete with:
27 Lever
28 Cable fixing screw
29 Washer for 26
30 Fixing screw for 26
31 Washer for same
32 Outer cable fixing screw
33 Secondary spindle operating lever
34 Distance bush
35 Secondary butterfly operating lever
36 Shutter
37 Spring for above
38 Primary spindle operating lever
39 Distance washer
40 Fast idling lever
41 Friction washer
42 Damper operating lever

43 Bush for lever
44 Fast idling operating rod
45 Spring for lever
46 Lever
47 Washer for secondary spindle
48 Spring washer
49 Nut
50 Butterfly operating lever
51 Locking washer
52 Nut
53 Spring for adjusting screw
54 Primary butterfly adjusting screw
55 Volume control screw
56 Cup
57 Gasket for 55
58 Fixing screw for overchoke device
59 Overchoke device
60 Pump cover
61 Pump diaphragm
62 Pump spring
63 pump cover fixing screw
64 Primary spindle return spring

65 Butterfly fixing screw
66 Butterfly valve
67 Primary spindle
67a Primary spindle, oversize
68 Secondary butterfly opening lever, complete with:
69 Screw
70 Nut
71 Lever
72 Secondary spindle
73 Pump discharge duct end screw
74 Float
75 Float chamber breather valve operating lever
76 Primary main jet
76a Secondary main jet
77 Needle valve
78 Float spindle
79 Filter plug
80 Gasket for 77
81 Filter element
82 Washer for rod
83 Overchoke device operating rod
84 Nut
85 Corrugated washer
86 Overchoke device operating lever
87 Carburettor cover fixing screw

6 Carburettor - dismantling, inspection and reassembly

1 Do not dismantle the carburettor unless it is absolutely necessary, when systematic diagnosis indicates that there is a fault with it. The internal mechanism is delicate and finely balanced and unnecessary tinkering will probably do more harm than good.
2 Although certain parts may be removed with the carburettor still attached to the engine, nevertheless it is considered safer to remove it and work over a bench.
3 With the carburettor off the car disassembly is as follows:- (this covers all carburettors fitted with slight variations, which, because of their simple and obvious nature, are not specifically mentioned each time, but it does not cover the automatic choke working of these carburettors so fitted).
    Remove the top of the carburettor by undoing its fixing screws with he correct size screwdriver. These screws are made of comparatively soft metal and will damage very easily. With the top should come the choke mechanism (both types)., fuel inlet pipe and filter (if fitted) and needle valve. The gasket and float itself should remain in the body of the carburettor, but relevant exploded diagrams will show you specifically what is fitted to each type.
4 Remove the gasket and the float placing bracket or spindle and then the float itself.
5 Remove any jets and their washers which have screwdriver cuts in their heads and are removable from the top and inside of the carburettor.
6 Unscrew all the external adjusting screws and the float chamber drain plugs. Retain all washers and springs and code for their relevant positions.
7 Turn the top of the carburettor upside down and remove the needle valve and its washer.
8 On Solex carburettors you will be able to remove the choke tube itself from the body of the carburettor.
9 There is no point under any circumstances in removing any more parts from the carburettor. If any of these parts are in need of attention, then a complete new carburettor is needed.It is safer and more efficient if you have reached this stage of need of repair to replace the complete unit.
10 All the parts which have been separated should be thoroughly cleaned in methylated spirits or clean petrol by hand and without the help of anything more than a soft non-fluffy rag. Do not use any scrapers, emery paper, wire wool, or hard pro jections such as a pin, on these parts. All have been machined to extremely fine tolerances. Blow all parts dry.
11 Inspect for blockages and scoring, the float for a puncture and the needle valve for easy operation. On plastic and metal floats make sure the metal tag is not bent or distorted. Replace if it is. There is no room for float level adjustment. Replace any parts which are obviously worn or damaged but also do so to any which you even suspect. (Make sure that parts are still available though before throwing away - you may need a temporary repair).
12 Replace all parts in the reverse of their removal using new copper washers and gaskets all the way through. Do not over-tighten anything, and do not use any gasket cement. If the top of the carburettor does not fit flat it needs replacing.
SPECIAL NOTE:
    No mention has been made of throttle, choke and butterfly spindles. As these spindles run directly in the body or top of the carburettor they are likely over a period of usage to wear. It is impractical to replace parts of these and any wear or failure in these parts must mean complete replacement.

7 Exhaust emission control

1 Crankcase fumes from the crankcase connection are fed to the oil separater mounted just above the fuel pump. From the oil separater the fumes pass along a pipe which feeds them to the air cleaner so that they are taken back into the engine and burnt. As a safety measure a wire brush flame arrester is inserted in the top

of the fume pipe just below the air cleaner union.
2 The emission control system built around the engine comprises the WEBER DHSA carburettor which has as mentioned earlier a inlet manifold vacuum actuated choke and primary barrel vacuum actuated secondary barrel accelerator valve.
    The inlet manifold now incorporates water heating and the cylinder head has been modified to provide the necessary water passages.
    The ignition coil has been changed to one that produces an increased energy spark to ensure better ignition.
    Lastly there is the 'fast idling system'. The object of this system is to maintain a flow of air through the carburettor which will allow the instruments to meter the fuel efficiently and **not** give a slightly rich mixture as normally occurs in engine over-run and slow idle condition.
3 The operation of the emission control system:
Carburettor: All carburettors have three vacuum diaphragm devices; one which monitors the vacuum in the primary barrel major venturi, actuates the secondary barrel accelerator valve. The second device monitors the inlet manifold vacuum or the vacuum near the minor primary venturi depending on the response of the WEBER electrovalve and overrides the manual control. The third device works in conjunction with the second but is connected to the primary barrel accelerator valve so as to prevent closure of that valve when the WEBER electrovalve is operated. The electrovalve is wired to three switches. The switches operated by the gearshift mechanism and the clutch pedal are in series, and when both are made, that is when either third or fourth gear is selected and the clutch released the WEBER electrovalve is energised. The vacuum from the inlet manifold now actuates the diaphragm device that prevents the accelerator valve in the primary barrel from closing, so that the engine idles faster. The third switch operates the electrovalve directly: it is a pushbutton switch mounted on the sidewall of the engine compartment. The switch is used when the fast idling speed needs adjustment; which can be achieved by pressing the pushbutton switch and turning the screw in the centre of the fast idle diaphragm device to move the actuating rod to obtain the fast idling speed desired - 1600 rpm.
4 Adjustment of fast idling speed:
a) Depress the pushbutton that operates the electrovalve so that the diaphragm device responds to the inlet manifold vacuum.
b) Bring the engine speed up to 2500 rpm, then allow the speed to drop onto the fast idle speed - which should be 1600 rpm. Adjust the screw in the middle of the diaphragm device as necessary.
c) Release the pushbutton and check that the engine speed falls to the normal idling speed within 1 to 3 seconds.
5 Adjusting idling mixture control on DHSA carburettors:
a) Adjust the idling speed to 800 - 900 rpm.
b) Adjust the mixture control screw to give a carbon monoxide reading (CO) of $2 \pm 5\%$ on a suitable monitor device.
6 The fuel evaporation control system:
    The purpose of this system is to vent the petrol vapour in the fuel tank through a carbon packed canister into the inlet manifold of the engine so that it is burnt. The petrol filler cap is a completely sealed type.
    The carbon filled canister is located in the engine compartment. There are three pipe connections, one from the fuel tank vent system, one from the inlet manifold, and the last one from the air cleaner.
    The fuel vapours 'soak' into the carbon, which is purged of the vapour when the engine is running. The vacuum in the inlet manifold draws warm air from the air cleaner through the carbon taking the vapour into the engine through the inlet manifold pipe. Between the petrol tank and carbon canister there is a three way valve. The valve allows air into the vent system to compensate for consumption of fuel. The valve also allows vapours to pass along the vent line to the carbon canister and engine.
    The third mode of operation of the valve allows vapour to vent directly to the atmosphere in the event of a blockage in the

vent pipe to the canister.

NOTE:
There is no means of repairing the emission control devices and in event of malfunction they should be replaced by approved spare units.

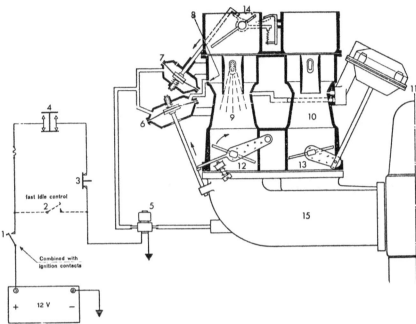

FIG.3.5. SCHEMATIC LAYOUT OF THE DHSA WEBER CARBURETTOR WITH EXHAUST EMISSION CONTROL DEVICES

1 Switch combined with ignition contacts
2 Pushbutton switch for fast idle adjustment
3 Micro-switch engaged when clutch pedal is released
4 Micro-switch engaged when 3rd, 4th (5th) gears are selected
5 WEBER electro-valve when power supplied connects vacuum devices to inlet manifold vacuum
6 Diaphragm vacuum device prevents primary accelerator valve from returning to slow idle when electro-valve is operating and vacuum in inlet manifold great enough
7 Diaphragm vacuum device actuated by barrel or inlet manifold vacuum, over-rides manual choke control
8 By-pass orifice neutralizes vacuum action on diaphragms 6 and 7 when electro-valve 5 closes, cutting off inlet vacuum
9 Primary barrel main venturi
10 Secondary barrel main venturi
11 Diaphragm vacuum device actuated by the vacuum in the primary barrel venturi - operates the secondary barrel accelerator valve
12 Primary barrel accelerator valve
13 Secondary barrel accelerator valve
14 Primary barrel choke valve
15 Inlet manifold

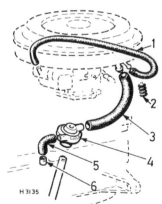

FIG.3.6. CRANKCASE VENTILATION SYSTEM

1 Vacuum line
2 Wire brush flame arrester
3 Vapour hose to carburettor inlet
4 Oil, vapour separator
5 Crankcase breather hose
6 Breather hose connection

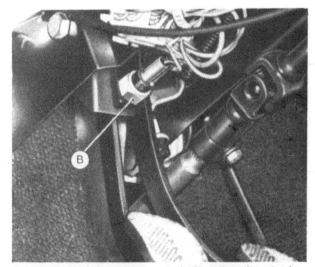

Fig.3.7. Electro-valve actuating switch 'B' closing when clutch pedal is released

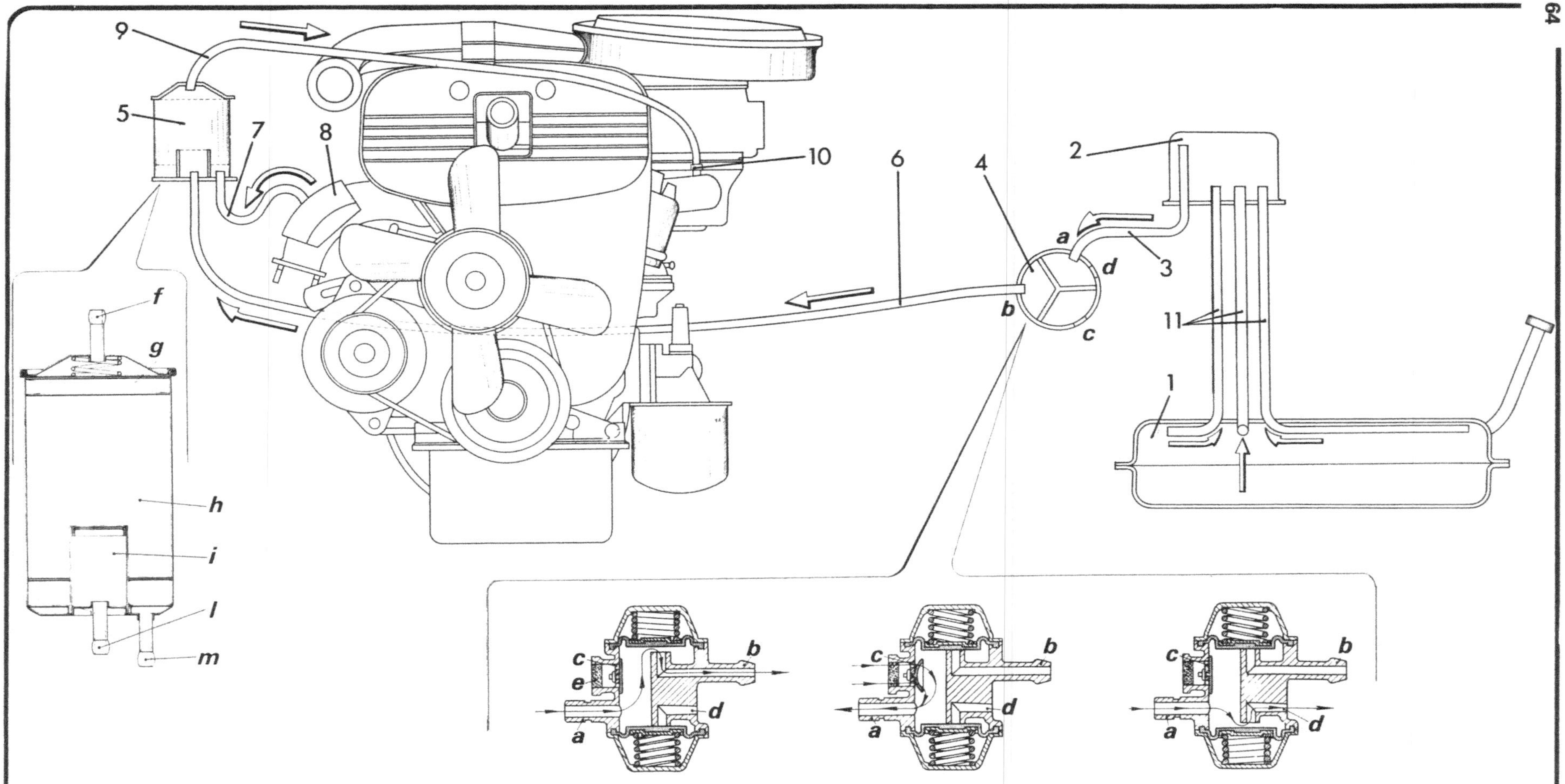

Position 1.

From fuel tank to activated carbon vapor filter.

Position 2.

Air from ambient into tank.

Position 3 (safety).

Vapor from tank to ambient (excess pressure in the tank).

FIG.3.8. FUEL VAPOUR RECIRCULATION SYSTEM

1 Fuel tank
2 Vapour liquid separator
3 Fuel vapour line from separator to three way valve
4 Three way valve
5 Carbon filter canister
6 Vapour line from three way valve to carbon canister
7 Clean warm air pipe
8 Exhaust manifold

9 Vapour line from filter to inlet manifold
10 Vacuum port
11 Line from tank to vapour liquid separator
a From fuel tank
b To carbon filter
c Fuel tank to air inlet
d Safety outlet

e Air filter
f To engine inlet manifold
g Synthetic filter
h Activated carbon filter
i Air purge paper filter
l Vapour inlet
m Warm air purge inlet

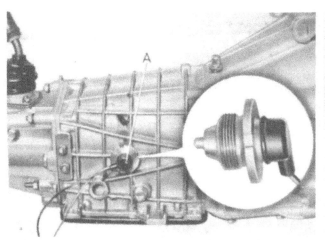

Fig.3.9. Electro-valve actuating switch 'A' closing when 3rd or 4th gears are selected

Fig. 3.10. Under bonnet actuating mechanism for emission control

1 Weber electro-valve, controlling fast idle vacuum device
2 Fast idle push button switch, used when fast idling speed needs adjustment

8 Fuel pump - removal and replacement

1 The fuel pump will need removing if it is to be dismantled for overhaul, but the filter can be cleaned in situ. Disconnect the fuel lines on the inlet and outlet sides by pulling off the connector pipes on both sides.

2 Undo the two nuts, holding the pump flange to the crank-case, and take the pump off. Keep the spacer and gaskets together and do not discard them. If necessary blank off the fuel line from the tank to prevent loss of fuel. (photo).

3 Replacement is a reversal of the removal procedure. Make sure that the total thickness of gaskets and spacer is the same as came off. Check that the fuel line connections are not leaking after starting the engine.

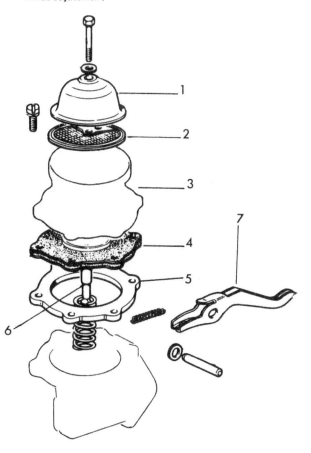

FIG.3.11. B.C.D. FUEL PUMP

1 Pump reservoir cover
2 Fuel filter
3 Pump body incorporating one way valves
4 Pump diaphragm
5 Gasket
6 Diaphragm actuating rod
7 Pump lever and pivot

8.2 Fuel pump removal

9 Fuel pump - inspection, dismantling and reassembly

1 First clean the pump exterior thoroughly and mark the edges of the two halves of the body.

2 Undo the cover retaining clip and lift off the cover. The gasket and gauze filter may then be removed.

3 Remove the eight screws and washers holding the two halves of the pump together and the top half may then be lifted off.

4 The diaphragm and pushrod should be removed next but you will have to remove the pump lever and its spindle to release it. This is done by releasing one of the spindles circlips. This can prove to be very fiddly but with patience and a strong blunt penknife blade the circlip can be 'peeled' off. Push the spindle through and pull out the lever. Retrieve the lever spring. Lift out the diaphragm and its rod carefully. This spindle is usually known as the rocker arm pivot pin.

5 If there are signs of wear in the rocker arm pivot, and rocker arm and link bushes then they should be renewed.

6 The valve assemblies should only be removed from the upper body if renewal is necessary. They are staked into the body and are destroyed when levered out.

7 Examine the diaphragm for signs of cracking or perforation and renew if necessary.

8 Overhaul kits are available for all pumps and are supplied with a new diaphragm, valves and sealing rings. Check the manufacture of the pump first.

9 When fitting new valve assemblies to the body, first fit the seating washers and then place the valves, making sure that they are the correct way up according to inlet and outlet. The body will have to be restaked at six (different) places round the edge so that the assemblies are firmly held in their positions. If this is not done properly and leakage occurs between the valve assembly and the seating ring the pump will not operate efficiently.

10 To replace the diaphragm and lever arm it will be necessary to place the diaphragm spring in the body of the pump. Then the diaphragm and its rod. Press the diaphragm spring down and fit the spring, push in the lever (the right way up) and connect over the top of the machined stop on the rod. Push in the rocker arm pivot pin and push through the lever. Replace the circlip on the pivot pin. Always use a new circlip.

11 Fit the upper half of the pump body and line up the mating marks. In order to assemble the two halves and the diaphragm properly push the rocker arm upwards so that the diaphragm is drawn level. Then place the eight screws in position lightly. It is best if the base of the pump is held in a vice whilst the rocker arm is pushed right up to bring the diaphragm to the bottom of its stroke. A short piece of tube over the rocker arm will provide easy leverage. In this position the eight screws should be tightened evenly and alternatively.

12 Fit a new filter bowl gasket carefully in the groove of the upper body, making sure that it does not twist or buckle in the process. Replace the cover and screw it tight.

13 When the pump is reassembled the suction and delivery pressure can be felt at the inlet and oulet ports when the rocker arm is operated. Be careful not to block the inlet port completely when testing suction. If the rocker arm were to be operated strongly and the inlet side was blocked the diaphragm could be damaged. NOTE: The dismantling sequence is not affected by the presence of a priming lever (if fitted). When replacing the pump in the engine block check its freedom of movement.

10 Fuel gauge sender unit

1 Before removing a suspected faulty fuel tank sender unit, check that all cable connections are clean and secure.

2 **For safety reasons disconnect the battery connections.**

3 Remove the fuel tank cover in the luggage compartment so that access is gained to the top of the sender unit.

4 Undo and remove the nuts securing the fuel sender unit to the tank. Recover the gasket.

5 Make a note of the electrical cable connections to the fuel tank sender unit and disconnect them.

6 Refitting the sender unit is the reverse sequence to removal. It is advisable to fit a new gasket to ensure a good petrol tight joint.

11 Exhaust system

1 The exhaust system is conventional in its arrangement and working. It is not too difficult to repair, having up to three silencers and supplied in three sections. The forward section consists of the twin pipes from the manifold and a convergent section.

The middle section includes the middle two silencers, and the rear section comprises the rear silencer and tail pipe.

2 It is wise to only use the original type exhaust clamps and proprietry made system. When any one section of the exhaust system needs renewal it often follows that the whole system is best renewed.

3 It is most important when fitting exhaust systems that the twists and contours are carefully followed and that each connecting joint overlaps the correct distance. Any stresses or strains imparted in order to force the system to fit the hanger rubbers will result in early fractures and failures.

4 When fitting a new part or a complete system it is well worth removing ALL the systems from the car and cleaning up all the joints so that they will fit together easily. The time spent struggling with obstinate joints whilst flat on your back under the car is eliminated and the likelihood of distorting or even breaking a section is greatly reduced. Do not waste time trying to undo rusted clamps and bolts. Cut them off. New ones will be required anyway if they are bad.

12 FAULT DIAGNOSIS

Unsatisfactory engine performance and excessive fuel consumption are not necessarily the fault of the fuel system or carburettor. In fact they more commonly occur as a result of ignition and timing faults. Before acting on the following it is necessary to check the ignition system first. Even though a fault may lie in the fuel system it will be difficult to trace unless the ignition is correct. The faults below, therefore, assume that this has been attended to first (where appropriate).

Symptom	Reason/s	Remedy
Smell of petrol when engine is stopped	Leaking fuel lines or unions	Repair or renew as necessary.
	Leaking fuel tank	Fill fuel tank to capacity and examine carefully at seams, unions and filler pipe connections. Repair as necessary.
Smell of petrol when engine is idling	Leaking fuel line unions between pump and carburettor	Check line and unions and tighten or repair.
	Overflow of fuel from float chamber due to wrong level setting, ineffective needle valve or punctured float	Check fuel level setting and condition of float and needle valve, and renew if necessary.
Excessive fuel consumption for reasons not covered by leaks or float chamber faults	Worn jets	Renew jets or carburettor body if not removable.
	Over-rich setting	Adjust jet.
	Sticking mechanism	Check correct movement of mechanism.
Difficult starting, uneven running, lack of power, cutting out	One or more jets blocked or restricted	Dismantle and clean out float chamber and jets.
	Float chamber fuel level too low or needle valve sticking	Dismantle and check fuel level and needle valve.
	Fuel pump not delivering sufficient fuel	Check pump delivery and clean or repair as required.

Chapter 4 Ignition system

Contents

Specifications

Spark plugs ... ... Marelli CW8LP, Champion N6Y or N9Y, Bosch W230T30
   Electrode gap ... ... 0.019 to 0.023 inches
   Tightening torque ... ... 29 lb f. ft

Coil
   Make ... ... Marelli BZR 202A; O.E.M. G37SU
   Resistance - Primary windings ... ... 1.64 to 1.76 ohms;   **2.**6 to 2.95 ohms
           - Secondary windings ... ... 7650 to 9350 ohms;   7000 to 8500 ohms

Electronic ignition system
   Available as special option pack only for series 132 vehicles:
   Spares part number 4249572
                           (Can be factory fitted to special order)

Distributor
   Make ... ... Marelli SI24A, S124B
   Rotation:
      Firing sequence 1 3 4 2
      Contact point gap ... ... 0.016 to 0.018 inch
   No vacuum advance

   Condenser capacity ... ... 0.20 to 0.25 microfarad, 50 to 1000 Hz

   Ignition timing ... ... 10º BTDC static   5º BTDC U.S.A. with exhaust
                                          emission devices

1 General description

In order that the internal combustion engine with spark ignition can operate properly, the spark which ignites the air-fuel charge in the combustion chamber must be delivered at precisely the correct moment. This correct moment is that which will allow time for the charge to burn sufficiently to create the highest pressure and temperature possible in the combustion chamber as the piston passes top dead centre and commences its power stroke. The distributor and ignition coil are the main devices which ensure that the spark plug ignites the charge as required.

Very high voltages need to be generated in the ignition system in order to produce the spark across the plug gap which ignites the fuel/air charge. The device in which these high voltages - several thousand volts - are generated is the coil (or electronic ignition pack which is an optional extra on recent models). The coil contains two sets of windings - the primary and the secondary windings. A current at 12 volts is fed through the primary windings via the contact breaker mechanism in the distributor. It is precisely when the flow is interrupted by the contact breaker that the huge voltage is momentarily induced in the secondary windings and that voltage is conveyed via HT leads and the rotor arm in the distributor cap to the appropriate spark plug.

It follows therefore that the contact breakers must part the instant a spark is required and the rotor arm must be aligned to the appropriate stud in the distributor cap which is connected to the spark plug which 'needs' the spark. The distributor shaft revolves at half crankshaft speed, and there are four rises on the distributor cam and four studs in the distributor cap, to cater for the four sparks the engine requires each two revolutions of the crankshaft. On this FIAT, the timing of the ignition is set by two means. The first is referred to as static timing and the second which is fully automatic is the centrifugal advance mechanism. This latter mechanism ensures that the spark arrives in that interval of time for combustion which corresponds to a greater angle of crankshaft movement when the engine is turning quickly than when it moves slowly. The mechanism comprises two weights on an arm on the distributor shaft. As the shaft speed increases the weights move outwards against their restaining springs. The contact breaker cam is attached to the weights so that as they move out it is rotated up to 20° relative to the distributor shaft, and therefore the contact breaker will open earlier relative to the distributor shaft and engine crankshaft as required.

The static ignition advance is that nominal amount which corresponds to the time for combustion at the idling speed of the engine. It is necessary therefore when carrying out ignition and carburation adjustments to ensure that the engine is turning at the speed appropriate to that test or adjustment.

2 Routine maintenance

a) Spark plugs
Remove the plugs and thoroughly clean away all traces of carbon. Examine the porcelain insulated round the central electrode inside the plug if damaged discard the plug. Reset the gap between the electrodes. Do not use a set of plugs for more than 9,000 miles. It is false economy.

b) Distributor
Every 9000 miles remove the cap and rotor arm and put one or two drops of engine oil into the centre of the cam recess. Smear the surfaces of the cam itself with petroleum jelly. Do not over-lubricate as any excess could get onto the contact point surfaces and cause ignition difficulties.

Every 9000 miles examine the contact point surfaces. If there is a build up of deposits on one face and a pit in the other it will be impossible to set the gap correctly and they should be refaced or renewed. Set the gap when the contact surfaces are in order.

c) General
Examine all leads and terminals for signs of broken or cracked insulation. Also check all terminal connections for slackness or signs of fracturing of some strands of wire. Partly broken wire should be renewed.

The HT leads are particularly important as any insulation faults will cause the high voltage to 'jump' to the nearest earth and this will prevent a spark at the plug. Check that no HT leads are loose or in a position where the insulation could wear due to rubbing against part of the engine.

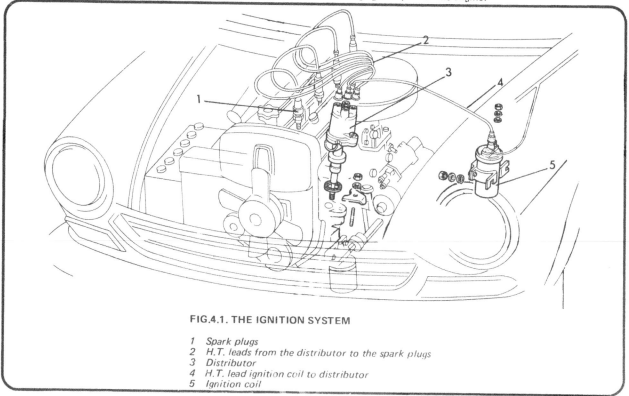

FIG.4.1. THE IGNITION SYSTEM

1 Spark plugs
2 H.T. leads from the distributor to the spark plugs
3 Distributor
4 H.T. lead ignition coil to distributor
5 Ignition coil

3 Distributor - contact points - gap adjustment

1 To adjust the contact breaker points so that the correct gap is obtained, first release the two clips or two screws securing the distributor body, and lift away the cap. Clean the inside and outside of the cap with a dry cloth. It is unlikely that the four segments will be badly burned or scored, but if they are, the cap must be renewed. If only a small deposit is on the segments it may be scraped away using a small screwdriver.
2 Push in the carbon brush located in the top of the cap several times to ensure that it moves freely. The brush should protrude by at least a quarter of an inch.
3 Gently prise open the contact breaker points to examine the condition of their faces. If they are rough, pitted or dirty, it will be necessary to remove them for resurfacing, or for replacement points to be fitted.
4 To make adjustment easier, note the position of the rotor arm relative to the automatic timing control table. Undo the two securing screws and lift away the rotor.
5 It will be noticed that it is not possible to fit the rotor the wrong way round due to the round and square pegs in the table.
6 Presuming the points are satisfactory, or they have been cleaned or replaced, measure the gap between the points by turning the engine over until the contact breaker arm is on the peak of one of the four cam lobes. A 0.0165 to 0.0189 in (0.42 to 0.48 mm) feeler gauge should now just fit between the points.
7 If the gap varies from this amount, slacken the contact plate seucring screw and adjust the contact gap by inserting a screwdriver in the notched hole of the stationary point and table and turn in the required direction to increase or decrease the gap. (photo).
8 Tighten the securing screw and check the gap again. (photo).
9 Replace the rotor and secure with the two screws. Refit the distributor cap and clip the spring blade retainers into position.

3.7 Turning the contact breaker gap adjusting screw

3.8 Checking the contact breaker gap with a feeler gauge

4 Distributor - contact points - removal and replacement

1 If the contact breaker points are burned, pitted or badly worn, they must be removed and either replaced or their faces must be filed smooth.
2 To remove the points, first unscrew the terminal nut and remove it together with the washer under its head. Remove the low tension cable from the terminal.
3 Unscrew and remove the contact breaker locking screw.
4 Detach the low tension and condenser cables and lift away the contact breaker points.
5 To reface the points, rub the faces on a fine carborundum stone or on fine emery paper. It is important that the faces are rubbed flat and parallel to each other so that there will be complete face to face contact when the points are closed. One of the points will be pitted and the other will have deposits on it.
6 It is necessary to remove completely the built up deposits, but not necessary to rub the pitted point right to the stage where all the pitting has disappeared, though obviously if this is done it will prolong the time before the operation of refacing the points has to be repeated.
7 Refitting the contact breaker points is the reverse sequence to removal. It will be necessary to adjust the points gap as described in Section 3.

5 Condenser - removal, testing and replacement

1 The purpose of the condenser (sometimes known as a capacitor), is to ensure that when the contact breaker points open there is no sparking across them which would waste voltage and cause wear.
2 The condenser is fitted in parallel with the contact breaker points. If it develops a short circuit, it will cause ignition failure as the points will be prevented from interrupting the low tension circuit.
3 If the engine becomes very difficult to start or begins to miss after several miles running, and the breaker points show signs of excessive burning, then the condition of the condenser must be suspect. A further test can be made by separating the points by hand with the ignition switched on. If this is accompanied by a flash it is indicative that the condenser has failed.
4 Without special test equipment the only sure way to diagnose condenser trouble is to replace a suspected unit with a new one and note if there is any improvement.
5 To remove the condenser from the distributor, first remove the distributor cap.
6 Detach the condenser cable from the terminal block and remove the condenser fixing to the body. Lift away the condenser.
7 Replacement of the condenser is the reversal of removal.

6 Distributor - removal and replacement

1 To remove the distributor from the engine, start by removing all the HT leads from the spark plug and remove the battery connections - a safety precaution.
2 Remove the HT lead from the coil, and disconnect the LT lead from the terminal on the side if the distributor.
3 Turn the crankshaft until the rotor arm is pointing to the insert in the distributor cap which is connected to the number 1 spark plug lead. Apply the handbrake, and select a gear to ensure that the engine does not turn subsequently.
4 Undo and remove the nut, washer and plate securing the distributor to the cylinder block (or exhaust cam assembly on 125 BC000, 1608 cc engines). (photo).
5 Lift away the distributor. Refitting the unit is simply the reversal of the removal procedure. Remember to hold the rotor arm and distributor shaft aligned to the No 1 cylinder insert in the cap, when refitting the distributor into the engine.
6 If the crankshaft has been rotated, before the distributor was refitted, it will be necessary to retime the ignition as described in Section 8 of this chapter.

6.4 The nut, washer and clamp plate
which secures the distributor to the
engine

7.3 Removing the contact breaker mounting plate

7.4 The centrifugal advance mechanism

8.5 The static timing marks on the timing
belt cover, and crankshaft pulley
'A'   10⁰ B T D C
'B'   5⁰ B T D C

7   Distributor - inspection and reassembly

1   Having removed the distributor cap, lift away the rotor arm
(on some distributors the rotor arm is on a disc secured by two
screws to the shaft assembly).
2   Remove the contact breaker points as described in Section 4
of this chapter.
3   The contact breaker mounting plate is removed after undoing
the two screws which retain it to the distributor body. Once this
plate is removed the centrifugal advance mechanism is exposed.
(photo).
4   If the centrifugal advance mechanism and or the distributor
shaft in the body, are found to be worn and sloppy, you should
check whether spares are available and consider an exchange
distributor. Very often in fine mechanisms as distributors, it will
not be a case of needing to renew just one component. (photo).
5   The centre shaft is removed after drifting out the parallel pin
which secures the shaft lower collar in position. Recover the
washer from under the collar and slide out the shaft.
6   Check the contact breaker points as described in Section 2.
Check the distributor cap for signs of tracking, indicated by a
thin black line between the segments. Replace the cap if any
signs of tracking are found.
7   If the metal portion of the rotor arm is badly burnt or loose
renew the arm. If slightly burnt clean the arm with a fine file.
Check that the carbon brush moves freely in the centre of the
distributor cover.
8   Examine the balance weights and pivot pins for wear, and
renew the weights or centre shaft if a degree of wear is found.
9   Examine the centre shaft and the fit of the cam assembly on
the shaft. If the clearance is excessive compare the items with
new units and renew either, or both, if they show excessive wear.
10 If the shaft is a loose fit in the distributor bushes and can be
seen to be worn it will be necessary to fit a new shaft and
bushes.
11 Examine the length of the balance weight springs and
compare them with new springs. If they have stretched, they
should be renewed.
12 Reassembly is a straightforward reversal of the dismantling
process, but there are several points which should be noted.
13 Lubricate the balance weights and other parts of the
mechanical advance mechanism, and the distributor centre shaft,
with Castrol GTX oil during assembly. Do not oil excessively but
ensure these parts are adequately lubricated.
14 Check the action of the weights in the fully advanced and
fully retarded positions and ensure they are not binding.
15 Finally, set the contact breaker gap to the correct clearance
of 0.0165 to 0.0189 in (0.42 to 0.48 mm).

8   Ignition timing

1   If the clamp plate pinch nut has been loosened on the
distributor and the static timing lost or if for any other reason,
it is wished to set the ignition timing proceed as follows:
2   Refer to Section 3 and check the contact breaker points.
Reset as necessary.
3   Assemble the clamp plate to the distributor body but do not
tighten the pinch nut fully.
4   It will be as well to remove the camshaft covers so that the
exact position of the engine assembly can be determined.
5   Slowly turn the crankshaft until the groove in the crankshaft
pulley lines up with the static ignition point on the timing belt
cover, and the camshaft lobes operating the No 1 cylinder valves
are pointing upwards and inclined towards the centre line of the
engine. The piston in the No 1 cylinder should then be just
approaching Top Dead Centre and just about to commence the
power stroke. (photo).
6   Provided the distributor shaft has not been disturbed or has
been refitted correctly the rotor arm should now point to the
position of the insert in the distributor which is connected to the
spark plug in the No 1 cylinder.

7   With the distributor body lightly clamped with the retaining plate; slowly rotate the distributor body until the contact breaker points are just beginning to open. Tighten the distributor clamp.

8   Difficulty is sometimes experienced in determining exactly when the contact breaker points open. This can be ascertained most accurately by connecting a 12v bulb in parallel with the contact breaker points (one lead to earth and the other from the distributor low tension terminal). Switch on the ignition and turn the distributor body until the bulb lightd up, indicating the points have just opened.

9   If it was not found possible to align the rotor arm correctly one of two things is wrong. Either the distributor drive shaft has been incorrectly fitted in which case the distributor must be removed and replaced as described in Section 6 of this chapter and chapter 1; or the distributor cam assembly has been

incorrectly fitted on the drive shaft. To rectify this, it will be necessary to partially dismantle the distributor and check the position of the cam assembly on the centrifugal advance mechanism: it may be 180° out of position.

10 It should be noted that this adjustment is nominal and the final adjustment should be made under running conditions.

11 First start the engine and allow to warm up to normal running temperature, then in a road test accelerate the car in top gear from 30 to 50 mph whilst listening for heavy pinking of the engine. If pinking is heard, the ignition needs to be slightly retarded until only the faintest trace of pinking can be heard under these operating conditions.

12 Since the ignition advance adjustment enables the firing point to be related correctly to the grade of fuel used, the fullest advantage of any change of fuel will only be obtained by re-adjustment of the ignition settings as described in paragraph 11.

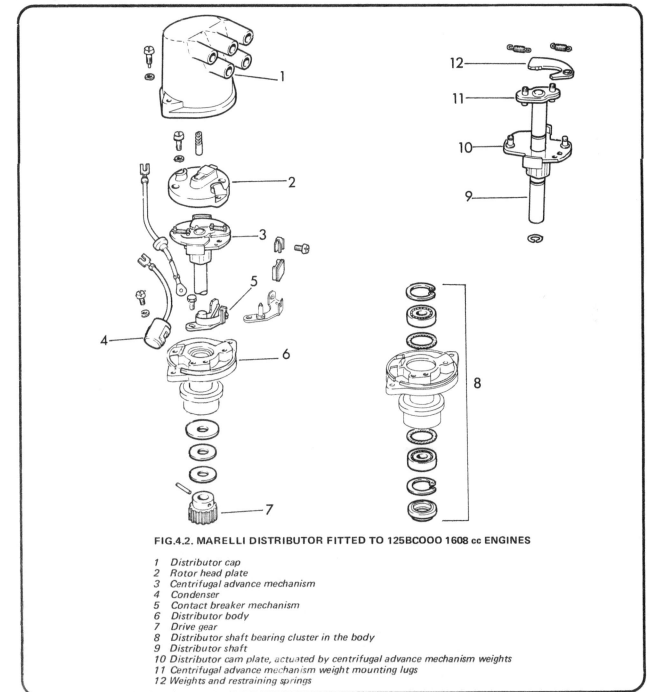

FIG.4.2. MARELLI DISTRIBUTOR FITTED TO 125BC000 1608 cc ENGINES

1   Distributor cap
2   Rotor head plate
3   Centrifugal advance mechanism
4   Condenser
5   Contact breaker mechanism
6   Distributor body
7   Drive gear
8   Distributor shaft bearing cluster in the body
9   Distributor shaft
10 Distributor cam plate, actuated by centrifugal advance mechanism weights
11 Centrifugal advance mechanism weight mounting lugs
12 Weights and restraining springs

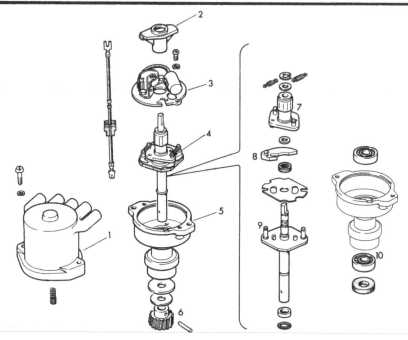

FIG.4.3. MARELLI DISTRIBUTOR ON 1608 cc ENGINES, TYPE SHOWN IN PHOTOGRAPHS

1   Distributor cap
2   Rotor arm
3   Contact breaker and condenser on mounting plate
4   Centrifugal advance mechanism
5   Distributor body
6   Distributor drive gear
7   Distributor cam plate - coupled to centrifugal advance mechanism weights
8   Weights
9   Shaft and advance mechanism drive plate
10 Shaft bearings

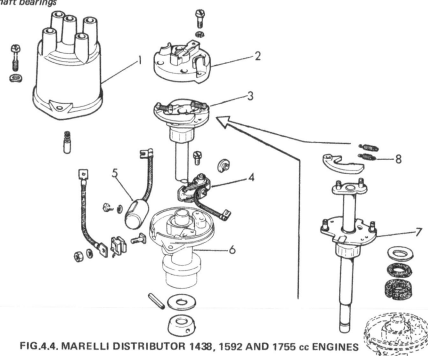

FIG.4.4. MARELLI DISTRIBUTOR 1438, 1592 AND 1755 cc ENGINES

1   Distributor cap
2   Rotor plate mounted on advance mechanism
3   Centrifugal advance mechanism
4   Contact breaker mechanism
5   Condenser
6   Distributor body
7   Distributor cam plate
8   Centrifugal advance mechanism weights and restraining springs

Common spark plug conditions

NORMAL

Symptoms: Brown to grayish-tan color and slight electrode wear. Correct heat range for engine and operating conditions.
Recommendation: When new spark plugs are installed, replace with plugs of the same heat range.

WORN

Symptoms: Rounded electrodes with a small amount of deposits on the firing end. Normal color. Causes hard starting in damp or cold weather and poor fuel economy.
Recommendation: Plugs have been left in the engine too long. Replace with new plugs of the same heat range. Follow the recommended maintenance schedule.

CARBON DEPOSITS

Symptoms: Dry sooty deposits indicate a rich mixture or weak ignition. Causes misfiring, hard starting and hesitation.
Recommendation: Make sure the plug has the correct heat range. Check for a clogged air filter or problem in the fuel system or engine management system. Also check for ignition system problems.

ASH DEPOSITS

Symptoms: Light brown deposits encrusted on the side or center electrodes or both. Derived from oil and/or fuel additives. Excessive amounts may mask the spark, causing misfiring and hesitation during acceleration.
Recommendation: If excessive deposits accumulate over a short time or low mileage, install new valve guide seals to prevent seepage of oil into the combustion chambers. Also try changing gasoline brands.

OIL DEPOSITS

Symptoms: Oily coating caused by poor oil control. Oil is leaking past worn valve guides or piston rings into the combustion chamber. Causes hard starting, misfiring and hesitation.
Recommendation: Correct the mechanical condition with necessary repairs and install new plugs.

GAP BRIDGING

Symptoms: Combustion deposits lodge between the electrodes. Heavy deposits accumulate and bridge the electrode gap. The plug ceases to fire, resulting in a dead cylinder.
Recommendation: Locate the faulty plug and remove the deposits from between the electrodes.

TOO HOT

Symptoms: Blistered, white insulator, eroded electrode and absence of deposits. Results in shortened plug life.
Recommendation: Check for the correct plug heat range, over-advanced ignition timing, lean fuel mixture, intake manifold vacuum leaks, sticking valves and insufficient engine cooling.

PREIGNITION

Symptoms: Melted electrodes. Insulators are white, but may be dirty due to misfiring or flying debris in the combustion chamber. Can lead to engine damage.
Recommendation: Check for the correct plug heat range, over-advanced ignition timing, lean fuel mixture, insufficient engine cooling and lack of lubrication.

HIGH SPEED GLAZING

Symptoms: Insulator has yellowish, glazed appearance. Indicates that combustion chamber temperatures have risen suddenly during hard acceleration. Normal deposits melt to form a conductive coating. Causes misfiring at high speeds.
Recommendation: Install new plugs. Consider using a colder plug if driving habits warrant.

DETONATION

Symptoms: Insulators may be cracked or chipped. Improper gap setting techniques can also result in a fractured insulator tip. Can lead to piston damage.
Recommendation: Make sure the fuel anti-knock values meet engine requirements. Use care when setting the gaps on new plugs. Avoid lugging the engine.

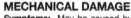

MECHANICAL DAMAGE

Symptoms: May be caused by a foreign object in the combustion chamber or the piston striking an incorrect reach (too long) plug. Causes a dead cylinder and could result in piston damage.
Recommendation: Repair the mechanical damage. Remove the foreign object from the engine and/or install the correct reach plug.

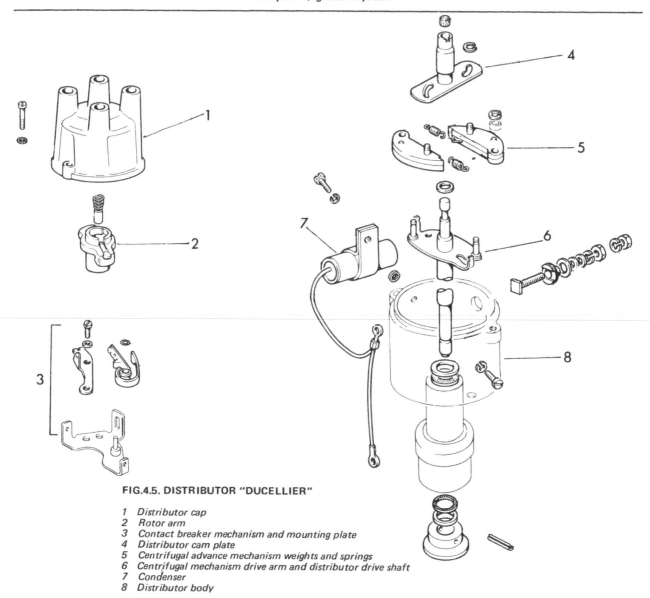

FIG.4.5. DISTRIBUTOR "DUCELLIER"

1   Distributor cap
2   Rotor arm
3   Contact breaker mechanism and mounting plate
4   Distributor cam plate
5   Centrifugal advance mechanism weights and springs
6   Centrifugal mechanism drive arm and distributor drive shaft
7   Condenser
8   Distributor body

9 Spark plugs and leads

1   The correct functioning of the spark plugs is vital for the correct running and efficiency of the engine.

2   At intervals of 6,000 miles (10,000 Km) the plugs should be removed, examined, cleaned, and if worn excessively, replaced. The condition of the spark plug will also tell much about the overall condition of the engine.

3   If the insulator nose of the spark plug is clean and white, with no deposits, this is indicative of a weak mixture, or too hot a plug. (A hot plug transfers heat away from the electrode slowly - a cold plug transfers heat away quickly).

4   The plugs fitted as standard are one of those as specified at the beginning of this Chapter. If the top and insulator nose is covered with hard black-looking deposits, then this is indicative that the mixture is too rich. Should the plug be black and oily, then it is likely that the engine is fairly worn, as well as the mixture being too rich.

5   If the insulator nose is covered with light tan to greyish brown deposits, then the mixture is correct and it is likely that the engine is in good condition.

6   If there are any traces of long brown tapering stains on the outside of the white portion of the plug, then the plug will have to be renewed, as this shows that there is a faulty joint between the plug body and the insulator, and compression is being allowed to leak away.

7   Plugs should be cleaned by a sand blasting machine, which will free them from carbon more than cleaning by hand. The machine will also test the condition of the plugs under compression. Any plug that fails to spark at the recommended pressure should be renewed.

8   The spark plug gap is of considerable importance, as, if it is too large or too small, the size of the spark and its efficiency will be seriously impaired. For the best results the spark plug gap should be set in accordance with the 'Specifications' at the beginning of this Chapter.

9   To set it, measure the gap with a feeler gauge, and then bend open, or close, the outer plug electrode until the correct gap is achieved. The centre electrode should never be bent as this may crack the insulation and cause plug failure if nothing worse.

10 When replacing the plugs, remember to use new washers, and replace the leads from the distributor in the correct firing order, which is 1,3,4,2, No 1 cylinder being the one nearest the distributor.

11 The plug leads require no routine attention other than being kept clean and wiped over regularly. At intervals of 6,000 miles (10,000 Km), however, pull each lead off the plug in turn and remove them from the distributor. Water can seep down into these joints giving rise to a white corrosive deposit which must be carefully removed from the end of each cable.

10 Coil and electronic ignition

1 The coil is a auto-transformer and has two sets of windings wound around a core of soft iron wires. The resistances of the two windings are given in the specification at the beginning of this chapter.

2 If the coil is suspect then these resistances may be checked and if faulty it may readily be replaced after undoing the mounting bolts.

3 The electronic ignition system comprises a transistorized switching circuit and a special ignition coil. The purpose of the switching circuit is to use a small current flowing through the contact breaker mechanism to trigger a larger current flowing through the primary windings of the coil. The interruption of the larger current in the primary winding will induce a greater voltage in the secondary, and hence the sparks will be more powerful. A transistor switching system is necessary because if the contact breaker mechanism were required to break - interrupt larger currents the wear and erosion would increase to impractical levels. The larger primary winding current is not only desired because of the more powerful sparks produced, but also because the larger current will offset the effect of the coils impedance which tends to reduce the energy of sparks at high engine speeds.

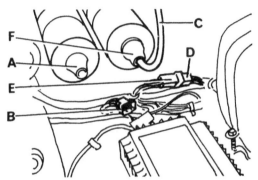

FIG. 4.7. ELECTRONIC IGNITION

A Coil socket (Auxiliary ignition circuit only)
B Connector (Auxiliary ignition circuit only)
C H.T. lead (Common) fitted to electronic ignition circuit coil
D Connector (Common)
E Connector (Electronic ignition circuit only)
F Coil socket (Electronic ignition circuit only)

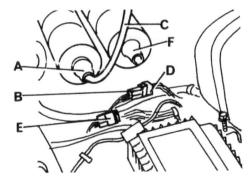

FIG. 4.8. AUXILIARY IGNITION

A Coil socket (Auxiliary ignition circuit only)
B Connector (Auxiliary ignition circuit only)
C H.T. lead (Common) fitted to auxiliary ignition circuit coil
D Connector (Common)
E Connector (Electronic ignition circuit only)
F Coil socket (Auxiliary ignition circuit only)

11 Ignition faults - symptoms, reasons and remedies

Engine troubles normally associated with, and usually caused by, faults in the ignition system are:
a) Failure to start when the engine is turned.
b) Uneven running due to misfiring or mistiming.
c) Smooth running at low engine revolutions but misfiring when under load or accelerating or at high constant revolutions.
d) Smooth running at higher revolutions and misfiring or cutting-out at low speeds.

a) First check that all wires are properly connected and dry. If the engine fails to catch when the starter is operated do not continue for more than 5 or 6 short burst attempts or the battery will start to get tired and the problem made worse. Remove the spark plug lead from a plug and turn the engine again holding the lead (by the insulation) about ¼ inch from the side of the engine block. A spark should jump the gap audibly and visibly. If it does then the plugs are at fault or the static timing is very seriously adrift. If both are good, however, then there must be a fuel supply fault, so go on to that.

If no spark is obtained at the end of a plug lead detach the coil HT lead from the centre of the distributor cap and hold that near the block to try and find a spark. If you now get one, then there is something wrong between the centre terminal of the distributor cap and the end of the plug lead. Check the cap itself for damage or damp, the 4 terminal lugs for signs of corrosion, the centre carbon brush in the top (is it jammed?) and the rotor arm.

If no spark comes from the coil HT lead check next that the contact breaker points are clean and that the gap is correct. A quick check can be made by turning the engine so that the points are closed. Then switch on the ignition and open the points with an insulated screwdriver. There should be a small visible spark and, once again, if the coil HT lead is held near the block at the same time a proper HT spark should occur. If there is a big fat spark at the points but none at the HT lead then the condenser is done for and should be renewed.

If neither of these things happen then the next step in this tale of woe is to see if there is any current (12 volts) reaching the coil (+ terminal). (One could check this at the distributor, but by going back to the input side of the coil a longer length of possible fault line is bracketed and could save time).

With a 12v bulb and piece of wire suitably connected (or of course a voltmeter if you have one handy) connect between the + or SW terminal of the coil and earth and switch on the ignition. No light means no volts so the fault is between the battery and the coil via the ignition switch. This is moving out of the realms of just ignition problems - the electrical system is becoming involved in general. So to get home to bed get a piece of wire and connect the + terminal of the coil to the + terminal on the battery and see if sparks occur at the HT leads once more.

If there is current reaching the coil then the coil itself or the wire from its - terminal to the distributor is at fault. Check the - or CB terminal with a bulb with the ignition switched on. If it fails to light then the coil is faulty in its LT windings and needs renewal.

b) Uneven running and misfiring should first be checked by seeing that all leads, particularly HT, are dry and connected properly. See that they are not shorting to earth through broken or cracked insulation. If they are, you should be able to see and hear it. If not, then check the plugs, contact points and condenser just as you would in case of total failure to start.

c) If misfiring occurs at high speed check the points gap, which may be too small, and the plugs in that order. Check also that the spring tension on the points is not too light this causing them to bounce. This requires a special pull balance so if in doubt it will be cheaper to buy a new set of contacts rather than go to a garage and get them to check it. If the trouble is still not cured then the fault lies in the carburation or engine itself.

d) If misfiring or stalling occurs only at low speeds the points gap is possibly too big. If not, then the slow running adjustment on the carburettor needs attention.

Chapter 5 Clutch

Contents

Specifications

	124ACOOO/132 series/1600		125BCOOO/132 series/1800	
Type	Single plate dry			
Throwout mechanism	Diaphragm spring			
Driven plate	With friction linings			
Lining outside diameter	7.87 in	7.87 in	8.464 in	8.307 in
Lining inside diameter	5.59 in	5.118 in	5.708 in	5.708 in
Clutch pedal free travel	1.00 in			

Torque wrench settings:	lb ft
Clutch mechanism to flywheel bolts	18
Clutch bellhousing to engine bolts	61.5

1 General description

The reason for the clutch unit being fitted between the engine and the transmission is so that there can be some form of connection between the engine and the tranmission. It enables the engine torque to be progressively applied to the transmission so enabling the car to move off gradually from rest, and then for the gear to be changed easily as the speed increases or decreases.

The main parts of the clutch assembly are: the clutch driven plate assembly, the cover assembly and the release bearing assembly. When the clutch is in use the driven plate assembly, being splined to the clutch shaft, is sandwiched between the flywheel and pressure plate by the diaphragm spring. Engine torque is therefore transferred from the flywheel to the clutch driven plate assembly and then to the transmission unit clutch shaft.

By depressing the clutch pedal, the clutch release bearing assembly is drawn against the diaphragm spring by a cable connecting the pedal to the release bearing throwout yoke. The pressure on the driven plate assembly by the diaphragm spring is released and therefore the drive between the engine and transmission is broken.

When the clutch pedal is released, the diaphragm spring forces the pressure plate into contact with the high friction linings on the clutch drive plate, at the same time forcing the clutch driven plate assembly against the flywheel and so taking the drive up.

As the friction linings on the clutch driven plate wear, the pressure plate automatically moves closer to the driven plate to compensate. This makes the centre of the diaphragm spring move nearer to the release bearing, so decreasing the release bearing clearance and therefore the clutch free pedal travel. The clearance will have to be adjusted as described in Section 2.

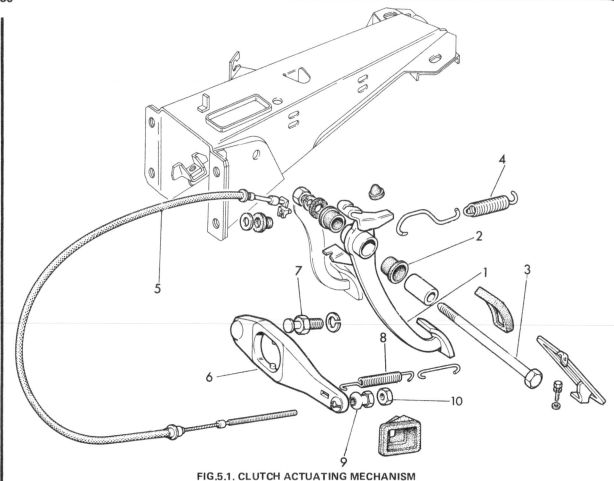

FIG.5.1. CLUTCH ACTUATING MECHANISM

1	Clutch pedal	4	Pedal return spring	6	Actuating lever	9 Pedal free travel adjusting
2	Pedal pivot bushes	5	Connecting cable and	7	Lever pivot ball pin	nut
3	Pivot bolt		sleeve	8	Lever return spring	10 Locking nut

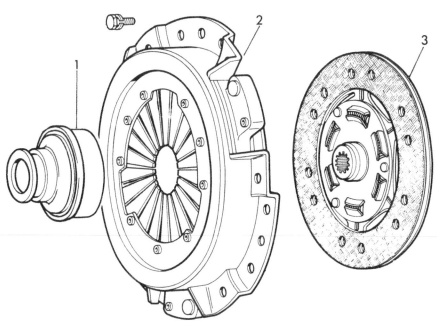

FIG.5.2 CLUTCH MECHANISM

1 Thrust ball race - to release clutch
2 Diaphragm spring assembly
3 Friction plate

2 Clutch pedal travel adjustment

1 When the clutch free pedal travel becomes less than about 1 inch (23 to 25 mm), the position of the control cable end in the clutch actuating lever should be adjusted.

2 Jack up the front of the car and support on firmly based chassis stands. Car ramps will also give sufficient car height to enable access to be gained underneath the car to the clutch bellhousing.

3 Working under the car hold the special actuating rod adjusting nut and release the smaller lock nut. (photo).

4 Now screw the special nut to increase pedal travel as necessary and once the correct free travel has been achieved lock the special nut with the lock nut.

5 Should adjustment or operation not be satisfactory inspect the cable grommets and guide sleeve to ensure that the movement of the inner cable is not inhibited. If any kink or warp in the sleeve or end grommets is noticed, this may account for faulty engagement of the clutch because the trapped cable would not allow the clutch mechanism to engage properly.

3 Clutch - removal and replacement

1 Refer to Chapter 6 and remove the tranmission unit from the car. (photo).

2 Mark the relative positions of the clutch assembly and flywheel and unscrew the six bolts securing the clutch asssembly to the flywheel. Turn the bolts half a turn at a time in a diagonal manner to prevent distortion to the cover flange.

3 As the bolts are being undone check that the cover flange is being released so sliding up the dowels, as it could fly off under the action of the diaphragm spring when all the bolts are removed if it is binding on one or more dowels.

4 With all the bolts and spring washers removed lift the clutch assembly off the ends of the locating dowels. The driven plate or clutch disc will fall out at this stage as it is not attached to either the clutch assembly or the flywheel (photo).

5 It is important that no oil or grease gets onto any clutch parts. It is advisable to replace the clutch with clean hands and to wipe down the pressure plate and flywheel faces with a clean dry rag before assembly begins.

6 Place the clutch disc against the flywheel with the longer end of the hub facing away from the flywheel. On no account should the clutch disc be replaced with the longer end of the centre hub facing towards the flywheel on reassembly as it will be found quite impossible to operate the clutch with the friction disc in this position. (photo).

7 Replace the clutch cover assembly loosely on the dowels, aligning the mating marks if the original parts are being refitted. Refit the six bolts and spring washers and tighten them finger tight so that the clutch disc is gripped but can still be moved.

8 The clutch disc must now be centralised so that when the engine and transmission unit are mated, the clutch shaft splines will pass through the splines in the centre of the driven plate hub.

9 Centralisation can be carried out quite easily by inserting a round bar or long screwdriver through the hole in the centre of the clutch, so that the end of the bar or screwdriver rests in the small hole in the end of the crankshaft containing the clutch shaft spigot bearing bush.

10 Using the clutch shaft spigot bearing as a fulcrum, moving the bar sideways or up and down will move the clutch disc in whichever direction is necessary to achieve centralisation.

11 Centralisation is easily judged by removing the bar and viewing the driven plate hub in relation to the hole in the release bearing. When the hub appears exactly in the centre of the release bearing hole, all is correct. (photo).

12 Tighten the clutch bolts firmly in a diagonal manner to ensure that the cover plate is pulled down evenly and without distortion of the flange. Finally tighten the bolts down to a

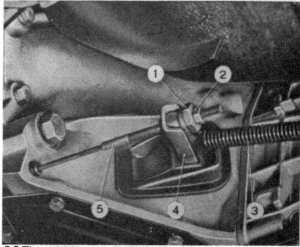

2.3 The pedal free travel adjusting special nut and locking nut
1 Adjusting nut
2 Locknut
3 Return spring
4 Release lever
5 Cable

3.1 The clutch exposed when the gearbox is removed

3.4 The clutch mechanism released from the flywheel

3.6 New clutch ready for assembly onto
the flywheel

3.11 The clutch friction disc centralized
on flywheel and release mechanism

6.1 The clutch release bearing and actuating lever in the bellhousing

torque wrench setting of 0.8 to 1 Kgm (5.8 to 7.2 lbs/ft)
(photo).
13 Lightly smear a little Castrol LM Grease onto the crankshaft
spigot bearing bush.

4   Clutch - dismantling and reassembly

1   It is not practical to dismantle the pressure plate assembly
and the term 'clutch dismantling and reassembly' is usually used
for simply fitting a new clutch plate assembly and clutch disc.
2   If a new clutch disc is being fitted it is false economy not to
renew the release bearing at the same time. This will preclude
having to replace it at a later date when wear on the clutch
linings is still very small.
3   If the pressure plate assembly requires renewal (see Section
5) an exchange unit must be purchased. This will have been
accurately set up and balanced to very fine limits.

5   Clutch - inspection

1   Examine the clutch disc friction linings for wear and loose
rivets, and the disc for rim distortion, cracks, broken hub springs
and worn splines. The surface of the friction linings may be
highly glazed, but as long as the clutch material pattern can be
clearly seen this is satisfactory. Compare the amount of lining
wear with a new clutch disc at the stores in your local garage,
and if the linings are more than three quarters worn replace the
disc.
2   Always renew the clutch driven plate as an assembly to
preclude further trouble, but, if it is wished to merely renew the
linings, the rivets should be drilled out and not knocked out with
a punch. The manufacturers do not advise that only the linings
are renewed and personal experience dictates that it is far more
satisfactory to renew the driven plate complete than try to
economise by fitting only new friction linings.
3   Check the machined faces of the flywheel and the pressure
plate. If either is grooved they should be machined until smooth
or renewed.
4   If the pressure plate is cracked or split or if the pressure of
the diaphragm spring is suspect it is essential that an exchange
unit is fitted.
5   Check the release bearing for smoothness of operation. There
should be no harshness and no slackness in it. It should spin
reasonably freely, allowing for the fact that it has been prepacked with grease.
6   Should it be considered necessary to carry out a full
dimensional check, first place the clutch cover assembly on a
base plate with a piece of suitable packing 7.9 mm (0.311 inch)
between the cover and base plate. Depress the release flange
several times to settle the clutch.
7   It should be observed that with a withdrawal travel of 8 mm
(0.315 inch) the pressure plate should be 1.8 to 1.9 mm (0.017
to 0.075 inch) out. The minimum permissible amount is 1.4 mm
(0.055 inch).
8   Measure the distance 'X' which should be 42 mm + 1.3/-1
mm (1.653 inch + 0.051/- 0.039 inch). If the measurement is
outside the limit measure the thickness of the friction ring (8)
which when new is 1.9 to 2 mm (0.075 to 0.079 inch) and make
any adjustment on the distance 'X' to compensate for friction
ring wear.

6   Clutch release bearing - removal and replacement

1   With the engine and transmission unit separated to provide
access to the clutch, attention can be given to the release bearing
located in the bellhousing over the clutch shaft. (photo).
2   The release bearing is a relatively inexpensive but important
component and, unless it is nearly new, it is a mistake not to
replace it during the overhaul of the clutch.
3   To remove the release bearing, first unhook the release

throwout arm spring from the throwout arm.

4 Extract the split pin securing the stretcher yoke to the actuating arm cotter pin. Lift away the plain washer and withdraw the cotter pin.

5 Remove the rubber gaiter and draw the release actuating arm and bearing assembly from the clutch shaft and actuating arm pivot.

6 Separate the release bearing assembly from the actuating arm.

7 Inspect the release bearing as described in Section 5, paragraph 5.

8 Reassembly is the reverse sequence to dismantling. Smear a little Castrol LM Grease onto the release bearing housing hub and check that the hub slides up and down freely.

7 Clutch shaft spigot bearing

To act as a support for the clutch shaft spigot a sealed ball race bearing is pressed into the end of the crankshaft. If this bearing is worn or dry it is removed with a hook shaped extractor. The hook passes through the centre bore and then is positioned behind the inner face so that the whole bearing may be pulled out of the bore in the crankshaft and flywheel. It should not be necessary to remove the flywheel.

The new bearing can be forced into place with a tubular drift which should only bear on the outer race.

8 Clutch pedal and Bowden cable - removal and refitting

1 As mentioned earlier the clutch on the FIAT 124 Sport is actuated by a wholly mechanical system. The clutch and brake pedals are mounted on a common fulcrum, attached to a bracket on the toe board. The procedure for removing the clutch pedal is identical to that for the brake pedal, details of which are given in Chapter 9.

2 The removal of the cable which links the clutch pedal to the actuating lever is straight forward. Begin by jacking up the front of the car so that access can be gained to the connection of the cable and actuating lever. It is wise to position chassis stands under strong points at the front of the car as a safety measure.

3 From underneath the car undo and remove first the locking nut and then the special nut on the threaded end of the inner cable. Once these nuts have been removed the inner cable may be pulled away from the actuating lever, and then the cable sleeve and inner cable away from the sleeve stop on the clutch bellhousing.

4 With the lever end of the cable free the pedal end may be disengaged from the fork in the top of the pedal arm, and the whole cable assembly extracted from the engine compartment of the car.

5 Refitting a new cable is simply the reversal of removal, remembering to adjust the pedal free travel as described in Section 2 of this chapter.

9 Clutch squeal - diagnosis and cure

1 If on taking up the drive or when changing gear, the clutch squeals, this is sure indication of a badly worn clutch release bearing. As well as regular wear due to normal use, wear of the clutch release bearing is much accentuated if the clutch is ridden, or held down for long periods in gear, with the engine running. To minimise wear of this component the car should always be taken out of gear at traffic lights and for similar hold-ups.

2 The clutch release bearing is not an expensive item.

10 Clutch slip - diagnosis and cure

1 Clutch slip is a self evident condition which occurs when the

clutch friction plate is badly worn, the release arm free travel is insufficient, oil or grease have got onto the flywheel or pressure plate faces, or the pressure plate itself is faulty.

2 The reason for clutch slip is that, due to one of the faults listed above, there is either insufficient pressure from the pressure plate, or insufficient friction from the friction plate to ensure solid drive.

3 If small amounts of oil get onto the clutch, they will be burnt off under the heat of clutch engagement, in the process gradually darkening the linings. Excessive oil on the clutch will burn off leaving a carbon deposit which can cause quite bad slip, or fierceness, spin and judder.

4 If clutch slip is suspected, and confirmation of this condition is required, there are several teste which can be made:

a) With the engine in second or third gear and pulling lightly up a moderate incline, sudden depression of the accelerator pedal may cause the engine to increase its speed without any increase in road speed. Easing off on the accelerator will then give a definite drop in engine speed without the car slowing.

b) Drive the car at a steady speed in top gear and braking with the left leg, try and maintain the same speed by pressing down on the accelerator. Providing the same speed is maintained a change in the speed of the engine confirms that slip is taking place.

c) In extreme cases of clutch slip the engine will race under normal acceleration conditions.

If slip is due to oil or grease on the linings a temporary cure can sometimes be effected by squirting carbon tetrachloride into the clutch. The permanent cure, of course, is to renew the clutch driven plate, and trace and rectify the oil leak.

11 Clutch spin - diagnosis and cure

1 Clutch spin is a condition which occurs when the release arm free travel is excessive; there is an obstruction in the clutch either on the clutch shaft splines or in the operating lever itself; or oil may have partially burnt off the clutch linings and have left a resinous deposit which is causing the clutch disc to stick to the pressure plate or flywheel.

2 The reasons for clutch spin is that due to any, or a combination of, the faults just listed, the clutch pressure plate is not completely freeing from the centre plate even with the clutch pedal fully depressed.

3 If clutch spin is suspected, the condition can be confirmed by extreme difficulty in engaging first gear from rest, difficulty in changing gear, and very sudden take up of the clutch drive at the full depressed end of the clutch pedal travel as the clutch is released.

4 If these points are checked and found to be in order then the fault lies internally in the clutch, and it will be necessary to remove it for examination.

12 Clutch judder - diagnosis and cure

1 Clutch judder is a self evident condition which occurs when the gearbox or engine mountings are loose or too flexible, when there is oil on the face of the clutch friction plate, or when the clutch pressure plate has been incorrectly adjusted.

2 The reason for clutch judder is that due to one of the faults just listed, the clutch pressure plate is not freeing smoothly from the friction disc, and is snatching.

3 Clutch judder normally occurs when the clutch pedal is released in first or reverse gears, and the whole car shudders as it moves backwards or forwards.

Chapter 6 Gearbox

Contents

Specifications

Synchronizers ... ... 1st, 2nd, 3rd and 4th cone, blocker type, 5th spring ring (Porsche) type

Gearbox ratios:

	Four speed Gearbox	Five speed Gearbox		
		1600/1800	1400	1600 option
Fifth	—	0.881	0.912	0.913
Fourth	1.00	1	1	1
Third	1.490	1.361	1.361	1.410
Second	2.300	2.100	2.100	2.175
First	3.750	3.667	3.422	3.797
Reverse	3.870	3.526	3.526	3.655

Gear backlash (measured at teeth) ... ... 0.004 inches

Ball bearings -	Radial play	0.002 inches maximum
-	End play	0.020 inches maximum

Clearance fit of 1st gear and 5th gear on their bushings ... 0.002 to 0.004 inches
Clearance fit of 2nd and 3rd gears on their mainshaft seating ... 0.002 to 0.004 inches
Clearance fit of reverse idler gear on its shaft ... ... 0.002 to 0.004 inches

Torque wrench settings:

	ft lbs
Gearbox to flexible mounting bolt	18
Flexible mounting to rear crossmember nut	18
Crossmember to chassis nut	11
Clutch bellhousing to engine block bolts	61
Bellhousing to gearbox stud nuts (large)	36
(small)	18
Front bearing to layshaft bolt	69
Rear bearing to layshaft nut	87
Transmission coupling spider to main gearshaft nut	58
Bottom cover gearbox nuts	7¼
Gear lever support nut	18
Selector bar locating spring cover nut	18
Selector fork fixing bolt	8.7
Speedometer drive bracket to gearbox nut	7¼

1 General description

FIAT have provided two gearboxes for this series of car, a four forward speed gearbox and one with five forward speeds. Both gearboxes have one reverse gear. The arrangement of components in the two gearboxes is broadly similar, and therefore only the dismantling of the five speed gearbox is described in Section 3 of this Chapter. The four speed gearbox differences are mentioned and in all cases tend to simplify the task of dismantling the gearbox.

The arrangement of the gear trains is that 1st and 4th gears are located in the gearbox casing, reverse and fifth are located in the rear extension of the gearbox. An input shaft from the clutch mechanism runs in ball race in the gearbox casing to a gear train that drives the layshaft.

The gearbox mainshaft runs in three bearings; a needle race in the end of the input shaft, a ball race set in rear of the gearbox main casing, and the third bearing in the rear extension of the gearbox. All the synchromesh systems are mounted on the mainshaft. As well as the two gearboxes there are two types of synchromesh mechanisms employed in the gearboxes. To begin with on the 4th, 3rd, 2nd and 1st gears there is the conventional sliding ring, spring loaded synchromesh cone mechanism. In this mechanism the sliding ring is impelled toward the gear to be engaged by the selector fork. As the ring approaches the gear it passes onto the synchronizing cone. The friction between the sliding ring and cone matches the rotational speeds of the cone (and hence gear to be engaged) and the sliding ring which is splined on a hub attached to the gearbox output main shaft. When the speeds are matched the sliding ring can engage the dog tooth wheel attached to the gear being selected.

The fifth gear synchromesh mechanism is based on a Porsche design, this mechanism can be found on other FIAT cars. It comprises a 'Porsche hub' which is splined onto the output shaft, a sliding sleeve which is dogged to the hub, a synchronizing ring, spring pieces, locking pieces, and finally a synchronizing ring gear.

The mechanism works as follows:-

The gear selector fork impels the sliding sleeve toward the gear to be engaged. As the sleeve approaches the gear it passes over the synchronizing ring, thereby compressing it. When the ring is contracted the ends abut a locking piece. The movement of this locking piece is restrained by two spring strips, which lie underneath the synchronizing ring and which themselves abut a stop keyed into the hub of the gear to be engaged.

The friction force between ring and the sleeve acts to match the rotational speeds of the sliding sleeve and the ring (and hence the gear being selected). Once the speeds are matched the sleeve can engage the ring gear attached to the gear train being engaged.

The selector forks are mounted on rods located in the gearbox casing. Smaller forked fittings - strikers - on the rear end of these rods are engaged by the pawl on the end of the gear lever mechanism.

Finally the speedometer drive is taken off the mainshaft just forward of the rubber flexible joint mounted on the end of the gearbox mainshaft which transmits the engine power from the mainshaft to the propeller shaft and rear wheels.

2 Gearbox - removal and replacement

The gearbox may be removed without disturbing the engine significantly. A special tool to undo the bolts securing the bellhousing to the engine is available, but the bolts can be undone with a long reach socket spanner quite satisfactorily. The procedure for gearbox removal is as follows:-

1 Raise the car so that there is approximately a 2 feet high space underneath the engine and gearbox region. Put chassis stands underneath the front of the car and chock the back wheels. It is essential to ensure that the car is safe in the raised position because when you will be working underneath it, the efforts made to loosen bolts and the like could easily topple an

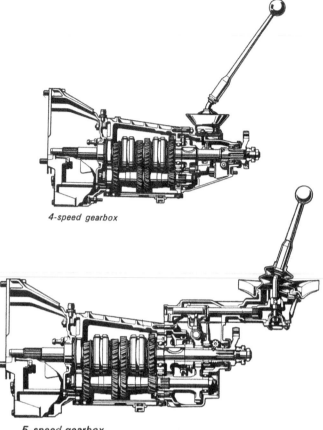

4-speed gearbox

5-speed gearbox

Fig. 6.1. Cross sections of the latest 4 and 5 speed gearboxes fitted to the 1600 and 1800 FIAT 124 Sports

inadequately supported vehicle.

2 The next task is to remove the gearlever extension; loosen the leather boot around the base of the gearlever and push down to reveal the lever joint. A sharp tug should be all that is required tp pull the upper lever from the lever stub. (Photo).

3 Returning to the underneath of the car, begin by removing the clutch cable from the actuating lever and remove the lever return spring. Pull the cable and sleeve clear of the bellhousing and tuck away beside the engine.

4 Next undo the nuts and bolts which secure the bottom flywheel shield on the clutch bellhousing and remove the shield.

5 Remove the electrical connections to the battery; this is an essential safety measure, taken whenever electrical systems in the car are going to be disturbed.

6 Disconnect the leads to the starter motor, and identify them as necessary to make certain that they are refitted to the correct terminals.

7 Undo the large bolts that secure the starter motor to the bellhousing and lift the starter motor away.

8 Continuing within the engine compartment, undo the nuts that retain the twin exhaust pipes to the manifold. Shake the exhaust pipe free of the manifold and then from underneath the car undo the clamp which secures the joint of the front twin pipes to the single pipe silencer system. Remove the forward twin pipe section of the exhaust system from the car.

9 Next undo the nuts and bolts which hold the rubber doughnut flexible joint to the transmission shaft spider at the rear end of the gearbox. The flanged metal strip bolted to the bodyshell across the transmission shaft tunnel, rearwards of the rear gearbox support cross member may now be removed. The

propeller shaft can then be moved aside tio provide more space for the actual removal of the gearbox. (photo).

10 The final tasks include the draining of the gearbox oil and the removal of the speedometer drive cable from the fitting on the rear extension of the gearbox. The cable is retained by a single knurled nut. (photo).

11 Some support for the engine should be devised now. A chassis stand placed underneath the sump and firmly wedged into position should serve adequately. Alternatively if a winch is available, a sling may be passed around the engine and attached to the winch overhead. The engine weight may then be taken by the winch.

12 A trolley jack should now be placed under the gearbox and raised to accept the weight of the box. Once the jack is in position, under the two nuts securing the rear support crossmember to the bodyshell.

13 The four main bolts that join the clutch bellhousing and engine block can now be undone - with FIAT tool No A55035 or a sufficiently long reach socket spanner.

14 The gearbox assembly is free to be moved away from the engine. Make sure that during its first movement backwards, to extract the gearbox drive input shaft from the clutch mechanism and flywheel, the gearbox is kept aligned to the engine. The gearbox input shaft and the clutch friction plate can easily be seriously damaged if the two are allowed to become misaligned. (Photo).

15 Having cleared the gearbox assembly from the engine, it may now be lowered from the car and wheeled away to a work bench ready for dismantling.

16 Refitting the gearbox is the reversal of the removal sequence. Remember that the major joint bolts have precise tightening torques and these are listed at the beginning of this Chapter. The clutch pedal free travel will need setting again too before the car is put on the road.

2.10 Speedometer cable connection

2.14 Gearbox ready for removal

3 Gearbox dismantling

1 Before commencing work, clean the exterior of the gearbox thoroughly using a solvent such as paraffin or Gunk. After the solvent has been applied and allowed to stand for a time, a vigorous jet of water will wash off the solvent together with all the oil and dirt. Finally wipe down the exterior of the unit with a dry non fluffy rag.

2 The first task in dismantling the gearbox is the removal of the gearlever assembly. The lever ball pivot is located in an aluminium alloy housing which is retained on the gearbox rear extension by four nuts on four studs. Recover the gasket that lay between the pivot housing and gearbox extension housing. (photo).

3 Following the removal of the gearlever assembly, the clutch bellhousing must be removed. It is secured by seven nuts on studs screwed into the aluminium alloy gearbox casing. Retrieve the bellhousing/gearbox joint gasket. (photo).

4 The rubber doughnut transmission coupling should be removed next. It is held to the spider on the gearbox output shaft by three nuts and three long bolts.

5 The rubber transmission coupling centering ring is now clearly seen next to the coupling spider. Slide off the dust shield from the end of the main shaft (output) and then remove the circlip ('snapring') retains the coupling centring fitting on the mainshaft. The centring fitting may now be pulled off the shaft with either a puller or judiciously placed screwdrivers.

6 Next flatten the tab on the mainshaft nut locking washer so that the nut may be undone and the spider pulled off the mainshaft. (photo).

7 The speedometer drive fitting should be removed next. It is retained by a single nut on a stud in the gearbox rear extension. (photo).

8 On five speed gearboxes there is a gearlever mechanism replay lever mounted on a spindle in the top of the rear extension to the gearbox. The spindle is extracted after first removing the small plate on the right hand side of the rear extension to the

2.2 Lever joint revealed

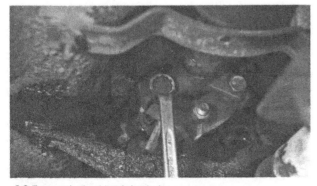

2.9 Removal of rubber joint bolts - mainshaft

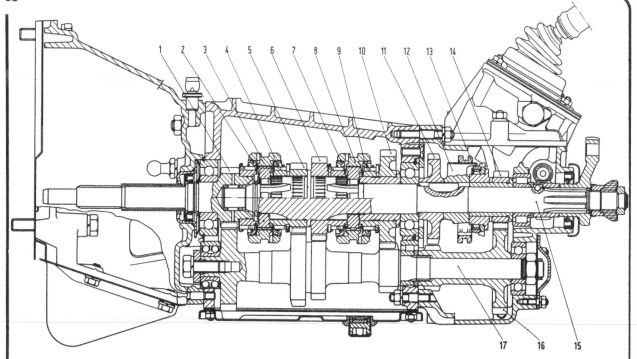

FIG. 6.2. CROSS SECTION OF AN EARLIER STANDARD OF 5 SPEED GEARBOX.
Later standards differ only in exact gear ratios and gearshift mechanism

1 Constant mesh gear pinion to layshaft and 4th gear direct drive
2 Synchronizing cone ring - 4th gear
3 Synchronizing sliding ring. 4th/3rd gears
4 Synchronizing cone ring - 3rd gear
5 3rd gear wheel on mainshaft
6 2nd gear wheel
7 Synchronizing cone ring. 2nd gear
8 Sliding sleeve. 2nd/1st gear selection
9 Synchronizing cone ring. 1st gear

10 1st gear wheel
11 Reverse gear wheel
12 Fifth gear synchronizer hub
13 Sliding sleeve. 5th gear selection
14 5th gear wheel and synchronizer
15 Mainshaft
16 Gear assembly on layshaft for reverse and fifth gears
17 Layshaft with 1st, 2nd and 3rd gear pinions and layshaft drive wheel

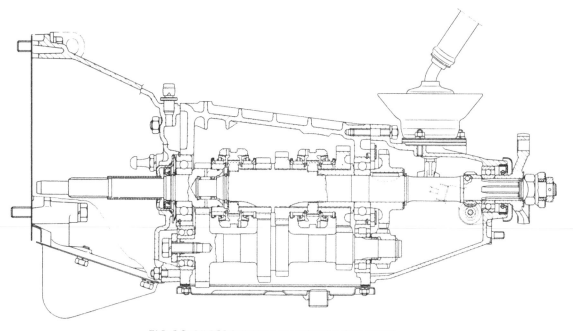

FIG. 6.3. CROSS SECTION OF 4 SPEED GEARBOX
Note the gear assembly in the centre main casing is the same in both 4 and 5 speed gearboxes.
Rear extension of gearbox much simplified, incorporating only the reverse gears which are secured on the respective shafts by circlips

gearbox. The plate is secured by two small nuts on studs screwed into the extension. (photo).

9 Extract the relay lever spindle and lift out the relay lever itself, together with the centring springs and spacers. (photo).

10 Next turn the gearbox upside down and remove the gearbox cover after unscrewing the ten nuts which secure it to the casing.

11 It will now be possible to undo the nuts inside the gearbox casing and the five nuts on the outside of the gearbox extension that hold the two gear housings together. (photo).

12 The gearbox rear extension should now be in a condition to be separated from the main casing. If dismantling a four speed gearbox or an early pattern five speed gearbox which does not have an extended gearlever mechanism in a special alloy housing, it will be necessary to disengage the gearlever pawl from the gear selector bar strikers before the rear extension can be removed from the main casing. In such cases remove the gear lever stop screw in the side of the rear extension so that the lever can be pushed to disengage the pawl on its lower end from the striker rods in the gearbox.

13 The rear extension can now be parted from the main gearbox casing. Do not use a chisel or a screwdriver to prise the extension off, both casings are of aluminium alloy and therefore the joint surface can too easily be damaged. A puller acting on the gearbox main shaft (output) and the rear support fitting which is bolted to the rear end of the extension should give the desired smooth force that will ease the extension off the main casing and in particular the mainshaft rear bearing. On five speed gearboxes the layshaft extension also runs in a bearing housed in the gearbox rear extension casing. (photo).

14 The gearbox is now ready for the final stages of dismantling. (photo).

15 Resume by removing the selector spring retainer plate, springs and balls. The plate is held on the rear right hand corner of the gearbox casing by two bolts. There are three springs and three inserts which register into the appropriate rods in the selector mechanism. Ensure that the springs and inserts are stored so that they are refitted into exactly the same hole from whence they came. (photo).

16 Remove the reverse and fifth gear selector rod and selector fork. (photo).

17 Push the remaining two selector forks so that they each engage a gear, thereby locking the mainshaft, layshaft and input drive shaft.

18 On five speed gearboxes undo the nut on the end of the extended layshaft and pull off the rear bearing race. Then slide off the layshaft extension together with the reverse idler gear (photo).

19 On four speed gearboxes, the layshaft is not extended and all that needs to be done at this stage is to remove the circlip on the rear end of the layshaft to permit the reverse gear pinion to be slid off the end of the layshaft.

20 On five speed gearboxes, working on the mainshaft, slide off the speedometer drive wheel, retrieving the ball bearing which registers it on the mainshaft. Then the roller race bearing may be pulled off the mainshaft, followed by the fifth gear, its synchromesh assembly and lastly the reverse gear wheel. Collect the locating key in the mainshaft.

21 Four speed gearboxes have a different arrangement on the rear part of the mainshaft. Begin by pulling off the rear most ball race, and then slide the speedometer drive wheel off the mainshaft. Again retrieve the single ball that registers the wheel on the shaft. Lastly remove the circlip that retains the reverse gear wheel in place so that it too may be removed together with the key which locates it on the mainshaft. It is advisable to use FIAT tool No 70158 to remove (and refit) the circlip, because the belleville washer behind the clip needs compressing when the circlip is to be removed or fitted.

22 From this point on the dismantling procedure for four and five speed gearboxes is the same.

23 With the mainshaft, layshaft and drive input shafts still locked remove the bolt that secures the forward end of the layshaft in the double ball race in that end of the casing.

24 Tap the layshaft forward just a little so there is sufficient space to get a pair of screwdrivers behind the circlip that is on the periphery of the forward layshaft bearing. Prise the bearing out of the casing and off the shaft. (photo).

25 Then drift the roller race at the rear end of the layshaft out of the gearbox casing and off the layshaft.

26 With both bearing races off the layshaft, it may be extracted from the box and placed aside for inspection.

27 Undo the locking bolts on the remaining two gear selector forks so that forks and their respective rods may be removed from the casing. Retrieve the three interlock plungers that act between the selector rods in a hole in the gearbox casing. (photo).

28 Turning your attention to the front of the gearbox; gently tap the rear side of this forward bearing so that there is sufficient space behind the circlip on its periphery to allow screwdrivers to be used to prise it out of its casing.

29 Pull the drive input shaft, with its bearing, layshaft drive gear and fourth gear synchronizing assembly away from the gearbox casing (photo).

30 The mainshaft bearing at the rear end of the gearbox casing is held in place by a plate and three screws. Remove the screws and the plate. The mainshaft may then be gently tapped rearwards so that as on the other bearings, there is space behind the circlip on the periphery of the bearing to allow a pair of screwdrivers to be used to prise the bearing out of the casing and off the seating on the mainshaft. (Photo).

31 The mainshaft can now be extracted from the gearbox casing.

32 The gearbox is now dismantled and the sub assemblies are ready for individual inspection and renovation.

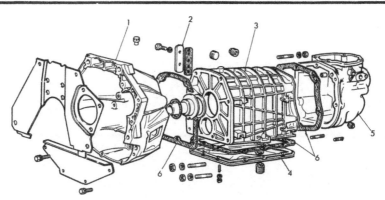

FIG. 6.4. MAIN GEARBOX HOUSINGS AND COMPONENTS

1 Clutch bellhousing
2 Detent ball spring retaining plate
3 Main gearbox casing

4 Main casing cover plate
5 Rear extension housing
6 Gaskets

3.2 Gearshift mechanism removal

3.3 Bellhousing/gearbox separation

3.6 The removal of the mainshaft spider

3.7 Removal of the speedometer drive

3.8 Relay lever spindle retaining plate

3.9 Removal of relay lever, spacers and springs

3.11 Removal of nut securing rear gearbox extension to main casing

3.13 Removal of rear extension casing

3.14 Gearbox ready for final dismantling tasks

3.15 Detent ball spring retaining plate removal

3.16 Removal of reverse/5th gear selector rod

3.18 Sliding of the rear gear trains

3.24 Prising out the forward layshaft bearing

3.27 3rd/4th selector rod removal

3.29 Removal of drive/input shaft

3.30 Removal of mainshaft intermediate bearing

4 Gearbox - examination

1 The gearbox has been stripped, presumably, because of wear or malfunction, possibly excessive noise, ineffective synchromesh or failure to stay in a selected gear. The cause of most gear box ailments is failure of the ball bearings on the input or mainshaft and wear on the synchro rings, cone surfaces and dogs. The nose of the mainshaft which runs in the needle roller bearing in the input shaft is also subject to wear. This can prove very expensive as the mainshaft would need replacement and this represents about 20% of the total cost of a new gearbox. (photo).
2 Examine the teeth of all gears for signs of uneven or excessive wear and, of course, chipping. If a gear on the mainshaft requires replacement check that the corresponding laygear is not equally damaged. If it is the whole laygear may need replacing also.
3 All gears should be a good running fit on the shaft with no signs of rocking. The hubs should not be a sloppy fit on the splines.
4 Selector forks shuold be examined for signs of wear or ridging on the faces which are in contact with the operating sleeve.
5 Check for wear on the selector rod and interlock spool.
6 The ball bearings may not be obviously worn but if one has gone to the trouble of dismantling the gearbox it would be short sighted not to renew them. The same applies to the four synchronizer rings although for these the mainshaft has to be competely dismantled for the new ones to be fitted.

4.1 Gear trains free for examination

7 The input shaft bearing retainer is fitted with an oil seal and this should be removed if these are any signs that oil has leaked past it into the clutch housing or, of course, if it is obviously damaged. The rear extension has an oil seal at the rear as well as a ball bearing race. If either have worn or oil has leaked past the seal the parts should be renewed.
8 Before finally deciding to dismantle the mainshaft and replace parts it is advisable to make enquiries regarding the availability of parts and their cost. It may still be worth considering an exchange gearbox even at this stage. You should reassemble it before exchange.

5 Input drive shaft - dismantling and reassembly

1 Place the input shaft in a vice, splined end upwards and with a pair of circlip pliers remove the circlip which retains the ball bearing in place. Lift away the spring spacer.
2 With the bearing resting on the top of open jaws of the vice and splined end upwards, tap the shaft through the bearing with a soft faced hammer. Note that the offset circlip groove in the periphery of the outer track of the bearing is towards the front of the input shaft.
3 Remove the oil caged needle roller bearing from the centre of the rear of the input shaft if it is still in place.
4 Remove the circlip from the old bearing outer track and transfer it to the new bearing.
5 Reassembly of the input shaft assembly follows the reversal of the dismantling procedure.

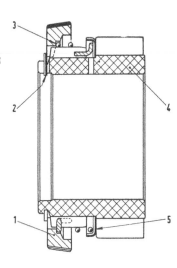

FIG. 6.5. SECTION AND EXPLODED VIEW OF CONE RING TYPE SYNCHRONIZER

1 *Synchronizing cone ring*
2 *Circlip*
3 *Thrust spring*
4 *Gear wheel*
5 *Thrust spring cup seat ring*
6 *Dog teeth on gear hub which sliding sleeves engage*
7 *Sliding sleeve*

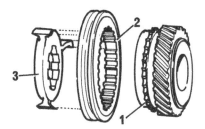

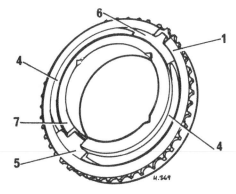

FIG. 6.6. SPRING RING (PORSCHE) TYPE SYNCHRONIZER FITTED TO 5TH GEAR

1 *Spring ring onto which sliding sleeve moves*
2 *Sliding sleeve*
3 *Sliding sleeve hub*
4 *Ring spreader springs*
5 *Spreader spring stop*
6 *Spring ring stop*
7 *Gear hub*

6 Mainshaft - dismantling and reassembly

1 The component parts of the mainshaft are shown in Fig. 6.7.
2 Remove the circlip from the forward end of the mainshaft, so that the spring spacer, synchronizing sleeve hub, 3rd gear wheel and synchromesh assembly can be slid off the main shaft. (photo).
3 The 1st gear wheel and its synchromesh cone assembly may be slid off the rear end of the mainshaft next, followed by the 1st and 2nd gear synchronizing sleeve and hub. Finally the 2nd gear wheel and its synchromesh cone assembly may be slid off the shaft.
4 It will be seen that each synchronizing cone assembly is located on a seating on the gear wheel itself. The assembly is retained by a wide circlip. Once this circlip is removed the synchronizing cone, thrust spring and spring seat ring can be removed from the gear hub.
5 On five speed gearboxes the fifth gear synchromesh system is, as mentioned previously, different form that employed on 1st, 2nd, 3rd and 4th gears. It is similar in one respect however, that it is retained on the gearhub by a wide circlip. Once this clip has been removed the components of this Porsche based mechanism can be easily separated.
6 Fig. 6.6 shows the layout of components of the fifth gear synchronizing mechanism.
7 Having separated all the parts of the mainshaft assembly they can be closely inspected for wear and tear.
8 The radial and end play on the ball races used in the gearbox can be checked with a clock gauge, but even if partially worn, it is advisable to renew the bearings if new ones are readily available. See the specification at the beginning of this Chapter. The radial play of the 1st, 2nd 3rd and reverse gears on their respective bushing and the seating on the mainshaft can be similarly checked against clearances given in the Specification.
9 Having collected all the new replacement parts, proceed to

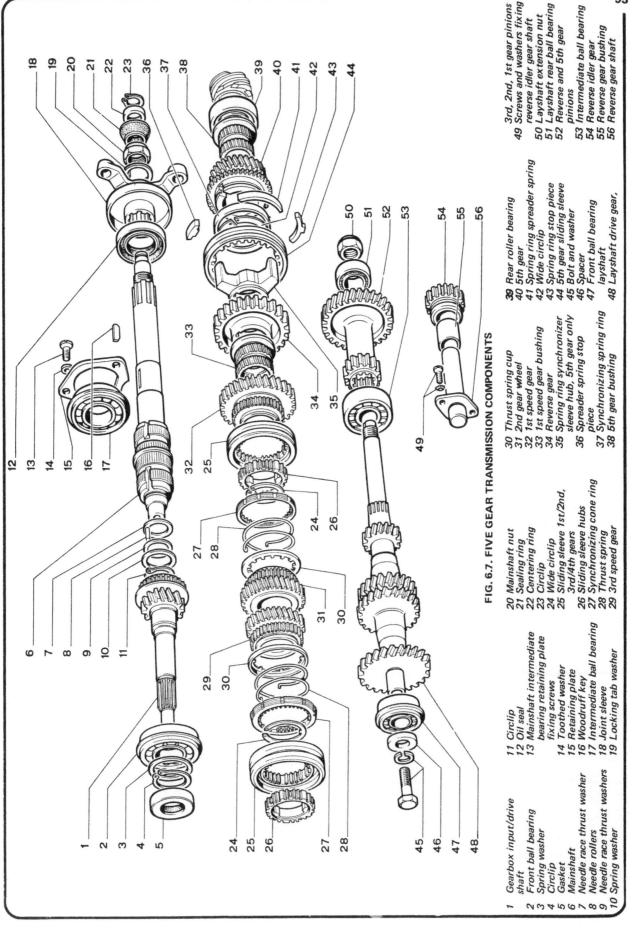

FIG. 6.7. FIVE GEAR TRANSMISSION COMPONENTS

1 Gearbox input/drive shaft
2 Front ball bearing
3 Spring washer
4 Circlip
5 Gasket
6 Mainshaft
7 Needle race thrust washer
8 Needle rollers
9 Needle race thrust washers
10 Spring washer
11 Circlip
12 Oil seal
13 Mainshaft intermediate bearing retaining plate fixing screws
14 Toothed washer
15 Retaining plate
16 Woodruff key
17 Intermediate ball bearing
18 Joint sleeve
19 Locking tab washer
20 Mainshaft nut
21 Sealing ring
22 Centering ring
23 Circlip
24 Wide circlip
25 Sliding sleeve 1st/2nd, 3rd/4th gears
26 Sliding sleeve hubs
27 Synchronizing cone ring
28 Thrust spring
29 3rd speed gear
30 Thrust spring cup
31 2nd gear wheel
32 1st speed gear
33 1st speed gear bushing
34 Reverse gear
35 Spring ring synchronizer sleeve hub, 5th gear only
36 Spreader spring stop piece
37 Synchronizing spring ring
38 5th gear bushing
39 Rear roller bearing
40 5th gear
41 Spring ring spreader spring
42 Wide circlip
43 Spring ring stop piece
44 5th gear sliding sleeve
45 Bolt and washer
46 Spacer
47 Front ball bearing layshaft
48 Layshaft drive gear,
49 Screws and washers fixing reverse idler gear shaft
50 Layshaft extension nut
51 Layshaft rear ball bearing
52 Reverse and 5th gear pinions
53 Intermediate ball bearing
54 Reverse idler gear
55 Reverse gear bushing
56 Reverse gear shaft
49 3rd, 2nd, 1st gear pinions

reassemble the mainshaft gear system as follows.

10 Begin by assembling the synchronizing cone components into their respective gear seating. Be particularly careful to ensure that the returned ends of the cone thrust spring are inserted into the slots in the blocker ring part of the gear seating, without the normal diameter of the spring being altered. Snap the wide circlips in position to retain each synchronizing system on their respective gears.

11 Figs. 6.5 and 6.6 indicate the correct placement of each component of the cone synchronizer and Porsche type synchronizers respectively.

12 Now that the individual gear sub-assemblies are complete they may be arranged onto the mainshaft to build the mainshaft gear assembly. (photo).

13 Start by sliding the 3rd gear and synchromesh assembly onto the forward end of the mainshaft, liberally coating the mainshaft bearing surface with gear oil as the gear is positioned.

14 Next slide on the synchromesh sliding sleeve and its hub onto the forward splined region on the mainshaft. This sliding sleeve engages the third and fourth gears.

15 Fit the spring spacer and circlip which retains the 3rd/4th gear synchromesh components onto the forward end of the mainshaft. Use a tubular drift to compress the spacer and push the circlip into its groove. (photo).

16 Then slide the 2nd gear wheel and cone synchromesh assembly onto the rear end of the mainshaft, right along to the shoulder onto which both it and the 3rd gear assembly abuts.

17 Follow this with the synchromesh sliding sleeve and hub which engage 1st and 2nd gears.

18 Slip the 1st gear and synchromesh assembly over their bush and slide them onto the mainshaft up to the Ist/2nd gear sliding sleeve hub.

19 Ensure that the 1st gear bush insert is correctly aligned, that is with the shoulder rearwards. The mainshaft is now ready to be refitted into the gearbox casing.

6.15 Refitting circlip to retain 3rd gear
and synchronizer on mainshaft

7 Gearbox - reassembly

1 Insert the mainshaft partly assembled as described in Section 6 into the gearbox casing. Slide the intermediate ball race over the mainshaft and gently drift into the casing bore. The drift should only contact the outer race, which should be driven in until the snap ring on its periphery is up against the face of the casing. (photo).

2 Lightly tap the forward end of the mainshaft with a soft face (wooden) drift to ensure that the intermediate bearing inner race is firmly against the shoulder of the first gear bush.

3 Check that the needle race in the centre of the rear end of the input shaft is intact and that the two spacer rings are in position, before pushing the input shaft into the gearbox casing. Do not tap the input shaft too sharply when fitting it into the gearbox casing, because the needles in the input/mainshaft bearing may be disturbed.

4 The inpur shaft bearing is driven into the casing until the ring in the periphery of the bearing abuts the front face of the gear box casing.

5 Fit the mainshaft intermediate bearing retaining plate and secure with the three set screws.

6 Fit the reverse idler gear spindle into position and secure with two screws.

7 Insert the 1st/2nd selector fork into the gearbox and push the selector rod through the casing to locate the fork piece. Insert the clamp bolt into the 1st/2nd selector fork boss and tighten the fork in position on the rod. (photo).

8 Insert the 1st/2nd selector rod, 3rd/4th gear selector rod interlock plunger into the hole in the rear end of the casing. (photo).

9 Fit the 3rd/4th gear selector rod relay plunger into the rod; insert the selector fork into the casing and onto the 3rd/4th sliding sleeve. (photo).

10 Push the 3rd/4th gear rod into the casing to locate the selector fork and interlock plungers. The 1st/2nd gear rod may need a push or pull to move the interlock plunger so as to allow the 3rd/4th gear rod to pass through the casing. (photo).

11 Do not tighten the 3rd/4th selector fork onto its rod at this stage and insert the third and last interlock plunger into position ready for the installation of the reverse (and 5th) gear selector rod later.

12 Next insert the layshaft and fit first the forward bearing and then the rear bearing. Drive the forward bearing into the casing until the ring in the outer race periphery abuts the forward face of the casing. The rear roller bearing outer race is not positively located in the gearbox casing but the inner race should abut a shoulder next to the 1st gear pinion on the layshaft.

13 Screw in and tighten the bolt and washer and spacer that secures the layshaft in the inner race of the forward bearing. Lock the layshaft, mainshaft and drive shaft by pushing both synchromesh sliding sleeves to engage gears. (photo)

14 The key which locates the reverse gear wheel on the slightly reduced diameter of the mainshaft should now be inserted into its slot and the reverse gear wheel slid onto the shaft, over the key and up against the bearing inner race. (photo).

6.2 Remove circlip on forward end of
mainshaft to allow removal of 3rd gear
and synchronizer

6.12 Mainshaft gear sub-assemblies ready
for mounting on shaft

15 The procedure for 4 and 5 speed gearbox differs from this point on. The following paragraph describes the finishing of the 4 speed gearbox mainshaft assembly, and the succeeding 4 paragraphs detail the last stages of the 5 speed gearbox mainshaft assembly.

16 4 speed gearbox :-

Slip the spring spacer over the rear end of the mainshaft and position it against the hub of the reverse gear wheel. Fit the circlip into the groove in the mainshaft adjacent to the spring spacer. The spacer may need compressing to allow the circlip to be fitted. FIAT supply a special tool No 70158 for this task. Next fit the single ball bearing which registers the speedometer drive pinion in position and slide the drive pinion onto the mainshaft onto the ball. Finally push the rear mainshaft bearing onto the shaft to abut against the speedometer drive pinion.

17 5 speed gearbox:-

Slide the fifth gear synchronizer hub and sliding sleeve onto the shaft, up to the reverse gear wheel and locate on the same key as the reverse gear. (photo).

18 Next slip the fifth gear, with its particular type of synchronizer, onto the special bush and slide both onto the rear end of the mainshaft. Note that the bush shoulder and helical gear should be rearwards.

19 Push the roller race bearing onto the mainshaft up to the fifth gear bush shoulder and slip the special spacer lip forwards onto the mainshaft to abut against the bearing inner track.

20 Fit the single ball bearing which registers the speedometer drive pinion on the mainshaft, into the hole provided and slide the drive pinion into position.

21 Considering the layshaft, again the procedure for four and five speed gearboxes are different.

22 Four speed gearbox:- fit the spacer onto the rear end of the layshaft next to the roller bearing inner race. Then slide the reverse gear pinion onto the splined end of the layshaft and secure in position with circlip.

23 Five speed gearbox:- fit the short spacer onto the rear end of the layshaft next to the roller bearing inner race. Then slide the reverse and fifth gear layshaft extension over the extended primary layshaft, together with the reverse idler gear onto its spindle. It will be necessary to temporarily remove the roller race from the rear end of the mainshaft, to allow the fifth gear pinion on the layshaft to pass into engagement with the fifth gear on the mainshaft. (photo).

24 Push on the rear ball bearing race onto the end of the layshaft and secure with a nut which when tightened is crimped to lock.

25 Both 4 and 5 speed gearboxes:- fit the reverse (and fifth) selector fork into position and push the selector rod through the fork and casing to locate the fork.

26 Drop the detent balls into their respective holes on the right hand side of casing and then insert the detent ball springs, (photo).

27 Refit the spring retaining plate, with a new gasket and tighten the two bolts which secure the plate to the casing.

28 Move the selector rods so that the detent balls register in the 'neutral' position on each rod. Move the selector forks and synchromesh sliding sleeves until the sleeve is in the 'neutral' niche position on the sleeve hub.

29 Insert the clamp bolts into the 3rd/4th selector fork boss and tighten the boss onto the rod.

Insert the reverse (and fifth) fork bolt and secure that fork into position.

30 The main gearbox casing assembly is now ready to receive the clutch bellhousing and the rear extension casing. (photo).

31 Beginning with the clutch bellhousing, insert a new oil seal into the input shaft sleeve fitting in the clutch bellhousing, and place the spring space in position on the rear face of the seal. (photo).

32 Ensure that the mating surfaces of the gearbox and bellhousing are clean and put a new gasket onto the studs on the gearbox casing. Mount the bellhousing onto the casing and secure with seven nuts. Tighten the nuts to the specified torque.

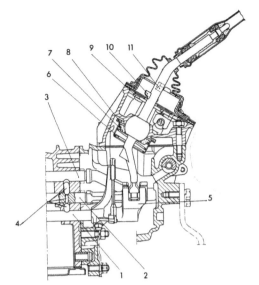

FIG. 6.8. SECTION THROUGH 5 SPEED GEARBOX GEARSHIFT MECHANISM

Note: Gear selector rod and striker blocks assemblies have remained basically unchanged

1 5th and reverse gear selector rod
2 3rd and 4th gear selector rod
3 1st and 2nd gear selector rod
4 Selector rod interlock plungers and relay plunger in the middle
5 Selector rod striker blocks
6 Reverse gear push spring on ball pivot
7 Gear lever ball pivot socket
8 Ball pivot socket retaining plate
9 Gear lever guide plate
10 Stop dog
11 Safety stop for reverse

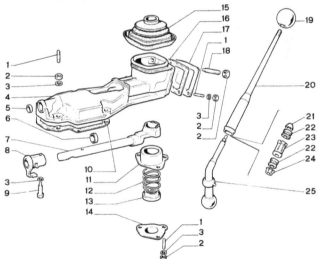

FIG. 6.9. EXPLODED VIEW OF LATER PATTERN GEARSHIFT MECHANISM

1 Studs	14 Lower ball socket
2 Nuts	15 Boot
3 Spring washers	16 Gasket
4 Gearshift mechanism housing	17 Cover plate
5 Plug	18 Lever stop pin
6 Gasket	19 Knob
7 Gear selector relay rod	20 Upper lever
8 Block and pawl	21 Shoulder block
9 Block retaining screw	22 Bushing
10 Gasket	23 Spacer
11 Ball pivot cover	24 Snap ring
12 Reverse stiffening spring	25 Pivot for relay rod
13 Upper ball socket	lever

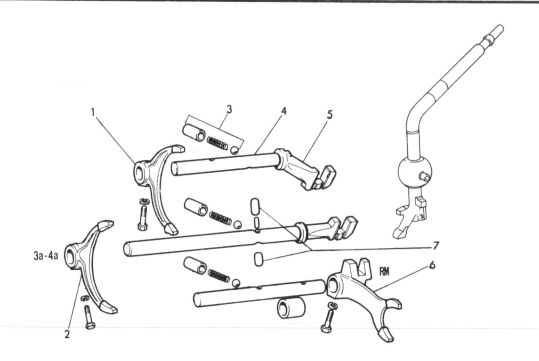

FIG. 6.10. EXPLODED VIEW OF FOUR SPEED GEARBOX GEAR SELECTOR COMPONENTS

1 1st/2nd gear selector fork
2 3rd/4th gear selector fork
3 Detent balls and springs
4 Selector rods

5 Striker blocks
6 Reverse gear selector fork and striker block
7 Interlock plunger and relay plunger

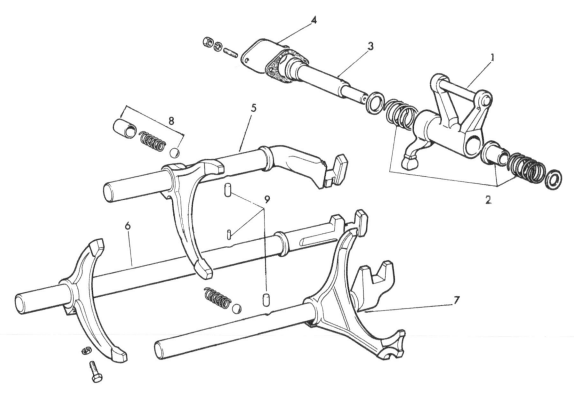

FIG. 6.11. EXPLODED VIEW OF FIVE SPEED GEARBOX GEAR SELECTOR COMPONENTS

1 Gear selector relay lever
2 Relay lever spacer and centering springs
3 Relay lever pivot spindle
4 Pivot spindle retaining plate
5 1st/2nd gear selector rod and fork

6 3rd/4th gear selector rod and fork
7 Reverse/5th gear selector rod and fork
8 Detent balls and springs
9 Interlock plungers

33 Having completed the assembly of the gears, bearings and gear selector rods on the rear lengths of layshaft and mainshaft, clean the mating surfaces of the gearbox casing and rear extension casing.

34 Place a new gasket on the gearbox joint face, and lower the extension casing into position. Lightly tap the extension home with a soft faced mallet. Tighten the six nuts that secure the extension to the main casing to the torque specified.

35 Press a new mainshaft oil seal into the rear of the extension casing, smearing the seal lip with a little gear oil.

36 Slip the flexible coupling spider onto the splines on the rear end of the mainshaft and tap home.

37 Place a new tag washer onto the mainshaft and then the nut which secures the spider on the shaft. Tighten the nut to the specified torque and tap a lip of the washer onto the nut flat to lock it in position. The spider will need to be locked when the nut is being tightened and an old bolt, or a piece of bar pushed through a hole in the spider so that it projects along the side of the extension casing will serve this purpose.

38 Once the spider nut has been tightened, slip the sealing ring spring, then the sealing ring and centering ring onto the end of the mainshaft and retain in position with the circlip. (photos).

39 Install the speedometer drive, using a new gasket, and secure in position with the single nut on the stud screwed into the gearbox casing.

40 Fit the rubber doughnut flexible coupling onto the mainshaft spider and tighten the nuts and bolts to the specified torque.

41 Check now that the input shaft, layshaft and mainshaft rotate smoothly and without any trace of impediment.

42 Fit the gearbox casing bottom cover in position; as always use a new gasket. Tighten the ten nuts to the specified torque.

43 On four speed gearboxes all that remains to do to complete the reassembly of the gearbox is to lower the gear lever and ball pivot assembly onto the extension casing. Ensure that the pawl on the lower end of the lever drop into engagement with the selector rod strikers and tighten the three nuts which secure the gear lever mechanism to the torque specified. Lastly screw in the gear lever travel stop bolt into the side of the extension casing until it prevents the lever end from disengaging the strikers and tighten the locknut.

44 Five speed gearboxes:- Some early pattern gearboxes have a gearlever installation of similar design to the four speed gearboxes; on later gearboxes the gear lever mechanism is mounted in a separate housing. The mechanism in this later installation incorporates a gearlever movement relay rod, and provides a gearshift with a shorter more upright gearlever than on the four speed and early five speed gearboxes.

45 The photo sequence shows the work necessary on the later type five speed gearbox gearshift mechanism.

46 Once the gearbox has been assembled as far as described to paragraph 42 of this Section, proceed to assemble the gearshift mechanism as follows:-

47 Place the relay lever, spacers and springs in the top aperture in the extension casing, and push the pivot spindle through the casing to locate the relay lever and associated components. (photo).

48 The pivot spindle is retained by a flat plate secured to the gearbox extension casing by two nuts on studs. As always use a new gasket. Ensure that the pawl on the lower end of the relay lever is correctly engaged in the striker forks on the selector rods in the gearbox.

49 After cleaning the joint surfaces of the gearshift mechanism housing and gearbox extension, fit a new gasket in position on the gearbox surface and lower the mechanism into place. Make sure that the bar across the top of the relay lever is positioned in the slot in the end fitting on the relay rod. The foot on the end fitting should fit into the base of the fork on the upper half of the relay lever (photo).

50 Tighten the nuts which secure the gearshift mechanism to the gearbox.

51 The installation of gearshift mechanism completes the gearbox assembly and it is now ready for refitting into the car.

7.1 Fitting intermediate bearing on mainshaft

7.7 Fastening the 1st/2nd gear selector fork to the selector rod

7.8 Fitting the 1st/2nd, 3rd/4th selector rod interlock plunger

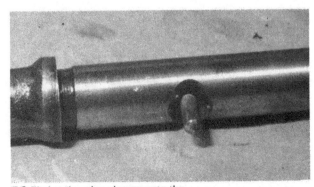
7.9 Fitting the relay plunger onto the 3rd/4th selector rod

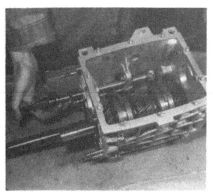

7.10 Inserting the 3rd/4th selector rod

7.13 Fitting bolt and spacer which locates layshaft in its forward bearing

7.14 Fitting the reverse gear woodruff key into the mainshaft

7.17 Slide the fifth gear synchronizer hub up to the reverse gear on the mainshaft

7.23 Sliding the reverse gear and fifth gear pinions onto the layshaft, together with the reverse idler gear and selector fork

7.26 Inserting detent ball springs

7.30 Rear gear assembly ready for the extension cover

7.31 Fitting the spring washer to the bellhousing gearbox drive shaft seal

7.38 Slipping the sealing ring, and centering spring onto the end of the mainshaft

7.39 Fitting circlip on end of mainshaft

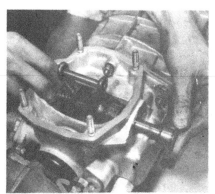

7.47 Pushing the relay lever spindle into position

7.49 Entering gear shift mechanism into place

8 Gearshift mechanisms and gearlever

1 There have been two basic types of gearshift mechanism employed by FIAT on their 4 and 5 speed gearboxes.

2 The first, fitted to all 4 speed transmissions, and some early 5 speed transmissions comprised a gear lever which extended below the ball pivot to a pawl which directly engaged the striker fittings on the ends of the selector rods. The sideways movement of the gearlever is restricted by stop bolts screwed into the rear gearbox extension casing, and there is a spring which returns the lever knob to the 1st/2nd gear neutral position. The ball pivot comprises two semi-spherical fittings in which the ball on the gearlever is sandwiched. A spring acting on the upper semi-spherical member maintains the bearing material in place on the ball.

Dismantling this type of mechanism is straightforward:-

Once the three nuts which secure it on the top of the gearbox casing have been undone and removed, the ball pivot assembly can simply be taken apart. The lower semi-spherical fitting is removed downwards. The upper fittings must be lifted upwards over the gearlever itself. It is essential to have previously pulled the upper part of the gearlever off the lower stub, and removed the circlip which holds the top bearing spring in position.

The reassembly follows the reversal of the dismantling procedure.

3 The 5 speed gearbox gearshift:- In this installation the gearbox is held in a ball pivot. Just above the pivot the lever is engaged by a relay rod. The sideways and forwards/rearwards movement of the gear lever knob is translated into rotational and axial movement of the relay rod.

On the end of the relay rod there is a block which engages the relay lever mounted in the aperture in the rear gearbox extension casing. The rod in the top of the relay lever fits into the slot cut in the underside of the block, and the pawl foot mounted on the block fits into the base of the forked upper half of the relay lever.

The rotational and axial movement of the relay rod is translated into sideways and tilting movement of the relay lever.

4 The gearshift mechanism is dismantled as follows:- Begin by removing the three nuts which secure the ball pivot members in position. The ball pivot, gear lever and pivot bearing components may then be extracted downwards out of the mechanism housing.

5 The block and pawl may then be unbolted from the forward end of the relay rod. The plate mounted on the rear end of the mechanism housing is removed next to allow the relay rod to be extracted from the housing.

6 Reassembly follows the reversal of the dismantling procedure.

7 Gearlever:- All the levers on the '124 Sport have a joint which enables the chromed top part of the lever to be pulled off the stub located in the gearbox. The joint is firm enough to transmit all the force and movement associated with gear change movements.

8 The joint comprises a shoulder block, rubber bushings, spacer and a nylon snap ring.

The position of the various components is shown in Fig. 6.13 and they can be dismantled once the upper lever has been removed. Refitting is equally straightforward.

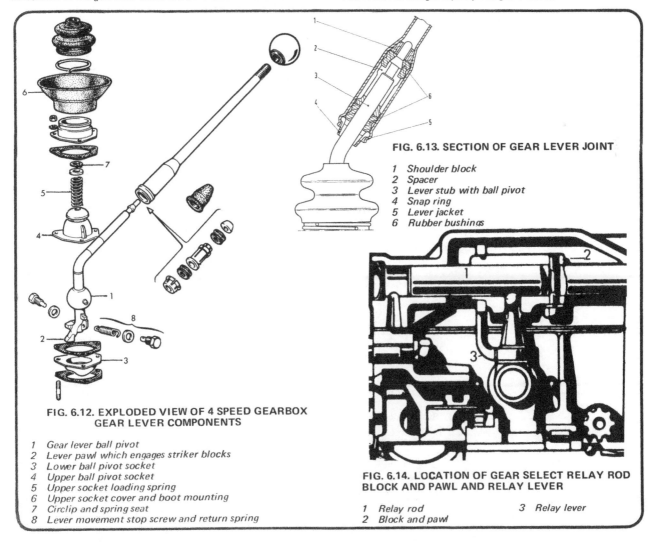

FIG. 6.12. EXPLODED VIEW OF 4 SPEED GEARBOX GEAR LEVER COMPONENTS

1 Gear lever ball pivot
2 Lever pawl which engages striker blocks
3 Lower ball pivot socket
4 Upper ball pivot socket
5 Upper socket loading spring
6 Upper socket cover and boot mounting
7 Circlip and spring seat
8 Lever movement stop screw and return spring

FIG. 6.13. SECTION OF GEAR LEVER JOINT

1 Shoulder block
2 Spacer
3 Lever stub with ball pivot
4 Snap ring
5 Lever jacket
6 Rubber bushings

FIG. 6.14. LOCATION OF GEAR SELECT RELAY ROD BLOCK AND PAWL AND RELAY LEVER

1 Relay rod   3 Relay lever
2 Block and pawl

9   Speedometer cable

1   The speedometer cable runs from the drive connection on the right hand side of the gearbox rear extension, through the rear of the engine compartment and then to the speedometer mounted on the instrument panel.

The cable is housed in a sleeve, the ends of which are held onto speedometer and drive connection by large knurled nuts. The cable has square ends which engage the spindles in the speedometer and drive connection.

2   Removal and replacement of the cable and sleeve is quite straightforward being a matter of simply undoing or refitting the knurled nuts as appropriate and threading the sleeve and cable through the holes in the engine compartment bulkhead.

3   The drive connection is renewable as a complete assembly, as well as the individual components. Frankly though if wear is suspected in the drive connection there is little point in renewing single components.

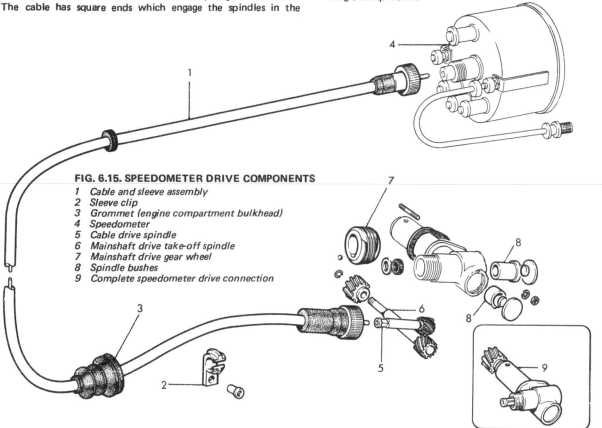

FIG. 6.15. SPEEDOMETER DRIVE COMPONENTS

1   *Cable and sleeve assembly*
2   *Sleeve clip*
3   *Grommet (engine compartment bulkhead)*
4   *Speedometer*
5   *Cable drive spindle*
6   *Mainshaft drive take-off spindle*
7   *Mainshaft drive gear wheel*
8   *Spindle bushes*
9   *Complete speedometer drive connection*

Fault diagnosis

Symptom	Reason/s	Remedy
WEAK OR INEFFECTIVE SYNCHROMESH		
General wear	Synchronising cones worn, split or damaged	Dismantle and overhaul gearbox. Fit new gear wheels and synchronising cones
	Synchromesh dogs worn, or damaged	Dismantle and overhaul gearbox. Fit new synchromesh unit.
JUMPS OUT OF GEAR		
General wear or damage	Broken gearchange fork rod spring	Dismantle and replace spring.
	Gearbox coupling dogs badly worn	Dismantle gearbox. Fit new coupling dogs.
	Selector fork rod groove badly worn	Fit new selector fork rod.
EXCESSIVE NOISE		
Lack of maintenance	Incorrect grade of oil in gearbox or oil level too low	Drain, refill, or top up gearbox with correct grade of oil.
	Bush or needle roller bearings worn or damaged	Dismantle and overhaul gearbox. Renew bearings.
	Gearteeth excessively worn or damaged	Dismantle and overhaul gearbox. Renew gear wheels.
	Laygear thrust washers worn allowing excessive end play	Dismantle and overhaul gearbox. Renew thrust washers.
EXCESSIVE DIFFICULTY IN ENGAGING GEAR		
Clutch not fully disengaging	Clutch pedal adjustment incorrect	Adjust clutch pedal correctly.

Chapter 7 Propeller shaft and universal joints

Contents

Specifications

Two patterns of shaft assembly fitted:

On AC and AS type bodyshells: A two part shaft comprising:-
a forward shaft running between a rubber flexible joint on the rear of the gearbox and a universal joint on the forward end of a rear shaft.
A rear shaft running in a tubular extension of the rear axle assembly.
The forward end of the extension is held in a rubber pillow block attached to the bodyshell.

On BC and BS and all current bodyshells: A two part shaft comprising:-
a forward shaft running between a rubber flexible joint on the rear of the gearbox and through a centre bearing held in a rubber pillow block mounted on the bodyshell.
The rear shaft is a conventional tubular unit coupling the rear end of the forward shaft with the flange fitting on the bevel pinion shaft on the rear axle.

Torque wrench settings:	ft lbs
Long bolts and nuts retaining rubber 'doughnut' onto forward shaft forward spider	72
Single nut securing universal joint yoke onto rear end of the forward shaft. Current pattern of shaft	87
Single nut securing universal joint yoke onto forward end of rear shaft (early pattern shaft assembly)	87
Nuts retaining pillow block support to bodyshell	18
Bolts securing pillow block to support crossmember (late pattern)	18
Nuts and bolts coupling shaft to flange fitting on bevel pinion shaft of rear axle (late pattern)	28
Bolts securing tubular extension/shaft cover to differential housing (early pattern)	51

Balanced pairs:
(These components must be assembled with their balance reference marks aligned).

EARLY PATTERN	LATER & CURRENT PATTERN
Shaft assembly Forward shaft forward spider and forward shaft	Rubber doughnut/forward shaft spider Forward shaft/mid universal joint yoke

1 General description

There have been two patterns of propeller shaft fitted to the FIAT 124 Sport series. When the car was first introduced and until chassis No. 1759186 for Spiders and chassis No. 0066059 for Coupes the propeller shaft consisted of a forward half running between a rubber flexible coupling on the gearbox, and a conventional universal joint on the forward end of the rear half of the shaft. The rear shaft was solid and ran in bearings mounted in a tubular extension fixed to the rear axle casting. The forward end of the extension was retained in a rubber pillow block mounted secured to the bodyshell by a small crossmember spanning the transmission tunnel.

The later type of propeller shaft is quite different in many respects. It still exists as two parts but the forward section now runs from the rubber flexible coupling on the gearbox, into a bearing mounted in a rubber pillow block attached to the body shell. The rear length of shaft is the usual pattern of shaft with a universal joint at both ends. There is no extension on the rear axle and the propeller shaft joins onto a flange fitted mounted on the pinion shaft of the differential assembly.

The propeller shaft requires periodic lubrication of the splined joint of the forward shaft into the flexible coupling on the gearbox output shaft. Some makes of replacement universal joint spiders have provision for grease lubrication of the needle roller bearings on the Spider.

It is essential to realise that the propeller shaft assembly is a finely balanced collection of components and that they are balanced only as they remain assembled in the alignment in which they were originally balanced.

Therefore whenever dismantling the propeller shaft pay particular attention to the reference alignment marks on the main shaft components so that they are correctly reassembled.

It is equally important to realise that if any single or small group of major components is renewed (except universal joint spider assemblies) the whole shaft assembly must be dynamically balanced professionally.

2 Propeller shaft - removal and replacement

EARLY TYPE:

This type of forward shaft is tubular and joins the mainshaft output from the gearbox to the universal joint on the forward end of the rear shaft housed in a tubular extension of the rear axle assembly. It is not possible to remove the two halves separately, therefore commence to remove the whole shaft assembly as follows:-

1 Raise the rear of the car onto chassis stands or drive the rear of the car onto car ramps. Make absolutely sure that the car is safely supported, because the efforts involved to undo and tighten nuts and bolts during repair tasks are sufficient to disturb an inadequately supported vehicle.

2 It might be an advantage to obtain FIAT tool A 70025 to remove the rubber 'doughnut' flexible coupling from the gearbox mainshaft.

3 Begin by removing the nuts and long bolts that secure the rubber 'doughnut' to the spider mounted on the forward transmission shaft.

4 Next remove the rear brake hose pipe clip from the tubular housing of the rear shaft.

5 Remove the brake reservoir caps and stretch a piece of polythene over the top of the reservoir and replace the caps. This measure will prevent hydraulic fluid draining from the brake system when the rear brake pipe is subsequently removed.

6 Disconnect the flexible hose from the metal brake pipe on the propeller shaft rear tubular housing and then remove the pipe from that housing.

7 Unhook the handbrake return spring from the crossmember which supports the centre rubber pillow b'ock.

8 Next remove the four bolts and washers which secure the rear shaft housing to the differential housing and then undo the four

nuts and washers which hold the pillow block support cross-member to the bodyshell.

9 The whole propeller shaft assembly is now free to be lowered from the car.

10 Lower and move the shaft assembly a little forwards to disengage the spline joint of the rear shaft to the bevel pinion shaft in the rear axle. Take care not to loose the spring which is located around the splined sleeve which joins the propeller shaft and bevel pinion shaft.

11 Take the whole assembly to a clean bench for examination and renovation.

12 Refitting the assembly is simply the reversal of the removal procedure, except that the following checks and tasks must be completed before the vehicle is taken onto the road.

a) Ensure all mounting nuts and bolts have been tightened to the correct torque given in the specification at the beginning of this Chapter.

b) Bleed the brake hydraulic system as detailed in Chapter 10 and check the operation of the handbrake.

c) Do not tighten the nuts securing the pillow block support to the bodyshell until the support position has been adjusted so that the gap between the left hand side (viewed from the front) of the tubular cover and the pillow block outer casing is about 0.040 inches (1 mm) less than the gap on the right hand side of the cover.

LATER TYPE:

This pattern of propeller shaft again does not permit the separate removal of the forward and rear sections of shaft. Therefore proceed to remove the whole shaft assembly as follows:

1 Begin by raising the rear of the car onto chassis stands or car ramps. Make sure that the car is safely supported because efforts involved in the removal and replacement of the shaft assembly could easily topple an inadequately supported vehicle.

2 Working underneath the car remove the three long bolts and nuts which hold the rubber 'doughnut' of the forward flexible coupling onto the spider fitting on the end of the forward shaft.

3 Next remove the handbrake return spring from the propeller shaft support crossmember and then remove the nuts and washers which hold the pillow block support crossmember and the idle crossmember to the bodyshell.

4 Some light support should be offered to the shaft at this point to prevent it from being grazed on the ground.

5 Finally remove the four nuts and bolts which hold the rear propeller shaft to the flange fitting on the bevel pinion shaft of the rear axle.

6 The propeller shaft assembly may now be lifted from the car and transfered to a bench ready for examination and repair.

7 The refitting procedure for the shaft assembly is simply the reversal of the removal procedure, except that the following checks must be made before the car is taken on the road.

a) Tighten all mounting nuts and bolts to the torque specified at the beginning of the Chapter.

b) Check that the handbrake operates properly.

3 Universal joints - inspection and repair

1 Wear in the needle roller bearings on the central spider is characterised by vibration in the transmission, 'clonks' on taking up drive and in extreme cases of lack of lubrication, metallic squeaking and ultimately grating and shrieking sounds as the bearings break up.

2 It is easy to check if the needle roller bearings are worn with the propeller shaft in position, by trying to turn the shaft with one hand, whilst holding the rear axle flange (later shaft assembly - rear joint) or the mating yoke (attached to the rear shaft on early design, or attached to the rear end of the forward shaft on the latest design) with the other hand. Any movement across the joint is indicative of considerable wear. A second check is to try to lift the shaft and noticing any movement in the joints.

3 If worn, the old bearings and spiders will have to be discarded and a repair kit purchased comprising new universal joint spiders,

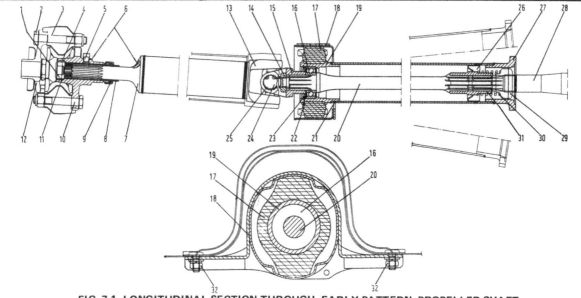

FIG. 7.1. LONGITUDINAL SECTION THROUGH EARLY PATTERN PROPELLER SHAFT AND PILLOW BLOCK

1 Gearbox mainshaft propeller shaft centering ring
2 Mainshaft spider bolts
3 Rubber 'doughnut' flexible unit
4 Propeller shaft spider fitting
5 Grease plug
6 Balance reference marks
7 Forward section of shaft
8 Seal
9 Cover
10 Propeller shaft spider bolts
11 Spider centre bush
12 Mainshaft spider
13 Universal joint yoke - forward shaft
14 Yoke retaining nut
15 Universal joint yoke - rear shaft
16 Ball bearing

17 Pillow block - rubber
18 Pillow block shell
19 Shaft cover end fitting
20 Rear propeller shaft
21 Tubular cover for shaft
22 Bearing shield
23 Circlip
24 Universal joint spider
25 Needle bearing on spider journals
26 Shield
27 Tubular cover end flange
28 Bevel pinion shaft
29 Bevel pinion shaft locknut
30 Spring
31 Pinion shaft/propeller shaft - coupling sleeve
32 Pillow block mounting studs

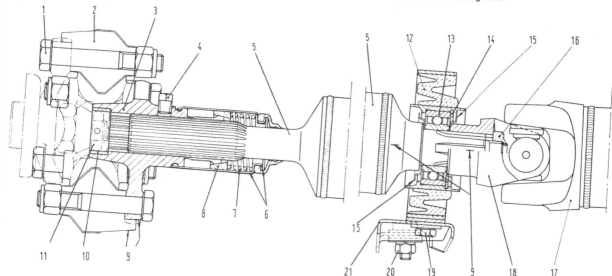

FIG. 7.2. LONGITUDINAL SECTION OF LATER/CURRENT PATTERN OF PROPELLER SHAFT

1 Mainshaft spider bolts
2 Rubber 'doughnut' flexible joint
3 Forward propeller shaft spider fitting
4 Grease plug
5 Propeller shaft - forward section
6 Seal and cover
7 Spring
8 Cone ring
9 Balance reference marks
10 Spider fitting centre bush
11 Centering ring on mainshaft

12 Centre pillow block - rubber
13 Ball bearing
14 Bearing circlip
15 Bearing shields
16 Nut retaining universal joint yoke
17 Propeller shaft - rear section
18 Universal joint yoke
19 Crossmember to bodyshell nuts
20 Pillow block to crossmember bolts
21 Crossmember

104

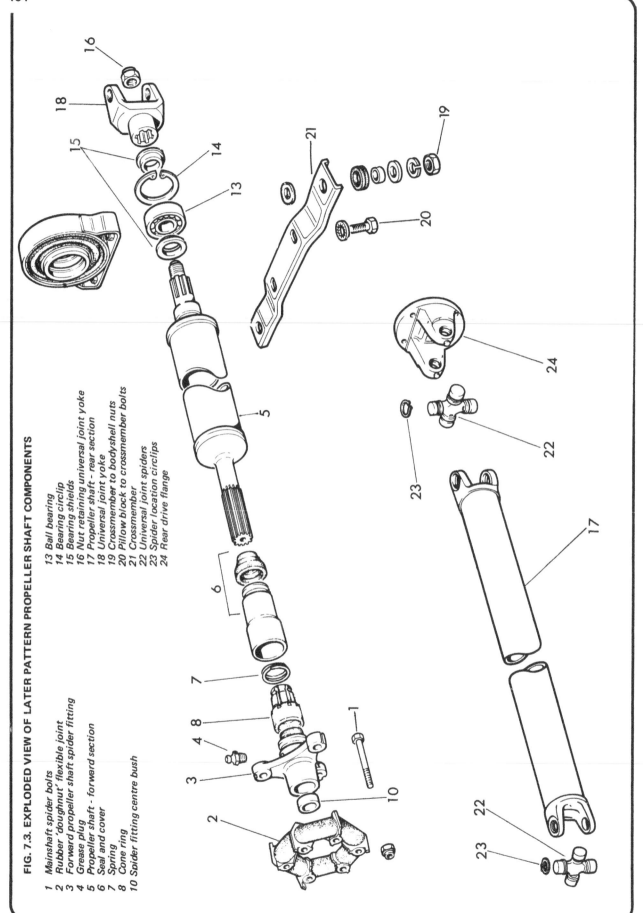

FIG. 7.3. EXPLODED VIEW OF LATER PATTERN PROPELLER SHAFT COMPONENTS

1 Mainshaft spider bolts
2 Rubber 'doughnut' flexible joint
3 Forward propeller shaft spider fitting
4 Grease plug
5 Propeller shaft - forward section
6 Seal and cover
7 Spring
8 Cone ring
10 Spider fitting centre bush

13 Ball bearing
14 Bearing circlip
15 Bearing shields
16 Nut retaining universal joint yoke
17 Propeller shaft - rear section
18 Universal joint yoke
19 Crossmember to bodyshell nuts
20 Pillow block to crossmember bolts
21 Crossmember
22 Universal joint spiders
23 Spider location circlips
24 Rear drive flange

bearings, seals and retainers.

Universal joints dismantling:-

4 Clean away all traces of dirt and grease from the circlips located on the bearing cups in the yokes. Remove the clips by pressing their open ends together with a pair of circlip pliers and lift them out with a screwdriver. If they are difficult to remove, tap the bearing cup top with a mallet to ease the pressure on the circlip.

5 Hold one side of the joint - normally the tubular shaft side to begin with and remove the bearing cups and needle rollers by tapping the yoke at each bearing with a copper or hide faced hammer. As soon as the bearing cups begin to emerge from their bores, they can be drawn out with either your fingers or a pair of pliers. If the bearing cups refuse to move then place a small drift against the inside of the bearing and tap it gently until the cup begins to move.

6 With all four cups removed together with their needle rollers, the spider can easily be extracted from the yokes. Once the spider is free the bearing faces may be wiped clean with a petrol damped cloth and the surfaces inspected. If any grazing scores or ridges are found the spider will need replacing. On some occasions when the universal joint has failed through lack of lubricant, the bores in the yokes in which the bearing cups fit can be worn. Again once the joint has been dismantled it is easy to check the condition of the yokes.

Universal joints - reassembly

7 Thoroughly clean the yokes and bores. Remember to check that the circlip grooves are clear.

8 Fit new grease seals and retainers on the new spider journals and place the spider into the shaft yoke. Assemble the needle rollers in the bearing cups and hold in place by smearing them with a medium lithum base grease. In new assemblies the needles should pack so well that they each retain the other in place.

9 Carefully ease the cups, packed with the needle rollers into the yoke bore and onto the appropriate spider journal. It is all too easy to hurry and a single roller might fall from place and prevent the cup from seating properly on the journal.

10 It may be necessary to tap the bearing cups home in the final stages and once the cup top face has placed the circlip groove in the yoke refit the circlips.

11 Once all the whole universal joint has been assembled there is provision on most makes of replacement universal joints for injecting extra grease into the spider bearings before the propeller shaft is refitted to the car. A small grub screw in the centre of the spider can be unscrewed and a grease nipple temporarily fitted to enable the joint to be greased.

4 Rubber pillow block - removal and refitting

EARLY PATTERN:-

1 The block is held in a housing fitted with attachments to join it to the bodyshell. It supports the forward end of the tubular cover of the rear half of the transmission. The block may be removed from the tubular cover once the shaft assembly has been removed from the car and the centre universal joint has been removed as described in Section 3. The joint yoke will then need to be removed from the rear half of the propeller shaft. It is retained on the solid shaft by a single nut, and once this nut is undone, the yoke may be pulled off the splined forward end of the rear propeller shaft.

2 The rubber pillow block seats in an annular recess on the forward end of the tubular cover and its simply pulled off.

3 The refitting of the pillow block is the reversal of the removal procedure, except that the following check must be included.

a) Ensure that the nut on the forward end of the rear shaft has been tightened to the torque specified at he beginning of this Chapter.

LATER TYPE:

4 The later type of pillow block supports a bearing in which the rear end of the forward propeller shaft turns. As with the early pattern of propeller shaft, it is necessary to remove the shaft assembly from the car and remove the central universal joint before the block can be removed.

Section 2 details the removal and refitting of the shaft assembly and Section 3 describes how to remove the central universal joint.

5 With the forward shaft separated from the rear, remove the single nuts from the rear end of the forward shaft and slip the universal joint yoke off the splined end of the shaft.

6 Support the inner ring of the rubber block on the shaft side. The propeller shaft may then be tapped gently out of the ball race bearing held in the inner ring of the rubber block. Use a hide or copper hammer on the shaft, a harder hammer will damage the shaft.

7 The bearing may be removed from the block inner ring after the circlip which retains it in the ring has been extracted.

8 Finally the block outer cover is separated from the support crossmember after the two securing bolts are undone.

9 The refitting of the pillow block onto the shaft follows the reverse procedure to removal, except the following checks and tasks must be included:-

a) Ensure that the centre bearing shields are in place on each side of the bearing.

b) Check that the nut on the rear end of the forward shaft is tightened to the specified torque at the beginning of this Chapter.

c) Make sure that the pillow block has been mounted the correct way around, with the bearing shoulder on the forward side of the assembly.

5 Forward shaft front spider - removal and refitting

On both types of propeller shaft the front flexible joint spider is splined onto the forward end of the forward shaft.

1 The spider can be removed once its rear cover is unscrewed and slipped away down the shaft. On later designs the cover is swaged into a groove machined on the rear of the spider fitting. A small drift will be required to force the cover off the spider.

2 The internal splines in the spider and the splines on the shaft may be cleaned and inspected.

3 If inspection reveals wear on either part, both should be renewed. It is false economy to renew just one of a finely matched pair of components.

4 It should be noted that the propeller shafts are finely balanced assemblies and it is essential to realise that the spider must be refitted on the front shaft so that the reference marks on the spider and shaft are aligned.

5 It is equally essential if parts are renewed singularly for the assembly to be balanced professionally before refitting on the car.

6 The propeller shaft pline cover and sealing components may be slid off the shaft when the spider has been removed.

7 The bush in the forward end of the spider fitting should be inspected for wear together with the centring ring on the end of the gearbox mainshaft on which it runs. If wear is found then both centring ring and bush should be renewed.

8 The bush can be drifted out of the spidering fitting and the centring ring removed from the mainshaft once the end nut has been removed. Check that the new components fit closely before assembling them onto the transmission.

9 Reassembly of the spider fitting onto the forward propeller shaft is the reversal of removal except that the following checks should be included:-

a) Grease the mating splines liberally with a lithum base grease such as Castrol LM Grease.

b) Ensure that the spider is fitted to the propeller shaft with the balance reference marks aligned properly.

6 Centre bearing - removal and replacement

1 Wear of this bearing is typified by transmission vibration and in the event of total loss of lubricant, metallic shrieking accompanied by excessive vibration. The condition of the bearing may be checked with the propeller shaft assembly in the car by shaking the shaft to discern radial play in the bearing.

2 Removal and replacement of the centre bearing in the early shaft assembly is straightforward. Begin by removing the propeller shaft assembly as described in Section 2.

3 Remove the mid-position universal joint as detailed in Section 3 of this Chapter.

4 Unscrew the nut on the forward end of the rear shaft and remove the universal joint yoke from the splined end of the shaft.

5 Remove the circlip which retains the centre bearing in the end of the rear shaft cover with a pair of circlip pliers.

6 Lightly tap the rear end of the rear shaft using a wooden drift to coax the shaft and bearing out of the forward end of the tubular cover.

7 With the rear propeller shaft out, tap the bearing off the forward end with a soft hide hammer.

8 Replacement of the centre bearing follows the reversal of the removal procedure.

9 Removal and replacement of the centre bearing fitted to the current pattern of propeller shaft proceeds as follows:-

10 Remove the propeller shaft assembly as described in Section 2 of this Chapter.

11 Remove the nut position universal joint as detailed in Section 3 of this Chapter.

12 Remove the centre pillow block assembly as described in Section 4 of this Chapter.

13 Tap the bearing out of the pillow block inner ring once the retaining circlip has been removed.

14 Replacement is the reversal of removal, but as with all tasks on the propeller shaft ensure that the reference marks are correctly aligned on reassembly. Make certain also that all the locating nuts and bolts have been tightened to the torques specified at the beginning of this Chapter.

Chapter 8 Rear axle

Contents

Specifications

Type	Semi floating
Bevel gears	Hypoid
Gear ratio	10/41 (1400) 10/43 (1600 & 1800)
Pinion bearing	2
Type of bearing	Taper roller
Pre-loading	By collapsible spacer and tightening with torque wrench
Pinion bearing pre-loading:	
(pinion shaft nut tightening torque)	108 to 166 ft lbs
Pinion running torque	1.16 to 1.44 ft lbs
Differential cage bearings	2
Type	Taper roller
Adjustment	By adjuster threaded rings
Bearing pre-loading: differential cap spread	0.0063 to 0.0078 in
Pinion and crownwheel	Matched pair
Adjusting differential side gears	Shims
Shim thicknesses for side gear axial adjustment	0.071 in to 0.083 in. in 0.002 in increments
Bevel pinion position adjustment	Shims
Shim thicknesses	0.100 in to 0.132 in. in 0.002 in increments
Backlash between pinion and crownwheel	0.0039 in to 0.0059 in.
Axle shaft bearings	Ball race
Rear track	51.97 in.
Axle lubrication	S.A.E. 90EP

Torque wrench settings:	ft lbs
Bevel pinion shaft nut (bearing preload)	181
Forward tubular extension to rear axle; securing bolts	51 AC and AS bodyshells
Nuts and bolts securing propeller shaft to flange fitting on	
bevel pinion shaft	28 BC, BS and current models
Rear wheel bolts	50
Panhard rod end bolts/nuts	72
Trailing arm end bolts	72
Rear brake caliper bracket retaining bolts	25
Crownwheel to differential cage	72
Differential cage bearing cap bolt	36
Differential cage bearing to axle housing	22

1   General description

There have been two methods of locating the rear axle onto the body shell. The first design of axle is located by a panhard rod, two trailing arms, two coil springs with shock absorbers and a tubular extension of the differential casing, the forward end of this extension being held in a rubber pillow block attached to the bodyshell. The extension also serves as a cover for the rear part of the propeller shaft. When the BC and BS series bodyshells were introduced at chassis numbers 1759184 for Spiders and 0066059 for Coupes, this method of locating the rear axle changed, the tubular extension /propeller shaft cover was abandoned and a more conventional two part shaft adopted. The axle is now located by four trailing arms, a panhard rod (transverse rod) and two coil springs with telescopic shock absorbers.

Apart from these location methods, the axle mechanics have remained basically unchanged. The bevel pinion input shaft runs in a pair of taper roller bearings, between which there is a 'collapsing cone' which maintains an end thrust on the bearing rollers. The pinion meshes with the crownwheel which is bolted onto the differential cage. Four straight bevel gears are housed in the differential cage and these complete the gear trains in the vehicles transmission. The whole differential cage runs in a pair of taper roller bearings and the axle shafts run through these bearings to the ends of axle casing where they emerge to support the rear wheels and brake discs.

It is very unlikely that the whole rear axle will need to be removed because virtually all maintenance and repair tasks can be readily performed with the axle casing in place. The procedure for axle removal has been included in this Chapter, but the sections which describe the renovation of the differential and axle shafts assume that the axle casing is in place on the car.

2   Rear axle removal and replacement

1   This section details the procedure for removal and refitment of the axle arrangement employed on current models of the FIAT 124 Sport. The additional tasks necessary if the early design of rear axle is being repaired are included.

2   Begin by preparing the vehicle and systems associated with the rear axle as follows.

3   Raise the rear of the car onto strong chassis stands situated adjacent to the major trailing arm anchorages on the bodyshell. FIAT do supply stands custom designed for the purpose of supporting the bodyshell so that enough space is available for removal of the axle assembly. The FIAT number for these stands is D15051.

4   With the car safely supported, remove the rear wheels and the special bolts which secure the brake discs onto the axle shaft flanged end.

5   Remove the brake fluid reservoir caps and stretch a sheet of polythene over the reservoir and refit the cap. This measure will prevent excessive loss of fluid from the brake system when pipes in the system are subsequently removed or disconnected.

6   Disconnect the flexible hose serving the rear brakes from the fixed metal pipe on the rear axle assembly. Tape over the ends of the pipes to prevent the ingress of dirt.

7   Remove the propeller shaft as detailed in Chapter 7.

8   Disconnect the handbrake cables from the two caliper assemblies on the ends of the axle casing. The cable and sleeve fit into forked recepticles on the caliper block and pad actuating lever.

9   Remove the trailing rods and transverse panhard rod. Long bolts secure the rubber bushed ends of these rods to the anchorage on the bodyshell and rear axle assembly.

10 Remove the smaller bolts now which connect the brake regulator actuator arms to the axle assembly.

11 Bring a jack underneath the centre of the axle assembly and raise to take the weight of the axle.

12 Then from within the boot undo the upper securing nuts for

the shock absorber assemblies. FIAT tool A57070 is especially designed for this purpose.

13 Check that all axle location rods and members have been removed or disconnected.

14 The axle is now free to be lowered from the car and once clear it should be lifted to a clean bench for dismantling.

15 Refer to Chapter 9 if it is desired to remove the brake assemblies.

16 The suspension springs can be lifted off the seating on the axle casing, together with the top seat caps. The telescopic shock absorber is retained by a single bolt to a bracket mounted on the axle casing.

17 The axle is now stripped of peripheral components and sub-assemblies and is ready for thorough renovation as described in Sections 3,4,5 and 6 of this Chapter.

18 Refitting the rear axle assemblies follows the reversal of the removal procedure, except for the addition of following tasks.

19 The rear shock absorber and spring are assembled as follows:- To begin with the lower spring seating members are positioned and then the telescopic shock absorbers are bolted to the brackets on the axle casing. The spring is then lowered down around the shock absorber, the upper end of which is pulled to project above the top level of the spring. The top spring caps are placed in position. The axle assembly is then moved to underneath the vehicle where it is raised into position with a jack. The tops of the shock absorbers must be guided through the holes in the bodyshell into the boot space and then the securing nuts screwed onto the absorber spindle.

20 It should be appreciated that the springs should be identical-in free length and stiffness, and that both rear shock absorbers should be at an identical point of wear. Never replace just one spring, or one absorber, these components must be renewed in pairs.

21 It is also interesting to note that rear springs are colour coded for stiffness. Chapter 11 gives detailed information for the springs and shock absorber used on the rear axle.

22 Check that all the nuts and bolts on the rear axle system have been tightened to the torque specified at the beginning of this Chapter.

23 Refer to Chapter 9 and bleed the brake system.

3   Axle shaft, bearing and oil seal - removal and replacement

1   The task of removing and refitting the axle shaft and associated components can be completed with the axle casing in position on the car. There is no advantage to be gained by removing the rear axle for this task.

2   On all axles fitted to the FIAT 124 Sport the bearing is retained in the end of the axle casing by a single circlip and is retained on the axle shaft by a ring which is shrunk into position.

3   Begin removal of the axle shaft components by raising the rear of the car onto chassis stands situated either under the axle itself or under the suspension anchorages. As always make sure that the vehicle is safely supported. Use at least two chassis stands and chock the front wheels.

4   Unbolt the brake caliper block (Chapter 9, Section 7) and position it out of the way; do not strain the flexible brake hose. Remove the caliper yoke.

5   Undo the two special bolts which retain the brake disc onto the end of the axle shaft and remove the disc.

6   The ends of the rear assembly is now clear for the removal of the axle shaft.

7   Next remove the circlip from its groove in the end of the axle casing, behind the flanged end of the axle shaft. Long reach circlip pliers or FIAT tool A81114 will be required to remove this circlip in the end of the axle casing.

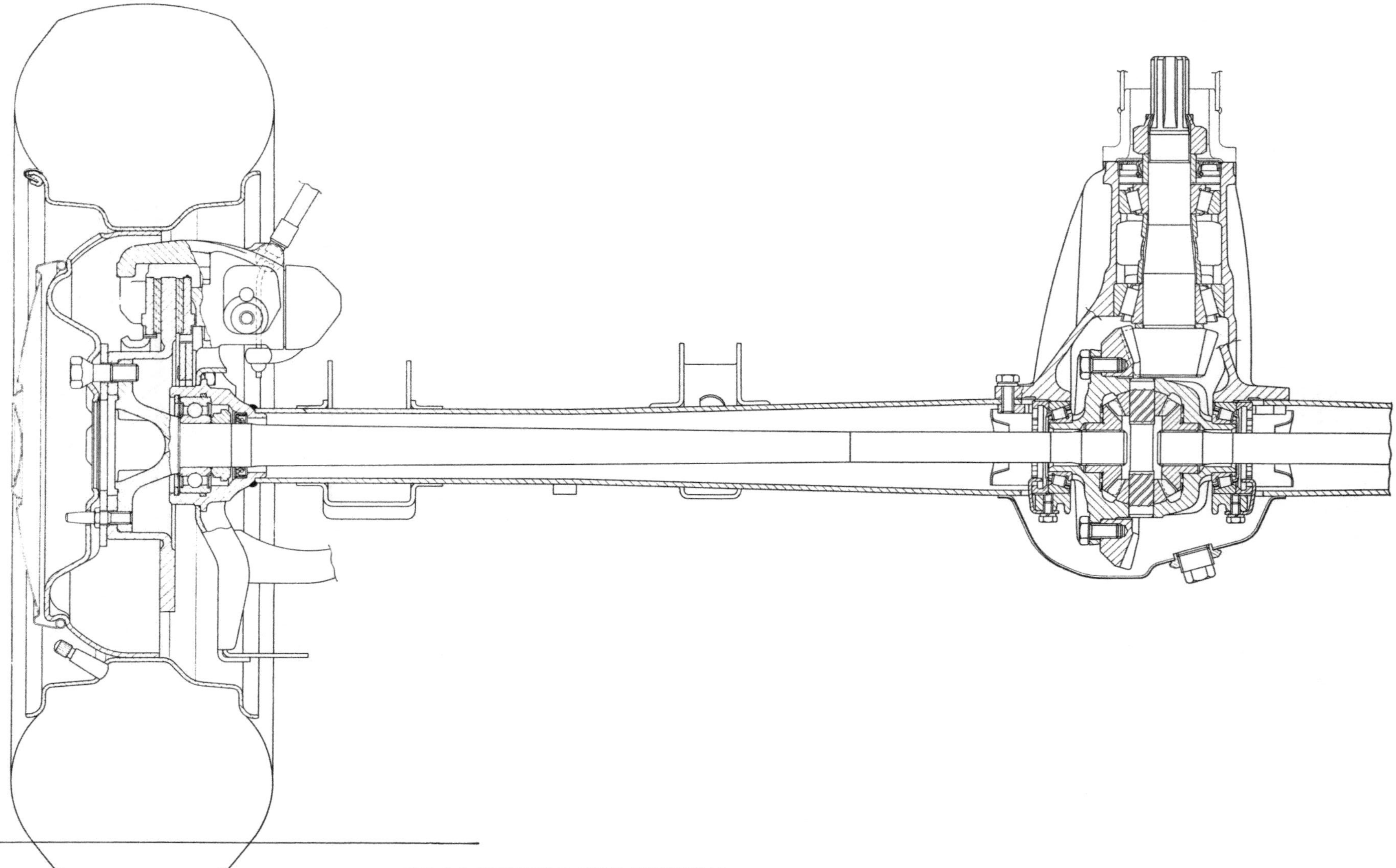

FIG. 8.1. THE REAR AXLE ASSEMBLY

A section through the final drive, differential assemblies and the L.H. rear wheel

8 The axle shaft together with the end bearings, locating ring and the seals can now be pulled out of the axle casing. FIAT supply a 'percussion puller' tool number A47017 for this task.

9 Inspect the shaft splines, the bearing and seal for signs of wear. It is also wise to check that the shaft is still straight by supporting the shaft on its centres and monitoring the surface of the shaft with a clock gauge as it is turned. If either shaft or bearing require replacement it will be necessary to remove the bearing from the shaft.

10 The bearing is retained on its seating on the shaft by a ring which is a shrink fit on the shaft. It will be a workshop task to remove and replace this ring because it will require a hydraulic press and FIAT tool 74108 to support the ring while the shaft is pushed through.

11 On no account reuse a bearing locating ring, once a ring has been removed, discard it and refit the bearing on the shaft with a new ring.

12 Reassembly of the axle shaft components proceeds as follows:-

13 Slide the circlip, bearing dust shield and bearing down the shaft to its flanged end. Place the shaft with those items loosely in place into a hydraulic press. FIAT supply tool number A 74017/1, which is a sole plate, to support the flanged end of the axle shaft whilst the shaft components are fitted.

14 Heat the new bearing locating ring in an oven to a temperature of approximately 300°C. Then slip the ring onto the shaft

making sure that the chamfered lip is away from the bearing. Again FIAT supply a special pair of pliers, tool number A60138, to handle the hot ring during the assembly operation.

15 Slide the tubular sleeve or FIAT tool number A 70417/2 over the shaft and down onto the hot ring which is ready to be forced home next to the bearing inner race.

16 Work the hydraulic press to force the ring against the bearing and the bearing against the shoulder on the flanged end of the axle shaft. Maintain the pressure on the ring until it has cooled down.

17 Next fit the axle oil seal onto its seating inside the end of the axle casing and carefully slide the axle shaft through the new seal into the axle casing and finally into the splines in the differential assembly.

18 Tap the shaft gently home with a soft faced mallet and fit the circlip (which retains the bearing into the axle casing) into its groove in the end of the casing.

19 Refit the brake disc and caliper assembly as instructed in Chapter 9. Reconnect the brake pipes and handbrake cable and bleed the brakes. Make sure that the bolts which secure the brake caliper block to the axle casing have been tightened to the torque specified at the beginning of this Chapter.

20 Finally refit the road wheels, tightening their retaining bolts to the specified torque and lower the vehicle off the chassis stands.

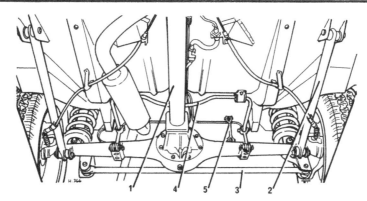

FIG. 8.2. THE REAR AXLE LOCATION ON EARLY MODELS OF THE FIAT 124 SPORT, AS SEEN FROM THE FRONT

1 Tubular extension on the rear axle which covers the rear half of the propeller shaft
2 Main trailing arms
3 Panhard rod
4 Anti-roll bar
5 Rear brake effort regulating mechanism

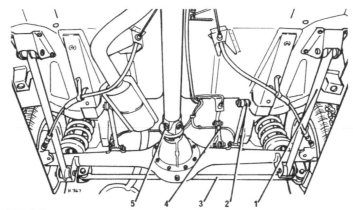

FIG. 8.3. LATER AND CURRENT LOCATION OF THE REAR AXLE AS SEEN FROM THE FRONT

1 Main trailing arms
2 Additional trailing arms for wheel torque reaction
3 Panhard rod
4 Rear brake effort regulating mechanism
5 Conventional propeller shaft attached to flanged fitting on the final drive input shaft

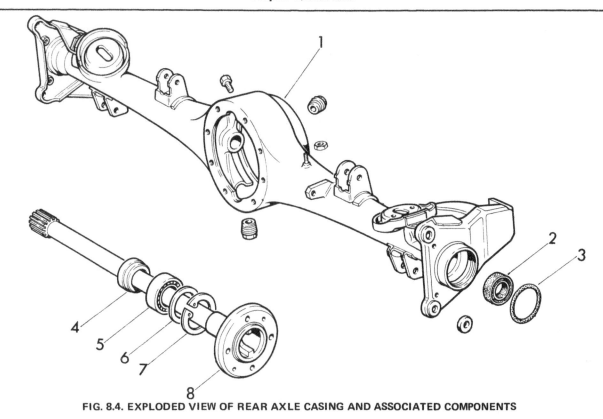

FIG. 8.4. EXPLODED VIEW OF REAR AXLE CASING AND ASSOCIATED COMPONENTS

1   Rear axle casing
2   Axle shaft oil seal
3   Axle shaft bearing 'O' ring
4   Axle shaft bearing locating ring

5   Axle shaft bearing
6   Bearing dust shield
7   Bearing locating circlip
8   Axle shaft

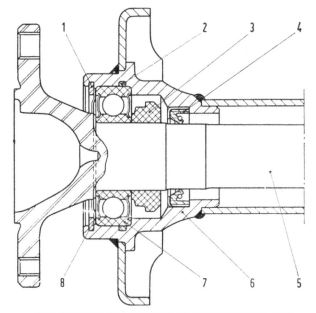

**FIG. 8.5. SECTIONAL VIEW OF LOCATION OF THE
AXLE SHAFT IN THE REAR AXLE CASING**

1   Circlip locating bearing in rear axle casing
2   'O' ring
3   Ring locating bearing on the axle shaft
4   Axle shaft oil seal
5   Axle shaft
6   Rear axle casing end fitting
7   Axle shaft bearing
8   Bearing dust shield

4   Pinion shaft oil seal

1   The renewal of the pinion shaft oils eal on this car is more involved than with many other cars. The complication arises from the use of a collapsible spacer between the inner races of the two taper roller bearings which locate the pinion shaft in the final drive housing. **This spacer must be renewed every time the pinion shaft end nut is loosened for one reason or another.**

2   It is possible if one is careful to remove the taper roller bearing inner race and roller assembly, either using the technique described later in paragraph 11 or using the special extractor tools available from Timkin Bearings.

3   The special extractors project into the bearing between the rollers to purchase on the rollers and cage so that the inner race and roller assembly can be pulled off the shaft.

4   Assuming that an appropriate tool is available or that you are going to try the technique described later proceed to remove the pinion shaft oil seal as follows:-

5   Raise the rear of the car onto car ramps or chassis stands. If the axle is of early design with the propeller shaft housed in a tubular extension fitted to the final drive housing, it will be necessary to remove the propeller shaft assembly as detailed in Chapter 7.

6   For the latest design of rear axle and propeller shaft it will only be necessary to undo the four nuts and bolts which retain the propeller shaft coupling on the flange fitting on the pinion shaft, to enable the rear half of the propeller shaft to be lowered away from the axle.

7   Once the propeller shaft is clear of the final drive housing, the pinion shaft end nut may be undone. It will be necessary to apply the handbrake really hard, because a considerable effort will be required to loosen the nut and it is easily sufficient to

turn a lightly braked wheel. It is also worth checking that the car is safely supported because the effort involved in undoing the pinion shaft nut could easily topple an inadequately supported vehicle.

8 FIAT supply tools A55075 and A70130 to remove the pinion shaft nut on the early design of axle where there is not a flanged fitting on the shaft.

9 After the nut has been removed the oil seal spacer or flanged fitting as appropriate may be slid off the shaft.

10 The pinion shaft oil seal may now be prised out of the final drive casing with a pair of screwdrivers. The oil flinger can then be slid off the shaft away from the first bearing.

11 The inner bearing race and roller assembly can be removed from the shaft by one of two methods:-

a) Using a speical TIMKIN extractor tool - specifically made for that bearing, insert the TIMKIN bearing extractor into the bearing and tighten the collar so that the pawls on the extractor purchase onto the bearing. Then screw in the central bolt in the extractor which will then pull the bearing off the shaft.

b) The second method does not assume costly special tools. Begin by gently tapping the pinion shaft into the final drive casing. It should move inwards a few tens of thousandths of an inch. Then pull the shaft outwards, the inner race of the first (outer) bearing should follow the shaft and a little space will be gained around that inner race. Next insert a pair of metal rods into the roller race to maintain jam the inner race in the position to which it has been brought. Repeat this procedure to move the inner race further and further along the shaft to a position from which it may be easily pulled off the shaft. The roller race will follow quite easily.

Once the inner race and roller race has been removed the spacer may be drawn off the shaft. Discard this old spacer, it must never be reused.

12 Reassembly of components can commence now with the fitment of the new collapsible spacer, followed by the roller assembly and inner race of the first bearing.

13 Slide the oil flinger into position and then fit the new pinion shaft oil seal.

14 The flange fitting or spacer are then fitted, followed by the pinion shaft nut and spring washer.

15 Tighten the pinion shaft nut to a torque of 160 ft lbs, and no more. Release the handbrake and check that the pinion shaft turns freely. Check also that there is no play of the shaft in the taper roller bearings. FIAT specify that the running torque for the pinion shaft should be between 14 and 17 inch lbs. It will not be possible to check this exactly without removing the final drive and differential from the rear axle casing. However, provided the nut has been tightened to or just less than the specified torque, the end load and therefore the running torque on the pinion shaft should be as specified by FIAT.

16 On the early design of axles the pinion shaft nut is locked by crimping the sleeve on the nut into one of the splines on the pinion shaft.

17 Refit the propeller shaft as detailed in Chapter 7 and bleed the brakes if necessary.

5 Final drive assembly - removal and replacement

This assembly has remained basically unchanged since the FIAT 124 Sports' introduction in 1967, therefore commence removal as follows:-

1 Raise the rear of the car onto chassis stands situated under the rear axle casing. Check the front wheels and ensure that the vehicle is safe.

2 Refer to Section 2 of this Chapter and slip the axle shafts out of the axle casing and differential in the middle of the casing.

3 Refer to Chapter 7 and either remove the whole propeller shaft in the case of the early design with a tubular extension on the final drive housing, or just remove the four nuts and bolts which secure the rear half of the propeller shaft to the flange on the final drive pinion shaft.

4 Remove the metal brake pipes from the rear axle casing in the vicinity of the central final drive housing. Remove the rear axle oil drain plug and drain the oil, which may be reused if fairly new.

5 Undo the eight bolts which secure the final drive assembly to the rear axle casing. The final drive may then be lifted out of the casing and transfered to a clean bench for dismantling and renovation.

6 Refitting the final drive follows the reversal of the removal procedure.

7 Ensure that all fixing nuts and bolts have been tightened to their correct torque and that the brake system has been bled, before taking the car off the stands and onto the road.

6 Final drive - dismantling, inspection, reassembly and adjustment

The final drive comprises the pinion shaft, crownwheel and differential assembly. The pinion and crownwheel are a matched pair of gears and should never be renewed individually.

Make sure before attempting to dismantle the final drive assembly, that it is both necessary and economic. Special tools are required in the reassembly of the final drive and it may well be cheaper to exchange the assembly as a complete unit at the outset.

1 With the final drive assembly removed from the rear axle casing and placed on a clean bench begin dismantling as follows:-

2 Dismantling.

Remove the bolts which secure the differential bearing adjuster locking tabs to the bearing caps. Lift away the locking tabs and unscrew the bearing adjusting rings. Make a note of the number of turns taken to remove the rings; the information will be useful when the differential is being reassembled.

3 Lock the differential in position by inserting a metal bar through the cage apertures; and loosen the input pinion shaft end nut. Undo the bolts which hold the bearing caps in place; do not remove the caps until identifying marks have been made on the caps and final drive casing to ensure they are replaced correctly.

4 Once the caps are removed, the differential, with the crownwheel bolted to it, can be lifted away from the final drive casing.

5 Begin dismantling the differential assembly by unbolting the crown wheel from the differential cage. Again make a mark on the crownwheel and differential cage so that they may be reassembled into the same position from which they were dismantled.

6 Then remove the differential cage bearing outer races roller assemblies, and using a conventional small sprocket puller, pull the bearing inner races off the seating on the differential cage sides.

7 Next with soft metal drift, tap the spindle on which the free bevel gears of the differential run out of the differential cage. Rotate the side bevel gears together to bring the free bevels into the apertures in the differential cage so that they may be lifted out of the assembly.

8 Once the free bevel gears are out, the two side bevels which drive the axle shaft may be withdrawn into the centre of the cage and then out through the apertures in the periphery. Retrieve the shims which are located behind the side bevels; these shims control the backlash between the side and free bevels.

9 The differential is now fully dismantled and the components may be cleaned ready for inspection.

10 The final drive input shaft and its associated components may be removed from the drive casing as follows:-

11 Remove the previously loosened end nut. If you omitted to loosen the nut, it will be necessary to bolt the drive flange to a length of rolled steel angle or similar so that the shaft may be held still whilst the nut is undone. If the early design of final drive assembly is being dismantled, there is not a flanged fitting on the shaft and therefore FIAT tool A70130 will be required to hold the splined end of the pinion shaft, whilst tool number A55075 or a ring spanner removes the end nut.

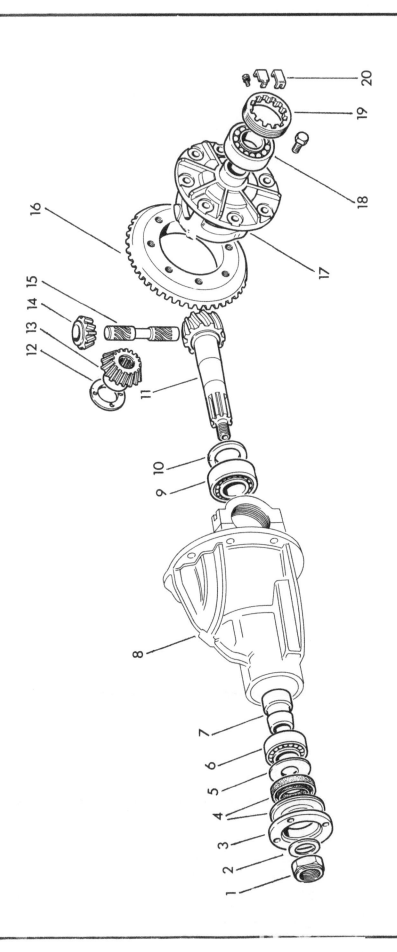

FIG. 8.6. EXPLODED VIEW OF THE FINAL DRIVE AND DIFFERENTIAL COMPONENTS

1 Pinion shaft end nut
2 Spring washer
3 Pinion shaft flanged fitting
4 Pinion shaft oil seal
5 Oil flinger
6 First (smaller) taper roller bearing

7 Collapsible spacer
8 Final drive casing
9 Second (larger) taper roller bearing
10 Shim spacer - (Change pinion
   position 0.002 inch increments)
11 Pinion shaft

12 Side gear shim washers
13 Differential side gear
14 Differential free gear
15 Free gear spindle
16 Crownwheel
17 Differential cage

18 Cage roller bearing
19 Cage position and bearing
   end load adjuster rings
20 Adjuster ring locking tabs

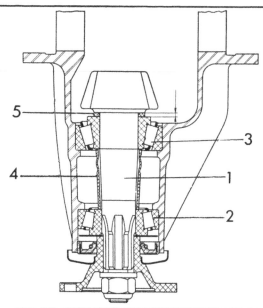

FIG. 8.7. ARRANGEMENT OF PINION SHAFT IN FINAL DRIVE CASING

1   *Pinion shaft*
2   *First (smaller) taper roller bearing*
3   *Second (larger) taper roller bearing*
4   *Collapsible spacer*
5   *Spacer shim - to set pinion gear position in the final drive casing*

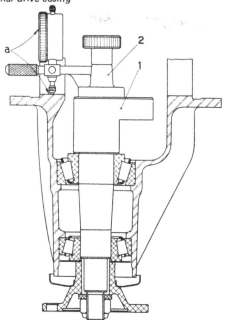

FIG. 8.8. SECTIONAL VIEW SHOWING FITMENT OF THE DUMMY PINION AND CLOCK GAUGE WHICH ENABLES THE THICKNESS OF THE PINION SPACER/SHIM TO BE DETERMINED

1   *Dummy pinion*   *FIAT tool No. 70184*
2   *Clock gauge*   *FIAT tool No. 95690*
Thickness of spacer/shim   =   dim. 'a' − (± b)

dimension 'a' is determined from the clock gauge, and is the average distance between the differential cage bearing seating and the top of the dummy pinion.

The gauge tool No. 95690 having been checked for zero on a surface plate.

dimension 'b' is stamped on the tapered section of the pinion shaft with its appropriate + or − sign; just below the production serial number of the shaft.

12 After the end nut has been removed, the flange fitting or spacer can be slid off the shaft. The pinion shaft oil seal can now be prised out of the final drive casing with a pair of screwdrivers.

13 Using a soft faced mallet, tap the pinion shaft down out of final drive casing. Collect the inner races and roller assemblies from the inside of the casing. The collapsible spacer can be removed too, but this must be discarded straight away because it must never be reused.

14 The shims directly behind pinion gear should be recovered and stored ready for reassembly of the final drive.

15 The outer races of the two pinion shaft bearings may be removed from the inside of the final drive casing using a soft metal drift. Tap the outer race rings carefully and be careful not to jam them in the seating.

16 All the components associated with the input shaft are now free and ready for inspection.

17 Inspection:-

Inspect the surfaces of the taper roller bearings inner races, outer races and the rollers themselves for pitting, scores and general surface deterioration. Renew complete bearings as necessary.

18 Inspect the surfaces of all the gear teeth for pitting, scoring and general surface deterioration. The pinion and crownwheel are a matched pair of gears and therefore must **never** be renewed individually.

The four bevel gears of the differential should be replaced in pairs too, the side gears and free gears.

19 The condition of the splines in the side gears and on the end of the pinion shaft should also be inspected. If wear is found, renew parts as necessary. Remember also to inspect the condition of the mating parts as well, that is the axle shafts and propeller shaft respectively since these too may be worn and merit replacement.

20 Lastly the condition of the bearing seatings both on the pinion shaft and in and on the casing and differential cage, should be inspected. If bearing failure necessitated repair of the final drive assembly, then it is possible that either the inner race or outer race spun on its seating. The seating would be damaged in that instant and therefore the appropriate casing will need replacement.

21 With all the components clean and ready for assembly commence by assembling the differential gears in the differential cage.

22 Assembly of differential gears:-

Begin by inserting the two side gears into the differential cage, together with their thrust washers. A nominal size of washer should be used if the side gears are new. Next fit the two free bevel gears into the cage and rotate the side gears to being the free gears into line with the holes through which the free gear spindle fits.

Insert the spindle and tap home with a soft drift. Now with either a clock gauge or feeler gauges determine the end play on the side gears. It should be 0.020 inches.

Select new shim thrust washers for the side gears to give the desired end play and reassemble the differential with the correct thrust washers. Again check the end play on the side gears to ensure that the new washers have had the desired effect.

23 The crown wheel may now be bolted onto the differential cage and the bolts tightened to a torque of 72 ft lbs. The two taper roller bearings in which the differential runs should now be assembled onto the cage. Gently tap the inner races of those bearings onto the seatings on the cage. Make sure the races are the correct way round with the highest ridge towards the centre of the cage. Then mount the roller assembly and the outer races onto the cage.

24 The differential is now ready for assembly onto the final drive, once the pinion shaft has been fitted.

25 Fitting the pinion shaft:-

Tap the outer races of the two taper roller bearings in which the shaft runs, into the final drive casing. Ensure that outer races are the correct way round. The high ridges should be towards each bearing.

26 There are special tools available from FIAT that enable the

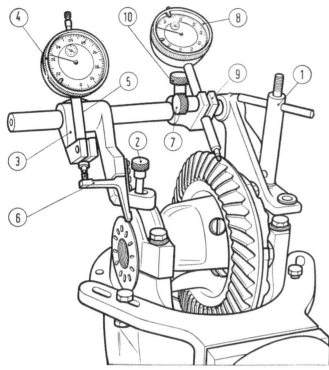

FIG. 8.9. CLOCK GAUGES AND FIXTURE TO MEASURE CROWNWHEEL/PINION BACKLASH AND THE DIFFEREN-TIAL GAUGE BEARING PRELOAD. FIAT TOOL NO. A95688

1   Column mounting
2   Knob support
3   Clock gauge support
4   Clock gauge reading differential cage bearing preload -
    bearing cap spread
5   Support mounting knob
6   Cranked lever
7   Clock gauge plunger knob
8   Clock gauge reading crownwheel/pinion gear backlash
9   Clock gauge support
10 Support mounting knob

pinion to be mounted in the correct position fairly quickly. The following paragraphs describe the technique necessary to match the relative positions of the pinion and crownwheel without special tools. (See Figs. 8.7 and 8.8).

27 Begin by assembling the pinion shaft bearings into the final drive casing, and then insert the pinion shafts into position, with the previously used spacer shim fitted behind the pinion gear. The pinion will then be very nearly in the same position as it was previously. Fit flange fitting or spacer onto the splined end of the shaft and then screw the nut onto the shaft. Tighten the nut a little so that the bearings are pinched together and the pinion gear is pulled back against the spacer shim and bearing.

28 Mount the differential onto the final drive, fit the bearing caps and tighten the cap bolts sufficiently to pinch the outer races of the differential cage bearings. Screw in the bearing adjusting rings so that they finish in the position from which they were removed. Paragraph 2 of this section instructed that a note be made of the number of turns it took to remove the adjuster rings.

29 Check that there is no end play on the differential and then tighten the bearing cap bolts to the specified torques.

30 Rotate the differential cage several times to settle the bearings and then, using feeler gauges or a clock gauge, deter-mine the backlash between the crownwheel and pinion. A clock gauge should ideally be used; place the probe on one of the crownwheel teeth and with the pinion gear held perfectly still rock the crownwheel to and fro. The correct backlash should be between 0.004 to 0.006 inches. (See Fig. 8.9).

31 Using a soft metal drift, tap the adjuster rings in the neces-

sary directions to move the differential cage in the required direction to give the required backlash. Should the adjusters be tight to move, slacken the differential cage bearing cap bolts slightly. After each adjustment tighten the cap bolts to the specified torque again.

32 Then clean the gear teeth on the pinion and crownwheel and brush a little 'mechanical blue' onto several teeth on both gears. Then roll the pinion and crownwheel together to mesh the gears coated with 'mechanics blue'.

33 Compare the patterns marked in the blue on the gear teeth with those shown in Fig. 8.10.

34 The illustration shows what remedial action should be taken if an incorrect mesh pattern is obtained.

The crown wheel may be removed by turning the adjusting rings in the appropriate direction.

The pinion may be moved by fitting a different thickness of washer shim between the gear and the main bearing inner race. The pinion washer shims are available in a number of thicknesses in 0.002 increments.

35 Once you are satisfied that the correct pinion washer has been fitted and that the correct position of the crownwheel and differential is known, temporarily remove the differential and crownwheel assembly from the final drive.

36 Undo the pinion shaft nut and remove the flange fitting or spacer. Pull the shaft out of the final drive casing.

37 The inner race and roller assembly of the smaller forward pinion shaft bearing should be removed. Then reinsert the pinion shaft into the final drive casing through the larger rear bearing. Slide a NEW collapsible spacer onto the pinion shaft and follow that by reassembling the forward bearing back onto the pinion shaft.

38 Slide the oil flinger disc into position next to the smaller forward bearing and then tap a new pinion shaft oil seal into its seating in the final drive casing.

39 The flange fitting or spacer should now be slipped onto the end of the shaft, and follow with the end nut and spring washer.

40 Tighten the end nut to just 15 ft lbs at this stage.

41 Reassemble the differential assembly onto the final drive, but do not fit the adjusting rings yet.

42 Lock the differential case with a bar inserted into the cage so that the pinion shaft nut may be tightened now to approxi-mately 140 ft lbs. Then with a dynamometer or more simply a weight pan and lever, check that the pinion shaft will turn with a running torque of between 14 in lbs and 17 in lbs.

If the shaft running torque has not reached that range, tighten the shaft nut further - but do not exceed a torque of 181 ft lbs.

If the torque exceeds that range check the conditions of the bearings closely before fitting a new collapsible spacer and repeating the assembly procedure.

43 With the pinion shaft satisfactorily installed in the final drive, reassemble the differential cage bearing adjusting rings into the final drive and screw into the positions determined previously.

Tighten the cap bolts to the specified torque.

44 Now that all the components have been assembled, recheck with mechanics blue on the gear teeth to be certain that the mesh pattern of the pinion and crownwheel is still correct.

The preload on the differential cage bearings can now be checked. Fit a clock gauge between the bearing caps, and then unscrew both adjuster rings equally by about ½ turn or slightly more. (Fig. 8.9).

45 Watch the clock gauge to check that the caps are still unloaded laterally. Set the clock dial to zero, and then screw in the adjuster rings **equally** until the caps have moved apart by 0.0063 to 0.0078 inches. The differential cage bearings are now correctly preloaded.

46 Check the torque on the bearing cap bolts and fit the adjuster ring locking tabs to the bearing caps.

47 The final drive assembly is now complete and ready for refitting into the axle casing as detailed earlier in this chapter.

48 Remember to use a new gasket between the final drive and the rear axle casing and do not forget to refill the axle with SAE. 90 EP oil.

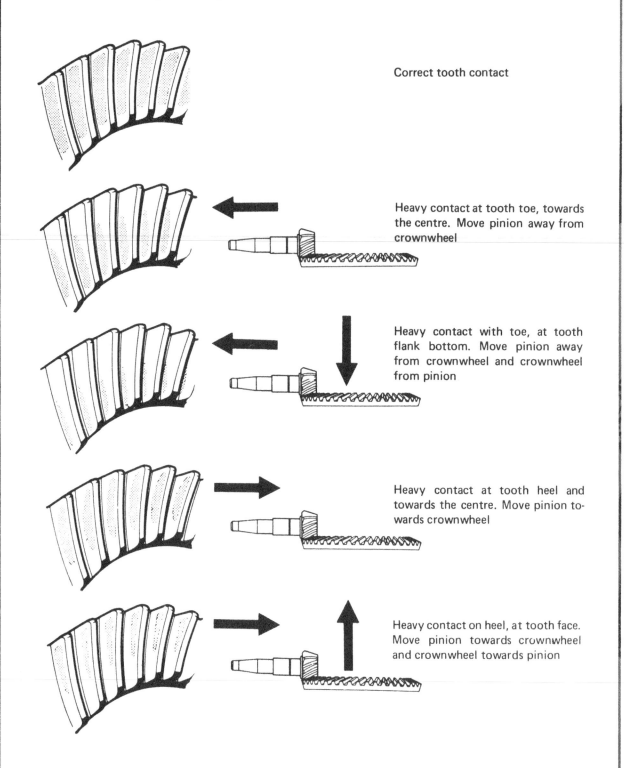

Correct tooth contact

Heavy contact at tooth toe, towards the centre. Move pinion away from crownwheel

Heavy contact with toe, at tooth flank bottom. Move pinion away from crownwheel and crownwheel from pinion

Heavy contact at tooth heel and towards the centre. Move pinion towards crownwheel

Heavy contact on heel, at tooth face. Move pinion towards crownwheel and crownwheel towards pinion

Fig. 8.10. Crownwheel/pinion gear mesh tooth patterns

Chapter 9   Braking system

Contents

Specifications

Pedal operated, servo assisted disc brakes on all four wheels. Disc brake calipers of floating type with single cylinder.

Front disc brake
Disc diameter	8.93 inches
Disc run-out	0.006 inches
Minimum pad thickness	2 mm (0.079 inches)
Minimum disc thickness after regrind	0.370 inches
Adjustment	Self adjusting
Caliper cylinder diameter	1.875 inches (48 mm)

Rear disc brake
Disc diameter	8.93 inches
Disc run-out	0.006 inches
Minimum pad thickness	2 mm (0.079 inches)
Minimum disc thickness after regrind	0.370 inches
Caliper cylinder diameter	1.375 inches (34 mm)

Servo unit
Manufacturer	'Master-Vac' Girling Benaldi
Vacuum cylinder bore	7 9/32 inches

Master cylinder
Bore	0.750 inches
Brake fluid	FIAT fluid blue label

Rear brake regulator
Distance from the bodyshell of the connection of the torsion bar and link to rear axle	3.74 inches ± 0.20 in (Sport Spider and Sport 1600 Spider); 5.787 inches ± 0.2 in (Sport Coupe and Sport 1600, 1800 Coupe) with piston of regulator just touching end of torsion bar.

Handbrake
Handbrake	Link and cable mechanical action to rear brake calipers: Self adjusting action

Torque wrench settings:
	lb ft
Bleed screw	4 to 6
Master cylinder retaining nuts	25
Servo retaining nuts	25
Brake caliper block retaining bolts	35
Brake disc securing bolts	15
Front caliper mounting plate and steering arm to kingpin member	43

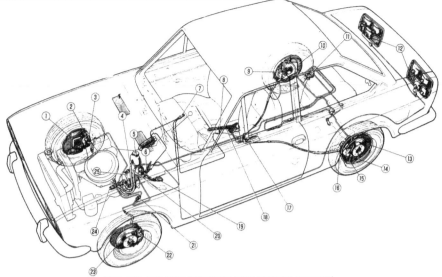

FIG. 9.1. BRAKE SYSTEM (124 SPORT COUPE)

1 Front brake disc shield
2 Front brake bleed screw
3 Front brake caliper block
4 Front brake circuit
5 Dual brake fluid reservoir(s)
6 Stop light switch
7 Brake circuit effectiveness and handbrake on indicator (US models)
8 Jam switch for handbrake ON signal and efficiency indicator (US models)
9 Rear brake bleed screw
10 Rear brake disc shield
11 Rear brake regulator device
12 Stop lights
13 Rear brake disc
14 Caliper yoke member
15 Rear brake caliper block
16 Mechanical brake on rear disc
17 Handbrake cable and harness
18 Handbrake operating lever
19 Brake pedal
20 Rear brake circuit
21 Pressure switch for indicator 7 (US models)
22 Brake pad and caliper yoke member
23 Front brake disc
24 Master cylinder with tandem pistons
25 Master-vac unit

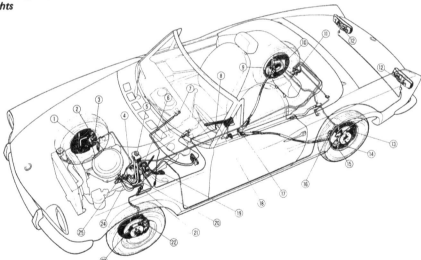

FIG. 9.2. BRAKE SYSTEM (124 SPORT SPIDER)

1 Front brake disc shield
2 Front brake bleed screw
3 Front brake caliper block
4 Master-vac unit
5 Dual brake fluid reservoirs
6 Stop lights switch
7 Brake circuit effectiveness and brake ON indicator (US models)
8 Jam switch for handbrake ON and brake indicator (7) (US models)
9 Rear brake bleed screw
10 Rear brake disc shield
11 Rear brake regulator
12 Stop lights
13 Rear brake disc
14 Caliper block yoke member
15 Rear brake caliper block
16 Mechanical rear brake
17 Handbrake cable harness
18 Handbrake operating lever
19 Brake pedal
20 Front brake disc circuit
21 Pressure switch for indicator (7)
22 Brake pad and caliper yoke member
23 Front brake disc
24 Rear brake circuit
25 Master cylinder. Tandem pistons

1 General description

FIAT have employed caliper disc brakes on all four wheels, which are hydraulically operated from a single or tandem master cylinder situated in the rear of the engine compartment. The master cylinder/brake pedal link passes through a Girling Bonaldi Master-Vac unit which uses inlet manifold vacuum to increase the effort exerted by the driver onto the master cylinder.

The handbrake is a conventional arrangement comprising a lever, linkage and cable which combine mechanically to act on specially adapted rear brake calipers. The adaption in the caliper assembly results in the handbrake system to be self-adjusting as far as brake pad wear is concerned. There is provision for adjustment to take up stretch and wear in the mechanical linkage to the rear brake assembly.

One final inovation employed by FIAT on the Sport series is a rear brake regulator. This device governs the braking effort of the rear brakes by monitoring the attitude of the vehicle during braking.

The regulator is positioned above the rear axle and a linkage connects the axle and regulator in such a manner as to close a valve in the regulator as the distance between the axle and body-shell increases - that is when the vehicle pitches front down during heavy braking. When the regulator valve closes the hydraulic action to the rear brakes is interrupted and the rear brakes have reduced effect. Rear braking is not lost entirely because there is fluid trapped under pressure in the brake lines to the calipers from the regulator unit.

2 Bleeding the hydraulic system

1 This without doubt will be one of the most frequent tasks to be performed on the brake system. You will require some small 1/8 or 3/16 inch bore rubber or clear plastic tubing and a clean dry glass jar. The tubing should be at least 15 inches long. You will also require a quantity of brake fluid (Castrol Girling Universal Brake and Clutch Fluid) probably between ¼ and ½ a pint.

2 It will be necessary to bleed the brakes whenever any part or all of the brake system has been overhauled or when a brake pipe connection has been undone in the course of performing tasks on other assemblies on the car. Bleeding of the brakes will also be necessary if the level of fluid in the brake system reservoirs has fallen too low and air has been taken into the system. During the task of bleeding the brakes, the level of fluid should not be allowed to fall below half way, or air will be drawn into the system again.

3 Although not necessary for the point of view of a success-fully completed task, it improves access to remove the road wheel adjacent to the brake to be bled. Beginning at the front of the car, place a jack underneath the lower arm of the suspension, Raise the road wheel off the ground and remove the wheel. Chock the other wheels and release the handbrake.

4 Remove the rubber dust cover from the bleed screw and wipe the screw head clean. Push the rubber/plastic tubing onto the screw head and drop the other end into the glass jar placed nearby on the floor. Remove the appropriate reservoir cap and ensure that the fluid level is near the top of the reservoir. Pour a little brake fluid into the glass jar, sufficient to cover the end of the tube lying in the jar. Take great care not to allow any brake fluid to come into contact with the paint on the bodywork, it is highly corrosive.

5 Use a suitable open ended spanner and unscrew the bleed screw about one half to a full turn.

6 An assistant should now pump the brake pedal by first depressing it in one full stroke followed by three shorter more rapid strokes, allowing the pedal to return of its own accord each time. Check the level of fluid in the reservoir and replenish if necessary with new fluid. NEVER re-use fluid.

7 Carefully watch the flow of fluid into the glass jar and when

the air bubbles cease to emerge from the bleed screw and braking system through the plastic pipe, tighten the bleed screw during a down stroke on the pedal. It may be necessary to repeat pumping detailed in paragraph 6 if there was a particularly large accumulation of air in the brake system.

8 Repeat the operations detailed in paragraph 3 to 7 on the other three brakes. When bleeding the rear brakes, place the jacks under the axle casing and remove the wheels. **Do not use chassis stands acting on the bodyshell otherwise the rear brake regulator system will be brought into operation** and prevent the flow of fluid to the rear brakes. A last additional point with regard the rear brakes, is to pump the pedal slowly and allow one or two seconds between each stroke.

9 Sometimes it may be found that the bleeding operation for one or more cylinders is taking a considerable time. The cause is probably due to air being drawn past the bleed screw threads, back into the system during the return stroke of the brake pedal and master cylinder, when the bleed screw is still loose. To counteract this occurance, it is recommended that at the end of the downward stroke the bleed screw be temporarily tightened and loosened only when another downstroke is about to commence.

10 Once all the brakes have been bled, recheck the level of fluid in the reservoir(s) and replenish as necessary. Always use new brake fluid — NEVER re-use fluid.

11 If after the bleed operation, the brake pedal operation still feels spongy, this is an indication that there is still some air in the system, or that the master cylinder is faulty.

3 Front disc brake pads - removal, inspection and refitting

1 Inspection of the disc pads on the brakes fitted to this series of car cannot be readily undertaken with the brakes in place. It will be necessary to remove the caliper block in order to expose the pads and enable an inspection to be made.

2 The caliper block seats in a yoke fitting which is bolted to the kingpin assembly and 'wraps around' part of the disc. The pads and their retaining springs, and the caliper block which acts on the pads, are held between the upper and lower parts of the enveloping yoke. The caliper is held so that it is free to move axially to centre itself on the pads and disc. It is retained radially by two wedges which are held axially to the caliper yoke by two cotter pins.

3 The procedure for removal of the disc brake pads is as follows:

Jack up the appropriate suspension and remove the road wheel. Chock the other wheels and release the handbrake if it is intended to remove the rear brake pads.

4 Pull out the cotter pins which retain the caliper block wedges. Wedges are fitted above and below the caliper block. Mark the wedges before removing them to be sure that you will be able to replace them in exactly the same position from which they were taken. (photos).

5 Once the wedges are out, the caliper block can be lifted away from the brake assembly. Be very careful not to strain the flexible brake pipe joining the caliper to the pipe system on the bodyshell. Rest the caliper on the lower suspension member. (photo).

6 Remove the brake pads. The pad anti-vibration springs may be removed if desired (photo).

7 Mark the pads so that they may be refitted into the exact positions from which they were taken.

8 Measure the pad material thickness, if less than 0.079 inches (2 mm) or quite close to that thickness, renew the pads.

9 Attention should be paid to the colour coding on the back plate of the pad. These codings indicate the type of material the pad is made of and as usual there is only one combination of pads which is safe to use.

10 The colour code (and hence pad material) on all eight pads used in the four brakes on the car MUST be the same. Make sure that the colour coding of the pads being discarded and the new pads is the same.

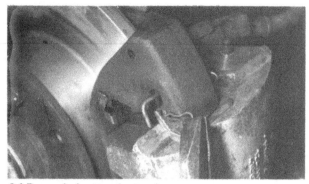

3.4 Removal of cotter pins to release
caliper locating wedge

3.4a Remove wedges

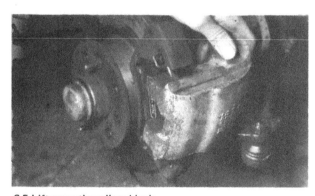

3.5 Lift away the caliper block

3.6 Pads removed. Anti-vibration springs
exposed

11 While you have the brake assembly of the disc it is a good
opportunity to inspect the disc for scores and excessive wear.
Check that its thickness is not less than 0.350 inches (9 mm).
Renew the disc immediately if it is too thin. You may only have
the disc reground if the thickness after machining will not be less
than 0.37 inches.
12 The refitting of the disc pads follows the reversal of the
removal sequence. See photo sequence.
13 Before the caliper block is refitted over the new pads, the
slave piston will need to be gently pushed back into its bore in
the caliper, to provide a sufficient gap between the piston and
the out claw of the caliper to accommodate the new pads and
the disc brake. It is as well to remember that as the piston is
moved back into the block brake fluid will be displaced and
returned to the reservoir. The reservoir will overflow unless fluid
is taken out with a device such as a pipette as the slave pistons
are moved.

4 Front brake caliper block and yoke — removal and refitting

1   There is no need to remove the calliper block yoke, if it is
only the caliper that is to be attended to.
2   The yoke needs little attention and the only occasions when
it is necessary to remove it, is when it is desired to remove the
brake disc which is mounted on the wheel hub.
3   Procedure to remove the brake caliper block as follows: Jack
up the appropriate lower suspension arm and remove the road
wheel.
4   Working inside the engine compartment, remove the cap on
the appropriate brake fluid reservoir, and stretch a thin sheet of
polythene over the top of the reservoir. Refit the cap.
   This measure will prevent excessive loss of brake fluid when
brake pipe connections are subsequently undone.
5   Undo the nut which is retaining the metal brake pipe to the
inner end of the flexible hose and lift the metal pipe free of the
hose end. Slacken the thinner nut on the hose inner end which
retains that end in the bracket on the bodyshell.
6   Unscrew the flexible hose from the caliper block and remove
the hose from the car. Tape both ends of the hose, the port in
the caliper block and the metal brake pipe on the car to prevent
the ingress of dirt.
7   Next remove the four cotter pins which hold the two caliper
wedges in position and withdraw the wedges from between the
caliper block and yoke. Mark them to ensure that you can refit
them into position from which they have been taken.
8   The caliper block can now be lifted free and taken to a clean
bench for overhaul.
9   Refitting the caliper block follows the exact reversal of the
removal procedure except the following tasks are added: Once
the caliper block has been refitted the brake system must be bled
to expel all the air from the system.
10 Caliper yoke removal. Proceed as directed in paragraphs 3 to
8 of this section and remove the caliper block. Then lift the
brake pads away. The two bolts which secure the yoke to its
mounting plate on the kingpin assembly can now be undone and
the yoke lifted free.
11 The brake anti rattle springs should be removed and the yoke
brushed clean. Renew the yoke only if cracks or serious wear is
found on the yoke.
12 Refitting is the reversal of removal and remember use new
lock washers and to tighten the retaining bolts to the torques
specified at the beginning of this chapter.

5 Front brake caliper overhaul

1   Once the caliper block has been removed as described in
section 4 the next problem is to extract the slave piston so that
the fluid seals may be removed and all components cleaned for
inspection. Some times the piston may be a smooth enough fit in
the bore in the caliper block to be pulled out directly. Very
often however the piston will need impelling out by force

exerted by a supply of high pressure air into the caliper block or a supply of hydraulic fluid under pressure. The latter is described here as it is a technique which does not require special equipment.

2   Temporarily reconnect the caliper block to the vehicle brake system and bleed the line to the block. Having bled the line, continue to pump the brake pedal to push the large piston in the caliper block out of its cylinder. It is as well to have an old tray beneath the area of work to catch any brake fluid spilt.

Once the piston protrudes a half inch or so it may be pulled out with your fingers. Once the piston has been removed, disconnect the caliper block and flexible hose and transfer all the components to a clean bench for cleaning, inspection and reassembly.

3   Remove the dust seal and 'O' ring seal from the bore in the caliper block using a plastic knitting needle or similar. Take care not to scratch the surface of the bore.

4   Wash all the components in Girling Cleaning Fluid or methylated spirit. Do not use any other cleaner fluid because traces will damage the seals and contaminate the hydraulic fluid when the block is reassembled.

5   Inspect the caliper bore and piston for scoring and wear, renew the whole caliper block assembly if such wear is is found.

6   To reassemble the caliper, wet the new 'O' ring seal with new brake fluid and carefully insert it into its groove near the rim of the bore in the caliper.

7   Then refit the new dust seal onto its seating at the rim of the bore in the caliper. Coat the side of piston with hydraulic fluid and carefully insert it into the bore in the caliper, until it protrudes by about one half inch. Fit the dust seal onto the top of the piston and then push the piston into the bore as far as it will go.

8   The caliper block is ready to be refitted to the brake assembly.

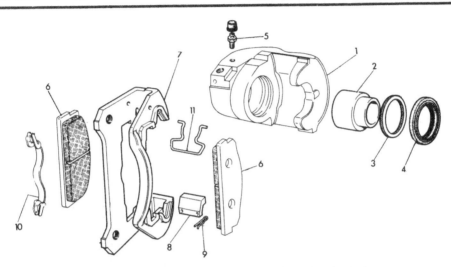

FIG. 9.3. EXPLODED VIEW OF FRONT BRAKE COMPONENTS

1   Caliper block
2   Piston
3   'O' ring piston seal
4   Dust seal
5   Bleed screw
6   Brake pad

7   Caliper block yoke member
8   Caliper block locating wedges
9   Cotter pins
10 Pad positioning spring
11 Caliper block positioning spring

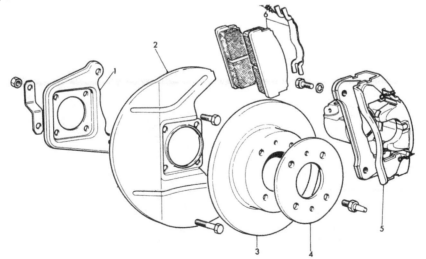

FIG. 9.4. EXPLODED VIEW OF FRONT BRAKE ASSEMBLY

1   Caliper yoke mounting plate (bolted to kingpin)
2   Disc shield
3   Disc

4   Spacer
5   Caliper block and yoke assembly
6   Pads and positioning spring

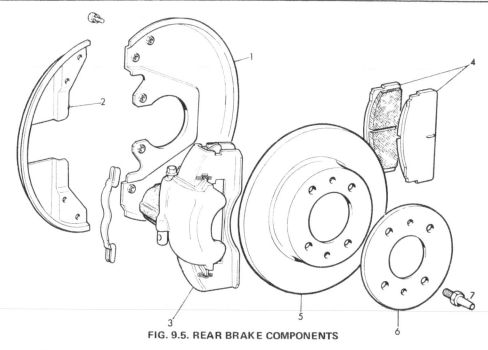

FIG. 9.5. REAR BRAKE COMPONENTS

1 Rear sector of disc shield
2 Forward sector of disc shield
3 Caliper block and yoke assembly
4 Pads

5 Brake disc
6 Spacer
7 Disc retaining spigot bolt

6 Rear disc brake pads - removal, inspection and refitting

1 As with the front brakes, the caliper block will need to be removed from the yoke member to expose the pads for inspection and removal when necessary.

2 The rear pads must be the same type and material as the front brakes. Colour stripes on the rear of the pad indicate type and material and it follows therefore that the colours on all pads must be the same.

3 The removal and refitting procedure for the caliper block and brake pads is the same as with the front brakes as detailed in Section 3.

4 The slave piston in the caliper will need screwing back into the caliper so there is a sufficient gap between the piston and outer claw of the caliper to pass over new pads and the disc brake.

5 Remember that as the slave piston is moved back brake fluid will be pushed out of the caliper cylinder along the brake lines and into the reservoir. The reservoir might easily overflow if, as the slave pistons are moved back, fluid is not drawn out of the reservoir with a device like a pipette.

6 Once the new pads have been fitted, try the brakes and pay particular attention to the feel of the brake pedal action: If it is at all spongy bleed the brakes, because you might have been unlucky and air might have been drawn into the brake system while the slave pistons were being moved to accomodate the new pads.

7 Rear brake caliper block - removal and refitting

1 This task is virtually the same as the removal and refitting of the front caliper block, except for the following additional tasks.

2 Release the handbrake and disengage the cable and sleeve from the lever and anchorage on the caliper block.

3 The forward sector of the disc brake shield will need to be unbolted from the rearward sector before the caliper can be separated from the yoke member (photo).

4 The rear brake caliper block is retained in exactly the same manner to the yoke member, as the front brake caliper block to its yoke member. (photo).

7.3 Remove forward sector of disc shield

7.4 Rear caliper block locating wedges

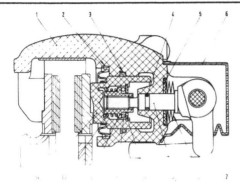

**FIG. 9.6. SECTION OF REAR BRAKE CALIPER
SHOWING MECHANICAL AND HYDRAULIC ACTUATION**

1 Caliper block   2 Dust shield
3 'O' ring piston seal   4 'O' ring plunger seal
5 Diaphragm spring thrust washer   6 Mechanical linkage
7 Mechanical brake actuating   protection boot
   lever   8 Diaphragm springs
9 Plunger with threaded end screwed into special nut in piston
10 Spun in disc   11 Ball thrust bearing
12 Special nut for self adjustment   13 Nut spring
14 Main piston

8 Rear brake caliper overhaul

1 Although the mode of operation of the rear brake caliper is similar to the front, the complication of the handbrake system incorporated in the block means that the overhaul procedure is rather different from that for the front brake.
2 Begin by removing the boot which covers the handbrake lever pivot assembly on the caliper block. Once the lever pivot is exposed, the pivot pin may be driven from the block to release the lever and cam.
3 Unscrew the plunger which follows the cam at the lever pivot to press the brake pads against the brake disc. Retrive the plunger return spring and the thrust washer. There will be an 'O' ring oil seal on the large shank of the plunger.
4 The slave piston can now be driven from the bore in the caliper block with a slim drift inserted through the hole which took the plunger and right in to contact the face of the piston. Be careful not to damage the threads in the special nut in which the plunger engaged. Remove the dust seal and 'O' ring seal from the bore in the caliper block with a plastic knitting needle.
5 Do not attempt to remove the special nut and its associated parts from inside the piston. You will certainly damage the spun in disc which retains the nut in position and FIAT do not supply the individual parts.
6 Once all the components have been separated they may be cleaned with Girling Cleaning Fluid or methylated spirits. Do not use any other solvent. Dry the parts and inspect for wear.
7 If any wear or scores are found on the piston or in the bore in the caliper block, the whole assembly should be renewed. You should not attemp to polish out scratches since the seals rely on a close fit to be effective.
8 Reassembly of the caliper block begins with the fitting of the new 'O' ring oil seal into its groove in the bore in the caliper block. The seal should be wetted in hydraulic fluid before fitment. Follow the fitting of the 'O' ring with the fitting of the dust seal.
9 Wet the sides of the slave piston assembly and push gently into the bore. When only one half inch is left protruding fit the dust cover onto the top of the piston and then push the piston into the caliper block as far as it will go.
10 Wet the small new 'O' ring that fits onto the plunger with brake fluid and slip over the plunger shank into its groove.
11 Place the plunger spring thrust washers and the springs themselves into position on the caliper block around the hole into which the plunger fits.
12 Insert the plunger into the caliper block and push and screw simultaneously to work it right into the special nut which is in the slave piston.

13 Place the lever pivot into position in the fork on the outside of the caliper block. You may need to slip a 'G' clamp over the lever pivot point and caliper block bearing onto the top face of the slave piston, to hold the lever in the position where the pivot pin may be driven into place.
14 The protective boot is now slipped over the handbrake lever and into place on the caliper block.
15 The caliper assembly is ready to be fitted to the yoke and wheel hub.

9 Master cylinder - removal and refitting

1 There have been two patterns of master cylinder fitted to the Sport Spider and Coupe cars, the early AC and AS bodyshell series were fitted with a master cylinder which comprised just one piston to push fluid to all four brakes. The BC, BS, CC and CS bodyshell series cars were fitted mainly with a tandem cylinder incorportating two pistons. One piston supplies the front brakes only, the other supplies the rear brakes only. To match the tandem master cylinder, there are two fluid reservoirs fitted, one for each brake system.
2 Both patterns of cylinder are readily removed for inspection and overhaul. They are retained to the Master-Vac assembly by two nuts on two bolts. Begin master cylinder removal as follows.
3 Remove the cap (s) from the brake fluid reservoir (s) and stretch a thin sheet of polythene over the reservoir top(s) and replace the cap(s). This measure will prevent excessive loss of fluid when the brake fluid pipes are disconnected from the master cylinder.
4 Undo the clips holding the fluid feed pipes to the master cylinder union (s) and pull the pipes clear. Next undo the connections of the metal brake pipes to the master cylinder and raise the pipe(s) a little clear of the cylinder.
5 Finally undo and remove the two nuts which secure the master cylinder to the Master-Vac and lift the master cylinder away to a clean bench for inspection and overhaul.
6 Refitting the master cylinder follows the reversal of the removal procedure except that the whole brake system must be bled before the car is taken on the road.

10 Master cylinder - overhaul

1 The overhaul of the tandem master cylinder will be described here, the single piston version is very much simpler but follows the same procedure.
2 With the cylinder out on a bench lightly jam the cylinder in a vice and undo the special bolt in the front end of the cylinder. Then undo the special bolts in the underside of the cylinder, these project into the bore to guide the pistons.
3 The pistons, seals, cups and springs may now be pushed out of the cylinder from the rear end.
4 Separate all the springs, seals and pistons and then, but only if absolutely necessary, remove the plastic inlet union and seal from the top of the cylinder.
5 Clean all the components with Girling Cleaning Fluid or methylated spirit and inspect the cylinder bore and piston sides for scores or wear. If scores are found replace the whole master cylinder assembly. Do not try to polish out scratches or score marks.
6 Once all the parts have been cleaned, gather them together for reassembly. Do not re-use seals always use new ones.
7 Begin reassembly by soaking the new seals in brake fluid before slipping the secondary seals onto the rear ends of each piston. Then slip the primary seal cups and the pimary seals themselves onto their seating on the forward end of each piston.
   Make sure the primary seals are correctly aligned with the fine lips towards the forward end of the piston.
8 Insert the rear piston (which is recessed to accept the pushrod from the Master-Vac assembly) making sure that both primary and secondary seals are not damaged. Position the piston with its groove lowermost and insert the special bolt to register into the piston. The bolt serves to limit the rearward movement of the piston.
9 Then drop the buffer and seal spreading springs down the

cylinder bore onto the first piston. Follow the springs with the second piston complete with both its secondary and primary seals. Position the groove lowermost. Screw in the guide bolt into the cylinder. The bolt registers into the groove of the second (forward most) piston and limits the rearward movement of that position.

10 Finish the assembly by dropping the second buffer and seal spreader springs and then the special end bolt. Screw in the end bolt tightly to seal the threads.

11 The single piston master cylinder design does not employ a spigot bolt to limit the rearward movement of the piston, instead a circlip is fitted in a groove in the rear end of the master cylinder bore.

12 If you have removed the inlet unions, they should be refitted now using new star washers and seal.

13 The cylinder is now ready to be refitted to the car.

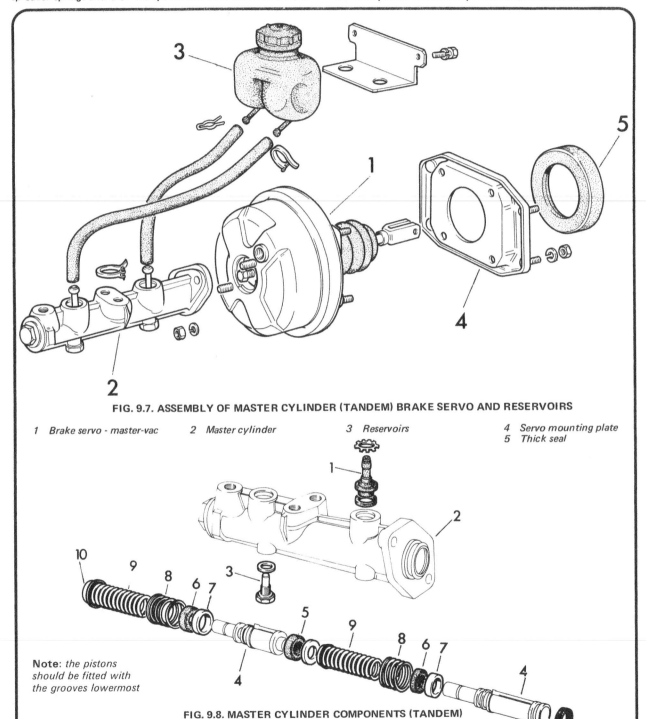

FIG. 9.7. ASSEMBLY OF MASTER CYLINDER (TANDEM) BRAKE SERVO AND RESERVOIRS

1   Brake servo - master-vac   2   Master cylinder   3   Reservoirs   4   Servo mounting plate
                                                                                       5   Thick seal

Note: *the pistons should be fitted with the grooves lowermost*

FIG. 9.8. MASTER CYLINDER COMPONENTS (TANDEM)

1   Plastic inlet union   6   Primary seals
2   Cylinder body   7   Primary seal retainers
3   Piston register bolts   8   Buffer spring
4   Piston   9   Primary seal spreader spring
5   Secondary seals   10   End plug

11 Brake discs - inspection, removal and refitting

1   The discs can easily be inspected by simply removing the roadwheels. If the swept surface of the disc is found to be scored or worn, check the thickness of the disc. The minimum thickness of the disc is 0.35 inches, and clearly if it its found to be thinner it should be renewed immediately.

The thickness of the discs when new is 0.39 inches and they can be reground to remove minor scores and scratches providing the thickness after machining is not less than 0.37 inches.

2   To remove the disc, begin by removing the brake caliper, block and disc brake pads as detailed in Section 3 and 6 of this chapter.

3   It should be appreciated that the caliper blocks must be carefully supported not to strain the flexible hose which connects it to the brake pipe system on the car.

4   The caliper yoke members may be unbolted from the support plate (front suspension) or axle casing bracket (rear suspension) as appropriate.

5   The two special spigot bolts which hold the disc to the wheel hub can now be removed and the disc lifted from the hub.

6   Refitting of the disc follows the reversal of the removal procedure. Remember to tighten the various nuts and bolts to the torques specified at the beginning of this chapter. Always use new spring locking washers as necessary.

12 Handbrake cable - adjustment

1   The rear brakes are self adjusting so the only adjustment

necessary on the handbrake is to accommodate cable stetch and wear in the mechanical linkage.

2   Adjustment will be necessary when the handbrake lever passes more than 4 to 5 notches on the pivot ratchet.

3   The adjustment mechanism can be found underneath the car just to the rear of the propeller shaft pillow block support member. (photo).

4   Simply loosen the locknut and turn the adjusting nut next to the cable harness to take up the slack in the cable, until the handbrake lever travel is as desired. Secure the adjusting nut in position by tightening the locknut against the adjusting nut.

12.3 The handbrake adjustment nuts

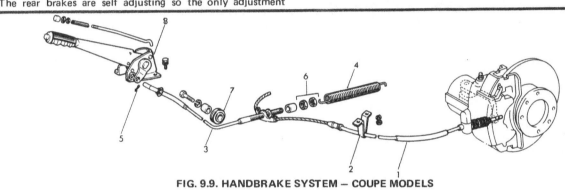

FIG. 9.9. HANDBRAKE SYSTEM — COUPE MODELS

1 Rear cable and sleeve assembly	6 System adjustment nuts
2 Sleeve support bracket	7 Link cable pulley
3 Cable harness	8 Handbrake lever assembly
4 System return spring	
5 Lever to rear cable link cable	

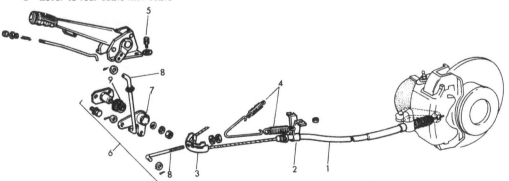

FIG. 9.10. HANDBRAKE SYSTEM — SPIDER MODELS

1 Rear cable and sleeve assembly	6 Operating lever to rear cable linkage
2 Sleeve support bracket	7 'L' shape lever
3 Cable harness	8 Link rods
4 System return springs	
5 Handbrake operating lever assembly	

13 Handbrake cable - inspection, removal and refitting

1 The cable will need renewal when it has been necessary to adjust it regularly to accommodate an increasing amount of stretch in the cable. Normally the handbrake linkage should only need adjustment once a year but if it becomes necessary to adjust at monthly intervals, then the cable merits renewal.
2 Begin the removal task by raising the rear of the car onto car ramps or chassis stands.
3 Undo the cable adjusting nut and its lock nut and unscrew them off the threaded rod from the handbrake lever.
4 Remove the cable harness from the threaded rod and lift the cable from the harness. Undo the nuts which retain the cable sleeve support brackets to the bodyshell and remove the brackets.
5 Finally pull the cable end from the slot in the end of the actuating lever on the rear brake caliper block. Lift the cable sleeve from its anchorage on the caliper block. The cable and sleeve is free from the car. Do not attempt to separate the cable and the sleeve. FIAT only supply the cable sleeve assembly as a spare. (photo).
6 Refitting is the exact reversal of removal, except that the new cable should be adjusted as described in Section 12 before the car is returned to the road.

13.5 The location of the handbrake rear cable on the rear caliper lever

14 Handbrake lever assembly - removal and refitting

1 The handbrake lever assembly exists as three sub assemblies. The lever and ratchet sub assembly is the largest, then the link cable sub assembly which joins the lever the the rear cable harness is next and finally there is a pulley sub assembly round which the link rod runs.
2 On the latest C5 bodyshell series Spider cars the handbrake lever is connected to the rear cable assembly by two link rods joined by an 'L' shape lever mounted on a pivot bolt in the transmission tunnel. The dismantling procedure for this system follows much the same pattern as described below for the earlier Spider and all Coupe models.
3 Remove the centre trim which covers the lever pivot assembly and heater controls as described in Chapter 12.
4 Then pull out the split pin which retains the link rod to the output lever from the handbrake lever pivot assembly. Remove the link rod.
5 Undo and remove the four bolts that secure the lever pivot assembly to the bodyshell, and lift away the lever assembly.
6 The link cable can now be removed once the adjusting lock nut and return spring are removed from the rear threaded end of the rod.
7 The link cable pulley runs on a bush on a bolt screwed into a bracket in the transmission tunnel.
8 Once the pulley bolt is removed the bush and pulley can be separated.

9 There is not any scope for repair work on the handbrake lever and its associated linkage; quite simply if it is worn or the cable is stretched, the parts should be renewed.
10 Refitting the handbrake lever assemblies follows the reversal of the removal procedure: remember to adjust the cable linkage as described in Section 12 before returning the car to the road.

15 Brake and clutch pedal assembly – removal, renovation and refitting

1 The brake and clutch pedals are mounted on a substantial channel member which at its lower end is bolted the the engine compartment bulkhead, and at the top end to the dashboard to support the upper section of the steering column.
2 The pedals themselves pivot on bushes running on a common long bolt which passes from one side of the channel to the other.
3 Thankfully it it not necessary to disturb this channel member when removing either clutch or brake pedal.
4 Begin removal of either pedal by removing the return spring acting on the clutch cable beside the gearbox underneath the car, and the return spring on the brake pedal situated immediately above the pedal.
5 Next undo and remove the nut on the end of the long bolt and then draw the bolt from the channel.
6 Remove the clutch pedal from the channel and disconnect the cable from the fork piece in the pedal lever.
7 The brake pedal can be removed once the clip that retains the servo push rod pin in place has been removed. The push rod can then be detached from the pedal assembly.
8 Once the pedals are free, an inspection can be made of the bolt, bushes and spacers. If the bushes are oval and worn, renew them.
9 The brake push rod and pivot are available as individual spares and therefore both should be inspected for wear.
10 The reassembly procedure is the reversal of the dismantling procedure, except that it is as well to check the clutch pedal free travel (Chapter 5) if the pedal assembly was repaired, also the operation of the brake light switch which is mounted adjacent to the brake pedal pivot.

16 Brake servo unit - description

The vacuum servo unit is fitted into the brake system in series with the master cylinder and brake pedal to provide power assistance to the driver when the brake pedal is depressed.

The unit operates by vacuum obtained from the induction manifold, and comprises basically a booster diaphragm and a non return valve.

The servo unit and hydraulic master cylinder are connected together so that the servo unit push rod acts as the master cylinder push rod. The driver's braking effort is transmitted through another push rod to the servo unit piston and its built in control system (photo).

The servo unit piston does not fit tightly into the cylinder but has a strong diaphragm to keep its periphery in contact with the cylinder wall so assuring an air tight seal between the two parts. The forward chamber is held under vacuum conditions created in the inlet manifold of the engine and during period when the engine is not in use the controls open a passage to the rear chamber so placing it under vacuum. When the brake pedal is depressed, the vacuum passage to the rear chamber is cut off and the chamber is opened to atmospheric pressure. The consequent rush of air into the rear chamber pushes the servo piston forward into the vacuum chamber and operates the push rod to the master cylinder. The controls are designed so that assistance is given under all conditions. When the brakes are not required, vacuum is re-established in the rear chamber when the brake pedal is released.

Air from the atmosphere passes through a small filter before entering the control valves and rear chamber and it is only the filter that will require periodic attention.

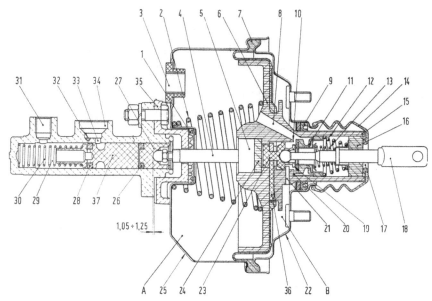

FIG. 9.11. SECTION OF MASTER VAC UNIT AND EARLY PATTERN SINGLE PISTON MASTER CYLINDER

1 Front seal
2 Piston return spring
3 Vacuum line union
4 Piston rod
5 Working piston
6 Actuating piston
7 Diaphragm
8 Vacuum duct
9 Rear seal
10 Vacuum port
11 Valve
12 Piston valve return spring
13 Retaining valve return spring

14 Guide tube gaiter
15 Actuating piston guide tube
16 Filter element
17 Servo unit air intake
18 Valve control rod
19 Atmospheric pressure air passage
20 Vacuum hole
21 Vacuum and air passage
22 Rear body
23 Piston valve
24 Reaction disc
25 Front body

26 Seal ring
27 Hydraulic piston
28 Floating ring valve
29 Master cylinder
30 Hydraulic piston return spring
31 Union for connection to 3-way brake distributor
32 Compensating hole
33 Master cylinder inlet port
34 Union for line from reservoir to master cylinder
35 Mounting flange
36 Piston retaining plate
37 Floating ring valve carrier

A Front chamber

B Rear chamber

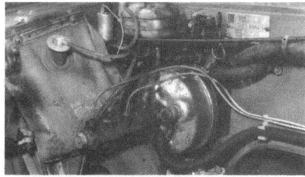

16.1 The location of the master-vac unit
and master cylinder

17 Brake servo unit - removal and refitting

1 Refer to Section 9 of this chapter and remove the brake master cylinder.
2 Slacken the hose clip and remove the vacuum hose from the inlet manifold from the union on the forward face of the servo unit.
3 Next remove the clip that retains the small pin that joins the servo push rod to the pedal lever. Remove the pin and the pedal return spring and separate the pushrod and pedal.
4 The servo unit is attached to the mounting plate by four nuts on studs in the servo unit. The mounting plate has another four bolts which pass through the rear engine compartment bulkhead and serve to join not only the servo unit to the bulkhead but also the pedal support channel on the rear side of the bulkhead.

5 Undo and remove the four nuts on those studs which join the pedal support channel to the compartment bulkhead. The servo unit and mounting plate can then be lifted away from the forward engine side of the bulkhead. Retrieve the thick seal sandwiched between the mounting plate and bulkhead.
6 Once the unit is free, transfer it to a clean bench and separate the mounting plate.
7 Refitting the brake servo unit follows the reversal of the removal procedure. Remember to use new spring lock washers and tighten nuts and bolts to their appropriate torques.

18 Brake servo unit - air filter renewal

1 Under normal operating conditions the servo unit is very reliable and does not require overhaul except possibly at very high mileages. In this case it is better to obtain a service exchange unit, rather than repair the original.
2 However, the air filter may need renewal and fitting details are given below.
3 Remove the brake and clutch pedals as detailed in Section 15 of this chapter.
4 Working inside the car in the driver's foot well, push back the dust cover from the push rod and control valve housing to expose the end cap and air filter element.
5 Using a screwdriver ease out the end cap and then with a pair of scissors cut out the air filter.
6 Make a diagonal cut through the air filter element and push over the push rod. Hold in position and refit the end cap.
7 Pull back the dust cover over the unit body and refit the pedal assembly.

19 Rear brake regulator - description

The regulator serves to reduce the braking effect of the rear brakes when the car body adopts a nose down attitude under heavy braking. (photo).

The reduction prevents the rear wheels from locking as they would do without the regulator because of the weight shift from the rear wheels to the front during braking.

The regulator operating mechanism comprises a torsion bar and link which connect the regulator to the rear axle casing. (photo).

As the distance between the axle casing and the regulator/bodyshell increases, due to pitching of the car during braking the linkage allows the plunger to move outwards from the regulator.

Inside the regulator the outward movement of the plunger closes the port admitting fluid to the rear brake pipes and thereby denies the rear brakes any more fluid.

When the regulator does close the rear brake line, the rear braking effort does not fall, because fluid remains trapped in the pipes between the regulator and rear caliper blocks. This fluid maintains the disc pads against the brake disc and braking effort. Once the regulator is closed however, any increased pressure of fluid in the main brake circuit will not be transmitted through the regulator to the rear brakes.

It follows that the regulator will need attention if the rear wheel begins locking regularly when the brakes are applied in earnest. It will be as well to check the condition of the rear brake pads or disc and the operation of the rear caliper blocks as well, to determine how many faults in the rear brake system have been causing the rear wheels to lock up.

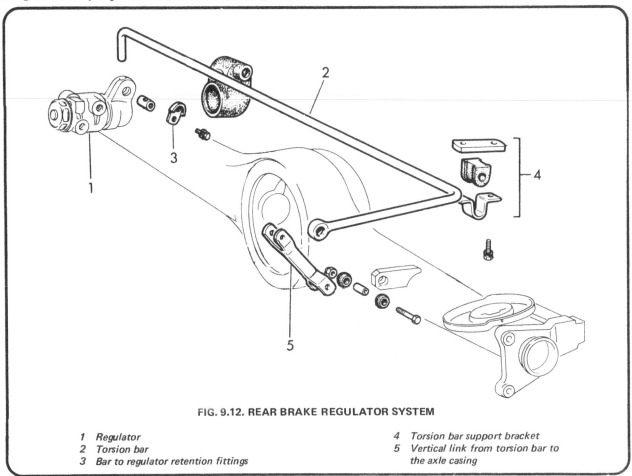

FIG. 9.12. REAR BRAKE REGULATOR SYSTEM

1 Regulator
2 Torsion bar
3 Bar to regulator retention fittings
4 Torsion bar support bracket
5 Vertical link from torsion bar to
   the axle casing

19.1 The regulator located above the rear axle

19.1a Vertical link between axle and torsion rod to the regulator

20.7 Regulator securing bolts which allow the regulator position to be adjusted

20 Rear brake regulator - removal, refitting and adjustment

1   Begin by raising the rear of the car onto car ramps and then from the engine compartment remove the brake fluid reservoir cap(s), stretch a sheet of polythene over the top(s) of reservoir(s) and replace the cap(s). This measure will prevent excessive loss of brake fluid when the pipes to the regulator are subsequently disconnected.

2   Undo the brake pipe connection to the regulator block and then undo and remove the two bolts which secure the block to the bracket on the bodyshell.

3   Undo and remove the nut and bolt which connect the torsion bar end to the vertical link to the rear axle cassing and finally undo and remove the nuts securing the torsion bar support cap to the bodyshell.

4   The regulator and torsion bar may now be lifted from the car and transfered to a clean bench for overhaul.

5   Refitting the brake regulator and torsion bar follows the reversal of the removal procedure except for the addition of one important task.

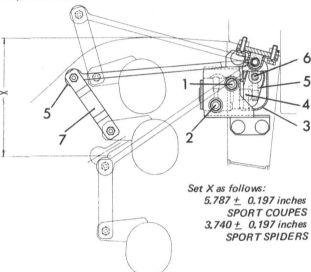

Set X as follows:
5.787 ± 0.197 inches
SPORT COUPES
3.740 ± 0.197 inches
SPORT SPIDERS

FIG. 9.13. FITTING AND ADJUSTING DIAGRAM OF REAR BRAKE REGULATOR

1 & 2   Regulator mounting screws
3   Protective boot
4   Plunger
5   Regulator torsion bar
6   Pin
7   Link from the axle to the torsion bar

6   It will be necessary to adjust the position of the regulator on its support bracket so that the linkage will operate the valve in the regulator at precisely the specified pitch attitude of the car.

7   As you will have noticed, two bolts secure the regulator to the bracket, one locates in a close fit hole, the other in a slotted hole. Once all the components have been assembled onto the car and before it is returned to the road, adjust the position of the regulator as follows:— (photo)

8   Slacken the two bolts that secure the regulator and with either a pair of jacks and/or passengers in the car arrange the ground height of the bodyshell so that the distance between the top of the vertical link to the torsion bar end and the underside of the bodyshell is 3.74 ± 0.2 inches for the SPIDER and 5.78 ± 0.20 inches for the COUPE.

9   Once the specified distance between the link and the bodyshell has been achieved, move the regulator so that the plunger just TOUCHES the acting end of the torsion bar. Tighten the regulator securing bolts and road test the car.

21 Brake regulator unit overhaul

1   The overhaul of the regulator is quite straight forward, with the regulator and torsion bar on a clean bench proceed as follos.

2   Slip the boot which protects the acting pivot end of the torsion bar and the regulator plunger off the regulator body and down the torsion bar. Undo and remove the bolt which secures the torsion bar retaining fitting and separate the torsion bar and regulator block.

3   Unscrew and remove the large end plug and retrieve the washer seal.

4   The spring, washer, slotted ring, plunger, primary seal, seal spring and rest ring and finally the secondary seal can all be extracted in that order.

5   Clean all components with Girling Cleaning Fluid or methylated spirits and wipe dry with a non-fluffy rag.

6   Inspect the seals, plunger shank and regulator bores for wear and surface deterioration. If only the seals are worn, renew the seals and reassemble the regulator. If regular plunger or bores are worn the regulator must be renewed as a whole assembly.

7   Reassembly of the regulator follows the reversal of the dismantling procedure, except that the primary and secondary seals should be wetted with clean brake fluid before they are refitted. It would be as well to use a new plug seal washer on each occasion the regulator is reassembled.

8   Refit the torsion bar and regulator block, slip the protective boot over the block and torsion bar end and commence refitment of the regulator assembly to the vehicle. (Section 20).

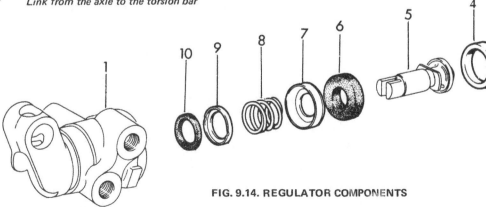

FIG. 9.14. REGULATOR COMPONENTS

1   Regulator block
2   End plug
3   Seal washer
4   Spacer ring slotted
5   Plunger
6   Primary seal
7   Seal cup
8   Seal spring
9   Rest ring
10 Secondary seal

22 FAULT DIAGNOSIS

Before diagnosing faults from the following chart, check that any braking irregularities are not caused by:-
1 Uneven and incorrect tyre pressures
2 Incorrect 'mix' of radial and cross-ply tyres
3 Wear in the steering mechanism
4 Defects in the suspension
5 Misalignment of the chassis

Symptom	Reason/s	Remedy
Stopping ability poor, even though pedal pressure is firm	Pads and/or discs badly worn or scored	Dismantle, inspect and renew as required.
	One or more wheel hydraulic cylinders seized, resulting in some brake pads not pressing against the discs	Dismantle and inspect wheel cylinders. Renew as necessary.
	Brake pads contaminated with oil	Renew pads and repair source of oil contamination.
	Wrong type of pads fitted (too hard)	Verify type of material which is correct for the car, and fit it.
	Brake pads wrongly assembled	Check for correct assembly.
Car veers to one side when the brakes are applied	Brake pads on one side are contaminated with oil	Renew pads and stop oil leak.
	Hydraulic wheel cylinder(s) on one side partially or fully seized	Inspect wheel cylinders for correct operation and renew as necessary.
	A mixture of pads materials fitted between sides	Standardise on types of pads fitted.
	Unequal wear between sides caused by partially seized wheel cylinders	Check wheel cylinders and renew pads and discs as required.
Pedal feels spongy when the brakes are applied	Air is present in the hydraulic system	Bleed the hydraulic system and check for any signs of leakage.
Pedal feels springy when the brakes are applied	Brake pads not bedded into the discs (after fitting new ones)	Allow time for new pads to bed in.
	Master cylinder or brake backplate mounting bolts loose	Retighten mounting bolts.
	Severe wear in brake discs causing distortion when brakes are applied	Renew discs and pads.
Pedal travels right down with little or no resistance and brakes are virtually non-operative	Leak in hydraulic systems resulting in lack of pressure for operating wheel cylinders	Examine the whole of the hydraulic system and locate and repair source of leaks. Test after repairing each and every leak source.
	If no signs of leakage are apparent all the master cylinder internal seals are failing to sustain pressure	Overhaul master cylinder. If indications are that seals have failed for reasons other than wear all the wheel cylinder seals should be checked also and the system completely replenished with the correct fluid.
Binding, juddering, overheating	One or a combination of causes given in the foregoing sections	Complete and systematic inspection of the whole braking system.

Chapter 10 Electrical system

Contents

Specifications

Both Coupe and Spiders equipped with a 12 volt electrical system with a negative earth.

Battery	48 amp hrs. (20 hrs. discharge) (60 amp hr. (20 hr. discharge) on Spiders with chassis numbers preceding No. 741)

Alternator	A12M 124/12/42M
Nominal voltage	12v
Output	770 watts
Maximum current	53 amps
Maximum speed	13,000 rpm
Drive ratio engine/alternator	1.8 : 1
Rotor coil windings measured across slip rings	4.5 ± 0.1 ohms
Rectifier diodes (6 off)	4AF2
Maximum forward current	25 amps
Nominal voltage	12 volts
Voltage regulator:	RC1/12B
Resistance between terminal 15 and earth	28.2 ohms
Resistance between terminals 15 and 67 open contacts ...	5.5 ohms

Spiders - preceding chassis No. 741

Dynamo	FIAT D115/12/28/4E - 400 watts
Nominal voltage	12 volts
Maximum steady output current	28 amps
Field winding	Shunt wound
Max. speed	10,200 rpm
Armature resistance	0.13 ohms
Field resistance	7.0 ohms

Regulator GN/2/12/28

Starter motor   ...   .   Type E100 - 1,3/12
    Operating voltage   ...   .   12 volts
    Nominal output   ...   .   1.3 kw
    Brushes   ...   .   Part No. 4045771

    Lubrication - Drive unit splines   ...   .   VS 10W Grease
              - Contact face, bearing disc, pinion carriage   ...   MR3 Grease

Spiders - preceding chassis No. 741

Starter motor   ...   .   E84 - 0.8/12 Var. 4
    Operating volts 12v
    Nominal output   ...   .   0.8 kw
    Brushes   ...   .   Part No. 4114432

    Lubrication as per E100 - 1, 3/12 starter motor.

Warning, control and indication system

    Cooling water temperature indication   ...   .   Electric
    Engine speed   ...   .   Electric tachometer
    Engine oil pressure:
        — low pressure warning   ...   .   Red light
        — pressure indication   ...   .   Pressure gauge
    No-charge warning   ...   .   Red light
    Fuel reserve indication   ...   .   Red light
    — amount of reserve supply   ...   .   5 to 7.5 litres (8 pts 16 ozs to 13 pts 4 ozs Imp.)

    Headlight main beam indication   ...   .   Blue light
    Sidelamps 'on' indication   ...   .   Green light
    Flashers 'on'   ...   .   Green light

Flashing direction lamps
    No. of flashes per min. at normal total load of 46 watts:
    — at nominal voltage, 12v and 20° C (68° F)   ...   .   85 ± 8
    — at 1.25 times nominal voltage (15v) and 40° C (104° F)   ...   ≤ 120
    — at 0.9 times nominal voltage (10.8v) and at −20° C (−4° F)   ≥ 60

Windscreen wiper   ...   .   Crank mechanism
    Wiper blades, strokes per min.   ...   .   52 to 70
    Bench test of motor and reducing gear:
        Feed voltage   ...   .   14 volts
        Resistance torque   ...   .   15 kg cm (1.08 ft lbs) 0.72 ft lbs
        Stator temperature rise   ...   .   ≤ 60° C (140° F)
        Speed, warmed up   ...   .   ≤ 70 rpm 68 rpm
        Current consumption, warmed up   ...   .   ≤ 3,5 amps   3.0 amps
        Speed, warmed up, with resistor in circuit   ...   .   ≤ 100 rpm   88 rpm
        Current consumption, warmed up with resistor in circuit   ...   ≤ 4 amps   3.5 amps
        Starting torque (with spindle held stationary), warmed up, at
        14v   ...   .   ≥ 140 kg cm   (10.13 ft lbs)
        Pressure of wiper blades on windscreen   ...   .   600 to 700 gr   (21 ozs to 24 ozs 8)
                       (400 to 500 gr)   (14 ozs to 17 ozs Spider)

Electric ventilating fan
    Speed in free air, with fan fitted, voltage 12v temp. 25° C (77° F):
    — 1st speed, resistor in circuit (1 ± 0.1 ohms)   ...   .   1900 to 2100 rpm
    — 2nd speed, resistor cut out   ...   .   2800 to 3000 rpm

    Nominal power   ...   .   20 watts
    Direction of rotation of motor (from fan side)   ...   .   Counter-clockwise

Lighting system:

Headlamps	Two
Double filament bulbs:	
— main beam	45 watts
— anti-dazzle beam	40 watts
Twin headlights - Dipped	40 watts
- Mains	60 watts
Front and direction lamps	Two
Side lamp bulbs	5 watts
Slashing direction lamp bulbs	20 watts
Side direction lamps	Two
Bulbs	4 watts
Tail lamps, rear direction and stop lamps and reflector	Two
Direction lamp bulb	20 watts
Double filament bulb:	
— tail lamp bulb	5 watts
— stop lamp bulb	20 watts
— stop back-up lamp bulb (Coupe only)	21 watts
Rear registration plate lamps	Two, (One Coupe)
Bulbs	5 watts
Outside lighting control	By switch on instrument panel
Headlamp dipping ,...	By lever under steering wheel
Interior lighting bulbs (two, under panel)	5 watts
Control:	
— by lever	On lampholder
— by push switch, automatic, operated by opening door ...	On door jambs
Instrument lighting:	
— five bulbs with switch on instrument panel	3 watts
Engine compartment:	
— two bulbs with automatic switch operated by opening bonnet	5 watts
Luggage compartment:	
— one bulb operated from automatic door switch	5 watts
Cigar lighter lighting:	
— tubular bulb	3 watts

Direction lamp repeaters:
- repeater bulb ... ...
Battery not charging ... ...
Low oil pressure ... ... } 7 bulbs 3 watts
Fuel reserve warning ... ...
Side lamps on ... ...
Headlamps main beams on ... ...
Cigar lighter spot light (Coupe only) ... ...

Electrical system - Fuses:

Fuses	Circuits Protected
A	— Dashboard lights. ⎫ Circuits not controlled by
(16 - Ampere)	— Electro-pneumatic horn ⎪
	— Inspection lamp socket ⎬
	— Cigar lighter ⎭ ignition switch
B	— Engine compartment lights
	— Instrument panel lights
	— Direction lamps and indicators
	— Stop lights
	— Windscreen wiper
	— Ventilation fan motor

C	...	...	...	...	...	...	— L.H. headlamp, main beam
							— Headlamp main beam indicator lamp
D	...	...	...	...	...	...	— R.H. headlamp, main beam
E	...	...	...	...	...	...	— L.H. anti-dazzle beam
F	...	...	...	...	...	...	— R.H. anti-dazzle beam
G		...	...	...	...	...	— L.H. side lamp
							— Side-lamp indicator lamp
							— R.H. tail lamp
							— L.H. number plate lamp
							— Cigar lighter spot lamp
							— Luggage compartment light
H	...	...	...	...	...	...	— R.H. front side lamp
							— L.H. tail lamp
							— R.H. number plate lamp
I	...	...	...	...	...	...	— Oil pressure gauge and low oil-pressure warning light
							— Temperature gauge
							— Fuel gauge and reserve warning light
							— Electro-magnetic fan-clutch
							— Engine speed indicator
L	...	...	...	...	...	...	— Voltage regulator
							— Alternator field winding

The following are not protected by fuses: ignition, starting and battery charge circuits (voltage regulator circuit excepted).

Torque wrench settings:							ft lbs
Alternator - upper bolt	...	...	...	...	...	...	32
- lower bolt	...	...	...	...	...	...	50
Starter motor to bellhousing bolts		...	...	...	...	...	25
Brake circuit back-up light switch		...	...	...	...	...	4.4

1 General description

1 The electrical systems on the 124 Sport models reflect the trend to ever more profusion as is apparent on so many new cars. Following this greater complexity comes the likelihood of failure of those systems, and therefore whereas the mechanics of your car will be more reliable than in years past, the electrical systems will most probably demand more attention than any other on the vehicle. Fortunately more often than not, it will be the electical lead connections and electrical device mountings that will give trouble, rather than the electrical devices themselves. It is from this feature that the need comes for cleanliness and neatness when dealing with electrical systems.

2 In this Chapter details are given of the maintenance and repair of each system, including any special precautions that may be taken to guard against premature failure of those systems.

3 As mentioned earlier, because you will probably have to deal with electrical faults more than anything else on this car it may be worth while investing in a few items of electrical test equipment. One very useful item of equipment is a voltmeter (0 to 20 volts) or at very often not much more expense, a multi-meter. This last device is really useful, not only for the car but also around the house. A multimeter can measure voltage, current or resistance on a variety of scales.

4 So much for the equipment and rationale. The electrical systems on the 124 Sports are conventional 12 volt systems. They are powered by an alternator driven off the engine. Some models have been fitted with a dynamo. The battery provides for starting, and reserve power should the demand from the systems exceed the output of the alternator. When the alternator is not fully loaded, the voltage regulator ensures that the battery is kept charged.

5 The potential power from the alternator or battery is sufficient to severely damage the electrical wiring, if faults occur which allow that power through the circuits. Therefore most are protected with fuses. Those not protected with fuses are associated with engine systems and include the ignition system and starter motor circuit.

2 Battery - removal and replacement

1 The battery is situated on the right hand side of the engine compartment, and is held in place by two tie rods and a pressed steel plate. It weighs approximately 43 lbs, with electrolyte.

2 To remove the battery begin by disconnecting the negative earth lead from the battery and bodyshell. Then disconnect the positive lead from the battery.

3 Once the leads have been removed, the two nuts which tension the tie rods onto the battery retaining plate, may be loosened and the plate moved aside.

4 Lift the battery from its seating in the bodyshell, taking great care not to spill any of the highly corrosive electrolyte.

5 Replacement is the reversal of this procedure. Replace the positive lead first and smear the clean terminal posts and lead clamp assembly before hand with petroleum jelly (Vaseline) in order to prevent corrosion. DO NOT USE ORDINARY GREASE.

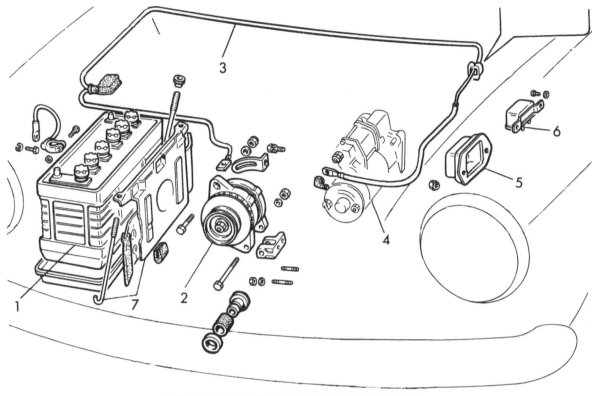

FIG.10.1. THE PRIMARY POWER SUPPLY SYSTEM

1 Battery	3 Major supply lead	5 Voltage regulator	7 Battery retaining tie rods
2 Alternator	4 Starter motor	6 Ignition light relay	and plate

3 Battery - maintenance and inspection

1 Check the battery electrolyte level weekly by lifting off the cover or removing the individual cell plugs. The tops of the plates should be just covered with the liquid. If not, add distilled water so that they are just covered. Do not add extra water with the idea of reducing the intervals of topping up. This will merely dilute the electrolyte and reduce charging and current retention efficiency. On batteries fitted with patent covers, troughs, glass balls and so on, follow the instructions marked on the cover of the battery to ensure correct addition of water.

2 Keep the battery clean and dry all over by wiping it with a dry cloth. A damp top surface could cause tracking between the two terminal posts with consequent draining of power.

3 Every three months remove the battery and check the support tray clamp and battery terminal connections for signs of corrosion - usually indicated by a whitish green crystalline deposit. Wash this off with clean water to which a little ammonia or washing soda has been added. Then treat the terminals with petroleum jelly and the battery mounting with suitable protective paint to prevent the metal being eaten away. Clean the battery thoroughly and repair any cracks with a proprietary sealer. If there has been any excessive leakage the appropriate cell may need an addition of electrolyte rather than just distilled water.

4 If the electrolyte level needs an excessive amount of replenishment but no leaks are apparent, it could be due to overcharging as a result of the battery having been run down and then left to recharge from the vehicle rather than an outside source. If the battery has been heavily discharged for one reason or another it is best to have it continuously charged at a low amperage for a period of many hours. If it is charged from the car's system under such conditions the charging will be intermittent and greatly varied in intensity. This does not do the battery any good at all. If the battery needs topping up

frequently, even when it is known to be in good condition and not too old, then the voltage regulator should be checked to ensure that the charging output is being correctly controlled. An elderly battery, however, may need topping up more than a new one because it needs to take in more charging current. Do not worry about this provided it gives satisfactory service.

5 When checking a battery's condition a hydrometer should be used. On some batteries where the terminals of each of the six cells are exposed, a discharge tester can be used to check the condition of any one cell also. On modern batteries the use of a discharge tester is no longer regarded as useful as the replacement or repair of cells is not an economic proposition. The tables following give the hydrometer readings for various states of charge. A further check can be made when the battery is under going a charge. If, towards the end of the charge, when the cells are meant to be 'gassing' (bubbling), one cell appears not to be, then it indicates that the cell or cells in question are probably breaking down and the life of the battery is limited.

4 Battery - charging and electrolyte replenishment

1 It is possible that in winter when the load on the battery cannot be recuperated during normal driving time (from a dynamo) external charging is desirable. This is best done overnight at a 'trickle' rate of 1-1.5 amps. Alternatively a 3-4 amp rate can be used over a period of four hours or so. Check the specific gravity in the latter case and stop the charge when the reading is correct. Most modern charging sets reduce the rate automatically when the fully charged state is neared. Rapid boost charges of 30-60 amps or more may get you out of trouble or can be used on a battery that has seen better days. They are not advisable for a good battery that may have run flat for some reason.

2 Electrolyte replenishment should not normally be necessary unless an accident or some other cause such as contamination

arises. If it is necessary then it is best first to discharge the battery completely and then tip out all the remaining liquid from all cells. Then acquire a quantity of mixed electrolyte from a battery shop or garage according to the specifications in the table. The quantity required will depend on the type of battery but three to four pints should be more than enough. When the electrolyte has been put into the battery a slow charge - not exceeding 1 amp - should be given for as long as is necessary to fully charge the battery. This could be up to 36 hours.

Specific gravities for hydrometer readings (check each cell) - 12 volt batteries.

Electrolyte temperature 60°F (15.6°C)

	Climate below 80°F (26.7°C)	Climate above 80°F (26.7°C)
Fully charged	1.270–1.290	1.210–1.230
Half charged	1.190–1.210	1.120–1.150
Discharged completely	1.110–1.130	1.050–1.070

NOTE: If the electrolyte temperature is significantly different from 60°F (15.6°C) then the specific gravity reading will be affected. For every 5°F (2.8°C) it will increase or decrease with the temperature by 0.002.

5 Alternator - general description

1 The FIAT A12M - 124/12/42M alternator is fitted as standard to both Spider and Coupe 124 Sports. The main advantage of the alternator lies in its ability to provide a relatively high power output at low revolutions. Driving slowly in traffic with a dynamo fitted invariably means a very small or even no charge at all reaching the battery. In similar conditions even with the wipers, heater, lights and perhaps radio switched on the FIAT A12M alternator will still ensure a charge reaches the battery. The alternator is of the rotating field ventilated design and comprises principally a laminated stator on which is wound 3 phase output winding and a twelve pole rotor carrying the field windings. Each end of the rotor shaft runs in ball race bearings which are lubricated for life. Aluminium end brackets hold the bearings and incorporate the alternator mounting lugs. The rear bracket supports the silicon diode rectifier pack which converts the AC output of the machine to DC for battery charging and output to the voltage regulator.

The rotor is belt driven from the engine through a pulley keyed to the rotor shaft. A special centrifugal action fan adjacent to the pulley draws air through the machine. This fan forms an integral part of the alternator specification. It has been designed to provide adequate flow of air with the minimum of noise and to withstand the stresses associated with the high rotational speeds of the rotor. Rotation is clockwise when viewed from the drive end.

The rectifier pack of silicone diodes is mounted on the inside of the rear end casing, the same mounting is used by the brushes which contact the slip rings on the rotor to supply the field current. The slip rings are carried on a small diameter moulded drum attached to the rotor. By keeping the circumference of the slip rings to a minimum, the contact speed and therefore the brush wear is minimised.

6 Alternator - testing, removal, refitting and maintenance

1 The alternator has been designed for the minimum amount of attention during service. The only items subject to wear are the brushes and bearings.

2 If the red ignition warning light on the instrument panel lights up indicating that the battery is no longer being charged, check the belt tension and the continuity of the leads to and from the alternator voltage regulator and battery.

3 Ensure all the connections of those leads are clean, and check for breaks in each cable by disconnecting both ends of a cable and reconnecting in series with a small battery and bulb. If the cable is complete the bulb will light up. If once the continuity checks have been completed and nothing found at fault, the alternator may be checked in situ as follows.

4 Accurate assessment of alternator output requires special equipment and a degree of skill. A rough idea of whether output is adequate can be gained by using a voltmeter (range 0 to 15 or 0 to 20 volts) as follows. Connect the voltmeter across the battery terminals. Switch on the headlights and note the voltage reading: it should be between 12 and 13 volts. Start the engine and run it at a fast idle (approx 1500 rpm). Read the voltmeter: it should indicate 13 to 14 volts.

5 With the engine still running at a fast idle, switch on as many electrical consumers as possible (heated rear window, heater blower etc). The voltage at the battery should be maintained at 13 to 14 volts. Increase the engine speed slightly if necessary to keep the voltage up. If alternator output is low or zero, check the brushes. If the brushes are OK, seek expert advice.

6 Begin removing the alternator by disconnecting the earth cable from the battery. Then remove the heavy black cable from the alternator followed by the yellow and grey leads. Identify the leads and terminals so that they may be refitted correctly later.

7 Next loosen the upper and lower attachment bolts and push the alternator toward the engine. Slip the drive belt off the pulley.

8 Move the alternator away from the engine and completely expose the lower attachment bolts. Undo and remove the nuts and bolts and lift the alternator from the engine.

9 Refitting is the reversal of the removal task, except that the fan belt must be tensioned as described in Chapter 2.

10 Before beginning to dismantle the alternator with the view to replacing the brushes and or bearings, check that the individual spares are available. The bearings are only listed as available either with the rotor or with the forward casing. The alternator is usually renewed as factory exchange unit and components are not readily available.

11 Assuming that the brushes or bearings are available, the rear end of the alternator which houses the brushes and rear bearing is retained by 4 long bolts and nuts which reach to the forward end of the machine. Undo and remove the four nuts and bolts and ease the rear casing from the machine. Retrieve the sealing ring which is fitted in the rear bearing seating.

12 The brushes are mounted in a plate screwed to the inside of the rear casing. Remove the plate and brushes and renew them if worn to less than 0.37 inches. Clean the slip rings on the rotor with a petrol damped cloth and reassemble the alternator. Take care not to damage the brushes when refitting the rear casing over the rotor.

13 As mentioned earlier the bearings are listed as available only with the rotor or front casing, their replacement follows similar lines to that for brushes, but the pulley and fan will need to be removed from the forward end of the rotor shaft to facilitate separation of the rotor and front casing.

6.4 Removal of main output leads from the alternator

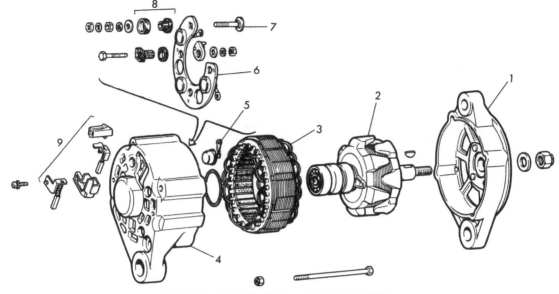

FIG.10.2. ALTERNATOR – EXPLODED VIEW

1 Forward end fitting	4 Rear end fitting	6 Frame on which terminals and	7 Terminal bolt
2 Rotor	5 Silicone diode (6 off)	rectifier bridge network mounted	8 Insulators
3 Stator			9 Rotor brushes

SECTION C-C

SECTION A-A

SECTION
(Scrap)
B-B

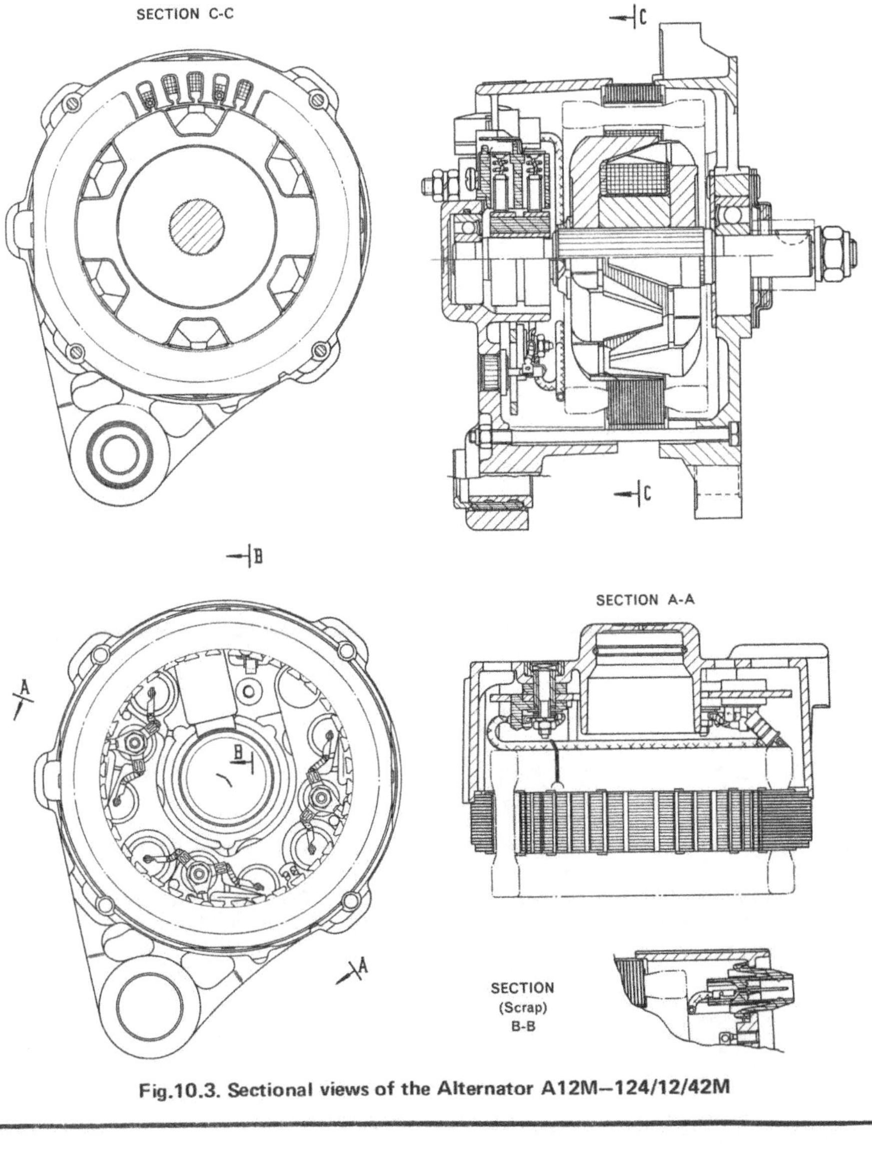

Fig.10.3. Sectional views of the Alternator A12M–124/12/42M

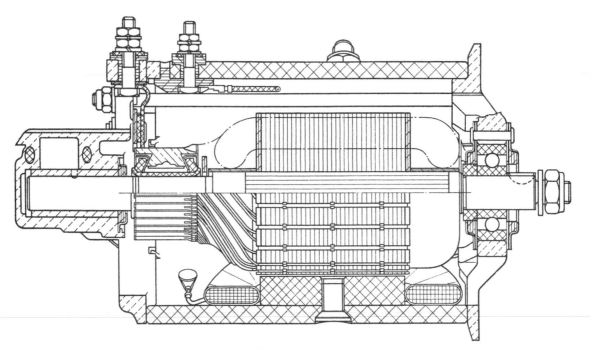

Longitudinal section through generator

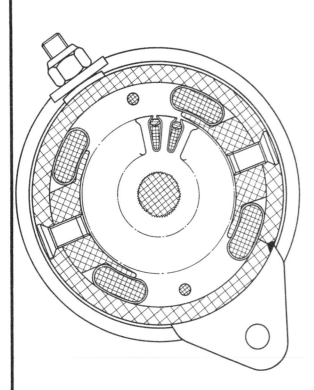

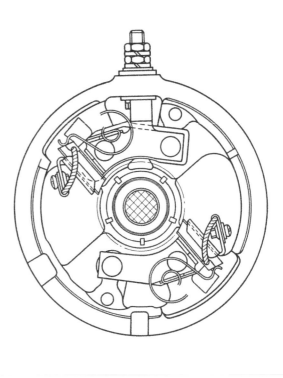

Cross section of generator through casing, pole-shoes and armature windings

Cross section of generator through armature shaft, and view of housing from commutator side

Fig.10.4. Sectional views of the D.C. Generator D115/12/28/4E

7 Dynamo - notes, inspection, removal, refitting and repair

1   A dynamo has only been fitted to 124 Sport Spider 1400 cars with chassis numbers preceding number 741 and in the event of suspected malfunction, it is of conventional design and is readily tested in situ. Maintenance involves injecting a few drops of engine oil into the rear bearing periodically.

2   Begin by checking the fan belt tension (Chapter 2) and the integrity of the various lead connections to the dynamo, battery and voltage regulator. Remove each connection in turn, thoroughly clean and then refit. Check for breaks in each cable by completing a continuity tests on each cable, disconnect both ends of the cable and reconnect into series with a small battery and bulb or voltmeter. If the lead is complete it will complete the test circuit and the bulb will come on.

3   If the red ignition light still remains on with the engine running, indicating a dynamo or voltage regulator fault, the dynamo performance may be checked as follows.

4   Pull off the leads from both terminals on the dynamo and join those terminals with a short length of bare wire. Using crocodile clips, attach a voltmeter with the positive lead to the centre of the bridge wire on the dynamo terminals and the negative to a good earth nearby. Switch off all ancilliary equipment. Start the engine and increase the revolutions smoothly to about 1000 rpm. The voltmeter should indicate an immediate smooth rise to approximately 15 volts. Do not increase engine speed in an attempt to increase the voltage if low. If the voltage is sub-normal the dynamo should be removed for closer examination.

5   Remove the dynamo by following the same procedure described in Section 6 for the alternator. Refitting the dynamo follows the reversal of the removal procedure.

6   Repairs of the dynamo detailed below cover the replacement of brushes and the testing of the commutator and armature.

7   With the dynamo removed from the car and placed on a clean bench, commence dismantling by undoing and removing the couple of nuts which retain the rear end fitting to the dynamo body.

8   Ease the rear end fitting off the armature shaft and away from the dynamo body. The brushes are mounted on this rear fitting and can now be slipped out of their retainers once the lead connections on each brush have been detached from the terminals in the end cover. Renew the brushes if they are less than 7/16 inches long.

9   Do not dismantle the dynamo any further, individual parts are not available as spares, and if it is more than just brushes which need attention, the dynamo should be taken to the nearest auto/electrician.

10 It will be as well to clean the commutator with a petrol dampened cloth and to pick out any accumulations of copper and/or carbon in between the commutator elements. Any traces of pitting, scoring or burring, if slight, can be cleaned off with fine glass paper. Do not use emery paper. Make sure there are no flat spots by drawing a strip of glass paper to and fro over the commutator. Do not try to clean off too much by this method as the commutator must retain a circular cross section.

11 The armature may be tested using a voltmeter or more simply a battery and bulb. The two tests determine whether there is a break or a short circuit in the armature. A battery, voltmeter and leads are connected in series, the ends of two leads are used as probes to check various circuits in the armature.

12 The first test determines if there is a broken wire in the armature:- Place the probes on adjacent segments of the commutator. All voltmeter readings should be similar on all adjacent segments and approach the voltage of the battery being used. If the meter does not register anything, that particular winding is broken, and if a couple of adjacent pairs of segments registering a voltage significantly different from all the other segments, there is a break down of insulation between adjacent windings. Ideally the resistance of each armature winding and the field winding should be checked with a multimeter. See the specification at the beginning of this Chapter.

13 The second test determines if there is a break in insulation between the windings and the frame of the armature. Hold one probe against the armature shaft and then place the other probe sequentially on each commutator segment. The voltmeter should register absolutely nothing if the insulation is intact.

14 Should either of the two tests outlined prove the armature faulty, the only course of action is to take it to the nearest auto/electrician to have it overhauled or the more 'cost effective' solution as far as you are concerned: purchase a factory exchange dynamo.

8 Starter motor - general description

1   The starter fitted to all Coupes and Spiders is the FIAT type E100, 1.3/12. Again there is an exception for the early Spiders with chassis numbers preceding 741; they were equipped with Type E84, 0.8/12 starters. Both starters are conceptually the same, only small dimensional differences distinguishing them.

2   The starter motors are of the pre-engaged type, in which the switch solenoid is also employed to physically move the drive pinion along the starter motor shaft, into contact with the ring gear on the flywheel before power is supplied to the motor for turning the engine.

3   There is a spring between the pinion and the actuating lever from the solenoid, so that in the event of an exact abutment of gearteeth as the pinion is impelled to engage with the flywheel ring gear; the solenoid switch will still continue and make power contact. The pinion will fall into engagement as soon as the motor shaft turns.

4   The location of the starter motor is as usual on the right hand side of the engine, and it is bolted to the clutch bell housing. The motor/pinion shaft projects into the bellhousing near the periphery of the flywheel.

9 Starter motor circuit - testing

1   If the starter motor fails to turn the engine when the switch is operated there are four possible reasons why:
a) The battery is no good.
b) The electrical connections between switch solenoid, battery and starter motor are somewhere failing to pass the necessary current from the battery through the starter to earth.
c) The solenoid switch is no good.
d) The starter motor is either jammed or electrically defective.

2   To check the battery, switch on the headlights. If they go dim after a few seconds the battery is definitely unwell. If the lamps glow brightly, next operate the starter switch and see what happens to the lights. If they go dim then you know that power is reaching the starter motor but failing to turn it. Therefore, check that it is not jammed by placing the car in gear and rocking it to and fro. If it is not jammed the starter will have to come out for examination. If the starter should turn very slowly go on to the next check.

3   If, when the starter switch is operated, the lights stay bright, then the power is not reaching the starter. Check all connections from battery to solenoid switch and starter for perfect cleanliness and tightness. With a good battery installed this is the most usual cause of starter motor problems. Check that the earth link cable between the engine and frame is also intact and cleanly connected. This can sometimes be overlooked when the engine has been taken out.

4   If no results have yet been achieved turn off the headlights, otherwise the battery will go flat. You will possibly have heard a clicking noise each time the switch was operated. This is the solenoid switch operating but it does not necessarily follow that the main contact is closing properly. (NB: if no clicking has been heard from the solenoid it is certainly defective). The solenoid contact can be checked by putting a voltmeter or bulb across the main cable connection on the starter side of the solenoid and earth. When the switch is operated, there should be a reading or lighted bulb.

If not, the solenoid switch is no good. (Do not put a bulb across the two solenoid terminals. If the motor is not faulty the bulb will blow). It, finally, it is established that the solenoid is not faulty and 12 volts are getting to the starter then the starter motor must be the culprit.

10 Starter motor - removal and refitting

1   Raise the front of the car onto car ramps and chock the rear wheels.
2   Continue by removing the positive connection from the battery and then stow it carefully aside. The negative earthing connection to the body may remain in place.
3   Then working on the starter motor assembly itself, remove the heavy leads to the starter and solenoid, identifying as necessary to ensure correct refitting later.
4   From underneath the car and from within the engine compartment, undo and remove the nuts and bolts securing the starter motor assembly to the bellhousing.
5   The starter motor is now free to be lifted from the engine and transferred to a bench for dismantling and repair. (photo).
6   Refitting the motor assembly to the engine follows the reversal of the removal procedure. Remember to use new locking washers as appropriate and tighten the starter motor retaining nuts and bolts to their specified torques at the beginning of this Chapter.

10.5 Lift the starter motor assembly from the engine

11 Starter motor - dismantling, repair and reassembly

1   The starter motor assembly comprises three sub-assemblies: The motor itself, the solenoid switch and the pinion actuator housing. The actuator housing forms the mechanical link between the motor and the solenoid switch.
2   Such is the inherent reliability and strength of the starter motors fitted to the Coupe and Spider that it is very unlikely that a motor will ever need dismantling until it is totally worn out in need of replacement as a whole.
3   The solenoid which is usually available individually as a spare is attached to the actuator housing by three nuts on three long bolts passing the length of the solenoid. Undo and remove these three end nuts and lift the solenoid from the starter motor assembly.
4   There is no possibility of repairing the solenoid and therefore if after reconnecting across the battery with two stout leads the unit remains lifeless or the switch part fails to work, the whole solenoid must be renewed.
5   Starter motor brushes:- On the forward end of the motor there is a wide strap, with a single screw to tighten it in position and it covers the aperture which allows access to the motor brushes.
6   The procedure for inspection and renewal of the brushes is straightforward. With the starter motor removed from the car and on a bench, slacken the single screw which clamps the end strap in position and slip the strap along the motor casing to uncover the brush access apertures.
7   The brushes are retained in their mountings by spiral springs. Move the ends of the spiral spring to allow the carbon brushes to be extracted from their mountings. Undo the small screw which secures the small lead from the brush to its terminal on the forward end fitting and remove the brush from the motor assembly. If the brush was worn to the extent when the spiral string applies little force, then the brush should be renewed.
8   Motor dismantling:- Having already removed the solenoid, the front end of the motor is the next unit to be separated from the motor assembly. The motor is held together by tie rods screwed into the actuator housing at the rear end and projecting through the front end fitting to accommodate nuts at the other.

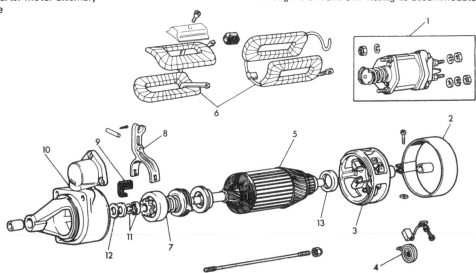

FIG.10.5. STARTER MOTOR — EXPLODED VIEW

1 Solenoid - for switching the motor power supply and engaging the pinion with the engine flywheel	7 Pinion and pinion carriage
2 Brush aperture cover strap	8 Pinion carriage actuating lever
3 Brush and terminal cage	9 Lever buffer, fitted to motor casing
4 Spiral spring and brushes	10 Pinion actuator housing
5 Motor armature	11 Pinion carriage retaining ring and sleeve
6 Field windings	12 Thrust washer
	13 Thrust washer front end

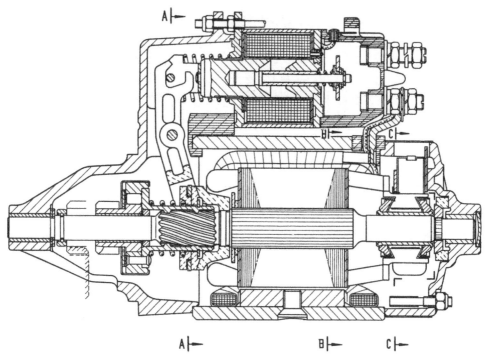

A B C

Longitudinal section of complete motor

SECTION A-A

SECTION B-B

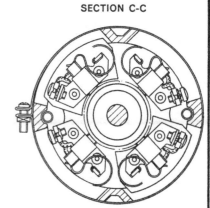

SECTION C-C

Section through pinion engaging device

Section through pole shoes and field and armature windings

Section through housing on commutator side, with view of brushes

Fig.10.6. Sectional views of the starter motor E100—1,3/12

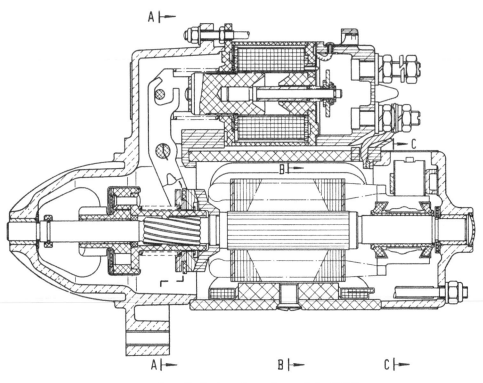

A ⊢→   B ⊢→   C ⊢→

A ⊢→

Longitudinal section of complete motor

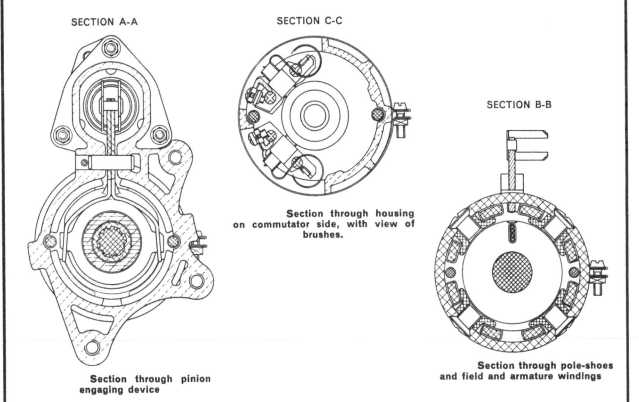

SECTION A-A

SECTION C-C

SECTION B-B

Section through housing
on commutator side, with view of
brushes.

Section through pinion
engaging device

Section through pole-shoes
and field and armature windings

Fig.10.7. Sectional view of the starter motor E84–0,8/12 Var 4

9 Before proceeding to separate the motor sub-assemblies it is necessary to disconnect the electrical connections between them. In particular the forward end cover strap should be removed and the brushes removed. The electrical leads running from the field windings in the motor, to terminals in the forward end cover, should be detached from those terminals.

10 Once the nuts on the tie rods have been removed, the forward end fitting, motor casing and the pinion actuator housing can be separated. The motor armature and drive pinion are mounted on a shaft which runs in bush bearings housed in the actuator housing and forward end fitting. It will be necessary to drive the pinion actuating pivot pin from the actuator housing to permit the separation of the actuator housing and motor armature assembly.

11 Be careful to retrieve the spacer shims and thrust bearings on each end of the motor shaft when the motor shaft is freed. You should refit the spacers and thrust components exactly in the positions which they occupied before dismantling.

12 The field windings are held to the inside of the motor main casing by special blocks which are in turn secured by screws passing through the casing.

13 The armature and pinion assembly will usually be separated from the motor components in order to gain access to the pinion assembly. Inspection and repair of the pinion assembly is described in Section 12.

14 Reassembly of the starter motor follows the reversal of the dismantling procedure. Fortunately there is little in the way of adjustments to make on the motor, the assemblies usually fit together to their correct relative positions.

12 Starter motor drive pinion - inspect and repair

1 Persistent jamming or reluctance to disengage may mean that the start pinion needs attention. The starter motor should be removed from the car first of all for general inspection.

2 With the starter motor removed, thoroughly clean all the grime and grease off with a petrol soaked rag. Take care to avoid any liquid running into the motor itself. If there is a lot of dirt, particularly on the pinion itself, this could be the trouble. The pinion should move freely along a spiral which is machined on the motor shaft. If the pinion motion is not smooth and easy remembering the springs which are fitted to the solenoid and pinion carriage to return it to its rest disengaged position - the motor should be dismantled and the armature/pinion assembly inspected and cleaned as follows.

3 Having removed armature/pinion assembly the commutator may be cleaned with a petrol dampened rag. The pinion carriage is retained on the motor shaft by a spring ring and sleeve. The sleeve should be driven off the end of the shaft exposing the spring ring which can now be slipped out of its groove seat and off the shaft.

4 Slide the pinion carriage off the rotor shaft and then clean the spiral which is exposed. Wipe the internal spiral in the pinion carriage clean. You should not dismantle the pinion carriage. Individual parts are not listed as available and if the pinion teeth are damaged then the pinion carriage or the whole starter should be renewed.

5 The spiral splines should be lubricated with VS 10W grease before reassembly of the carriage onto the shaft. The intermediate disc that forms the thrust bearing between the actuating lever ring and the pinion carriage sleeve, should be lubricated with FIAT MR3 grease or its equivalent.

6 Reassembly of the pinion carriage onto the motor shaft follows the reversal of the removal procedure.

13 Voltage regulator - general description

1 The regulator fitted to all Coupes and Spiders is the FIAT RC1/12B - charge indicator relay. Again Spiders with chassis numbers preceding 741 were equipped with a three core regulator, FIAT GN 2/12/28.

2 The RC1/12B regulator is specially designed to be compatible with the rectified output from the alternator. The alternator is self limiting as far as the current output is concerned and the regulator's function is to control the output voltage of the alternator. It is of the dual vibrating contact type.

3 The GN2/12/28 is the usual pattern of regulator employed with DC generators which need voltage and current regulation. The incorporation of a cut out device is necessary to isolate the dynamo from the battery when the voltage developed in the dynamo falls below the 12 v from the battery. This capability is essential to prevent the battery from discharging through the dynamo when the engine is running slowly or is stopped.

4 The RC1/12B regulator does not incorporate a cut out to prevent the discharge of the battery through the alternator because the main terminal on the alternator is connected into the diode bridge which apart from rectifying the alternator output also prevents current from flowing into the device from the battery.

5 There is no point in acquiring the equipment necessary to check the regulators, so if they are suspect it will be a job for a qualified auto electrician.

14 Voltage regulator - maintenance, checking and replacement

1 The voltage regulators are sealed units without provision for easy adjustment, and the only maintenance that need and can be done whilst the regulator is operating satisfactorily is to check the wiring contacts to the regulator every 9,000 miles.

2 The leads should be disconnected one at a time and their ends cleaned with methylated spirit. They may be rubbed clean with emery paper. The contacts on the regulator should receive the same treatment.

3 It will only be necessary to check that the regulator is working properly when checks have been completed on the alternator, cables, battery and terminals and no fault found.

4 When you are satisfied that it must be the regulator at fault, take the car to an auto-electrician and he will have the equipment to enable him to tell you very quickly where the malfunction is.

5 It is as well to remember that the RC1/12B regulator is used in conjunction with a relay which is used to switch on/off the red ignition warning light.

   If your only indication that something is faulty is the glowing of the red ignition light, then it might be the special relay which switches this light. Should the actuating coil in the relay fail, the relay switch will remain closed and the ignition light will glow.

6 The ignition light relay switch is readily checked by an auto-electrician, although if you have a 14.5 volt source and connect ot across the terminals Number 85 +ve and 86 -ve (yellow lead and black/white lead) on the relay the ignition light should go out. If it does not, renew the relay.

15 Fuses

1 There are 10 fuses fitted in the electrical system on both Coupes and Spiders. They are mounted on a holder mounted underneath the dashboard.

2 The symptom of fuse failure is the simultaneous 'failure' of a number of electrical systems. The fuse which has broken can then be identified by which combination of electrical systems do not operate.

3 Nine of the fuses are 8 amp and one is 16 amp. The 16 amp fuse protects those systems not controlled by the ignition switch.

4 The table specification at the beginning of this Chapter details the identification of each fuse and the circuits protected by them.

5 Never think you can leave fuses out or by-pass them; or substitute a fuse with a piece of tin foil or similar. A fuse blows for a reason and if the fault is not righted immediately, you will do serious damage to the wiring on the circuit involved and even

adjacent wiring.

The plastic electric insulation material is also a heat insulation material and if excessive currents flow through the wires, they will soon heat up and melt the insulation.

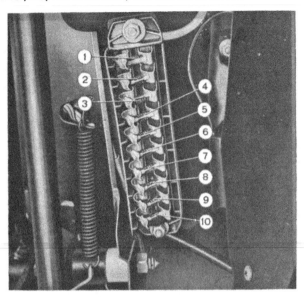

FIG.10.8. THE FUSE BOX, SITUATED UNDERNEATH THE DASHBOARD, ADJACENT TO THE STEERING COLUMN

1. Fuse A - 2. Fuse B - 3. Fuse C - 4. Fuse D - 5. Fuse E - 6. Fuse F - 7. Fuse G - 8. Fuse H - 9. Fuse I - 10. Fuse L.

16 Windscreen wipers - fault diagnosis

1 Two windscreen wiper control systems have been employed on the Coupes and Spiders. Earlier cars had an on/off switch, a wiper speed control and a foot operated push switch which operates the wipers and the electric screenwash pump.

2 Later models retained the on/off switch for continuous operation, a wiper speed control but now includes a switch operated intermittent operation. The period of the intermittent operation is also variable. The screen wash is a separate system.

3 Switch and lead failure will be indicated when only a particular mode of operation of the wiper is malfunctioning. The duplication of wiper motor activating circuits helps with fault diagnosis. If none of the circuits manage to operate the motor, it is unlikely that all activating routes have failed simultaneously; the motor is suspect and should be removed.

4 When the wipers work in one or two modes only then the suspect activiation circuit should be tested. Check the continuity of the leads in that circuit and the operation of the switch with a small battery, a bulb and some leads. Note the intermittent operation control unit is a separate piece of equipment, mounted adjacent to the wiper motor.

5 If the wipers run too slowly it will be due to something restricting the free operation of the linkage or a fault in the motor. In such cases it is well to check the current used by connecting an ammeter in the circuit. If it exceeds three amps something is restricting free movement. If less than three amps then the commutator and brush gear in the motor are suspect.

6 The wiper motor and gearbox are mounted behind the dashboard and operates the individual wipers through a crank linkage which is sufficiently exposed for occasionally a foreign object to interfere with the linkage movement.

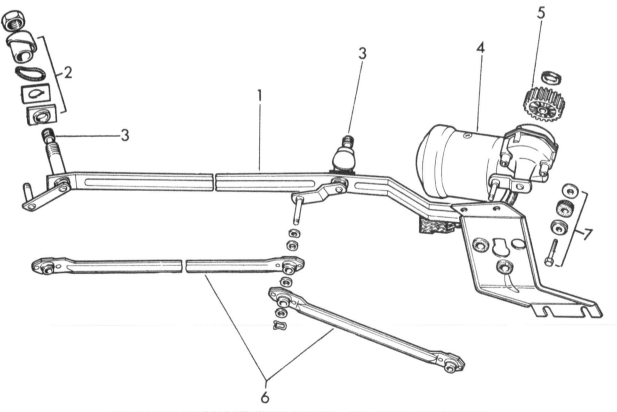

FIG.10.9. THE WINDSCREEN WIPER SYSTEM – EXPLODED VIEW (SPIDER)

1 *Wiper linkage frame*
2 *Wiper spindle housing retaining nuts and trim*
3 *Wiper spindle*
4 *Wiper motor*
5 *Main gear wheel*
6 *Wiper linkage rods*
7 *Motor mounting bolts and bushes*

17 Windscreen wiper motor and mechanism - removal and replacement

1 Although it is pssible to separate the motor from the wiper linkage in situ; it is advisable to remove the linkage and motor together, it is a fiddly business disconnecting the motor from the mounting and linkage, particularly when the task concerns the SPIDER, as the whole wiper system is behind the dashboard. The COUPE wiper installation, being behind the engine compartment rear bulkhead, is better as far as access is concerned; but it still will be as well to remove the whole wiper system and separate the parts on the workbench.

When all is said and done, it is probable if the motor has worn to an extent to merit replacement; the linkage too will merit inspection and renovation.

2 Two patterns of wiper have been fitted to the 124 Sport Series, one for each of the Coupe and Spider varients. The differences are of dimensions and sizes of gear wheel in the motor gearbox. Conceptually they are the same and therefore the same removal and refitting procedures apply.

3 Begin by removing the windscreen wiper arms from the splined hub. Then undo and remove the nut and chromed trim with gasket which retains the wiper spindle housing on the bodyshell. (photo).

4 Once the wiper spindle housings have been freed; the bolts which secure the motor mounting to the bodyshell bracket can be removed to allow the motor/wiper assembly to be lifted clear of the car.

5 The electrical lead connectors to the motor fly leads should be separated as the wiper assembly is removed. It will probably require some manipulation to extract the wiper assembly from crowded area behind the SPIDER dashboard.

6 Refitting the system follows the reversal of the removal procedure.

18 Windscreen wiper motor and mechanism - dismantling and reassembly

1 The motor and gearbox unit is held to its mounting plate by three bolts which pass through rubber grommets to provide a 'soft' mounting for the motor.

2 If the motor is known to be faulty, there is no point in trying to effect a repair, a new motor/gearbox assembly should be purchased.

3 If the problem was one of sloppy operation with a great deal of slack in the operating mechanism, the bearing bushes at each end of both link rods should be closely inspected. The bearings are not renewable individually and it will be a matter of replacing the appropriate link rods. It should be remembered that if the bearings were worn, then so the pins on which the bearings run could also be worn.

4 Still with the investigation into sloppy operation, it is worth removing the top of the gearbox so that the condition of the worm driven gear wheel can be inspected. This main gear wheel is listed as available as spare and may be replaced, if found to be worn.

5 The wiper spindles are not replaceable individually and if the spindles are a shakey fit in their housings, it will be necessary to replace the whole main framework. This framework comprises both spindles, the housings and spacing member and finally the motor mounting.

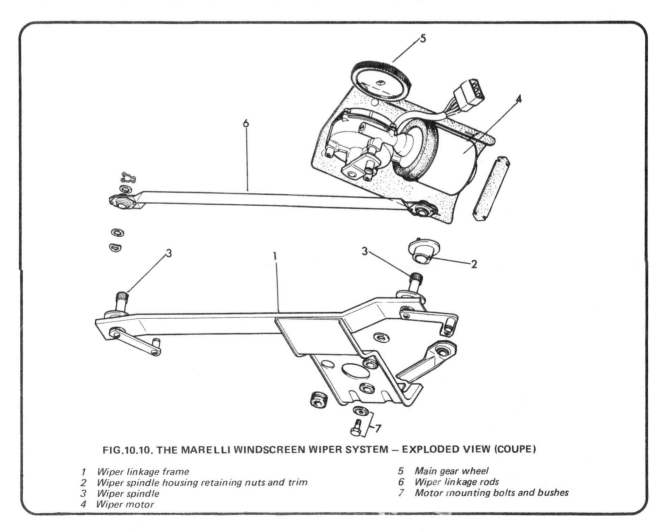

FIG.10.10. THE MARELLI WINDSCREEN WIPER SYSTEM – EXPLODED VIEW (COUPE)

1 Wiper linkage frame
2 Wiper spindle housing retaining nuts and trim
3 Wiper spindle
4 Wiper motor
5 Main gear wheel
6 Wiper linkage rods
7 Motor mounting bolts and bushes

17.3 Removal of windscreen wiper arm
boss from the splined spindle

19 Windscreen wiper switches - removal and refitting

1   Several types of switch have been employed to operate the
wiper system. Currently the wiper switch is mounted on the
steering column and is stork operated. In order to remove switch
units mounted on the steering column proceed as follows:
2   Disconnect the positive terminal from the battery.
3   Remove the steering wheel as directed in Chapter 11.
4   Undo and remove the screws securing the column switch half
covers to the switch frame, and remove the covers.
5   Slacken the switch unit retaining strap which secures the
lower end of the unit to the upper steering column support
bracket.
6   Disconnect the block cable connectors that join the switch
units to the vehicle electrical loom.
7   Slide the switch unit off the steering column.
8   The steering column switch unit is renewable only as a
complete assembly.
9   Refitting the switch unit follows the reversal of the removal
procedure.
10 Dashboard mounted switches and controls:- Removal of
switches from the dashboard is quite straightforward since they
are either a push in spring fit in the apertures in the dashboard,
or are retained in the aperture by a chrome ring which screws
onto the switch housing.
11 Testing is straightforward again, requiring the switch to be
connected in series with a bulb and small battery. The light
should come on when the switch is turned on.

20 Direction indicator switch - removal and replacement

1   The direction indicator switch has always been incorporated
in the steering column group of switches. It is stork operated.
The various switches are not replaceable individually and it will
be necessary therefore to renew the whole column switch
assembly if any one switch has broken.
2   To remove and refit the steering column switch assembly
follow the operations detailed in paragraphs two to nine of the
preceding Section, Number 19.

21 Direction indicator - circuit fault diagnosis

1   The direction indicator circuit comprises a flasher unit, the
indicator switch and indicator lights connected in series in that
order.
2   The most common faults in the indicator circuit are either
the failure of the flasher unit or the failure of the indicator light
bulbs or earth to those bulbs. If when the indicator switch is
operated there is completely no response, continue as directed in
paragraphs 3 to 5. If on the other hand the lights do work too

rapidly, or only one works, proceed as directed in the paragraphs
6 and 7.
3   The flasher unit is situated behind the dashboard and can be
readily checked in situ. There are four terminals on the unit.
Two terminals are the supply and earth. Black and yellow lead(s)
connect to the positive terminal on the flasher unit and a black
and black/white leads connect to the earth connection. The blue
lead goes to the green repeater light in the speedometer. The
fourth lead, which is either violet or black/white, connects the
flasher unit to the switch on the steering column.
4   The test procedure for the flasher is to push the indicator
switch to either position and bridge the positive supply terminal
(black/yellow leads) to the switch terminal (violet or black/
white) on the flasher. If the flasher was at fault, the appropriate
indicator lights should now be bright.
5   Should the lights still fail to come on, even when the flasher
unit is bridged, the supply lead to the flasher should be checked.
6   Should the fault be confined to one light, or that the lights
flash too rapidly then begin by removing the indicator lamp
glass, the bulb itself and then thoroughly clean the bulb holder,
connections and bulb. Check that the bulb filament is intact.
7   Another fault that results in erratic or fast operation of the
flasher circuit is a deterioration of the earthing joint between the
indicator light holder and bodyshell. Very often when the earth
is at fault the adjacent bulb to the indicator bulb in the side light
cluster will flash instead. In this instance remove the light cluster
assembly from the bodyshell, thoroughly clean the mating
surfaces to ensure electrical contact when they are reassembled.

22 Horn - maintenance and repair

1   Two types of horn have been fitted to the 124 Sport series.
In all probability you will find a pair of air trumpet horns; but
the more usual twin tone diaphragm horns have been fitted by
FIAT to many cars.
2   Maintenance of the air horns involve periodic oiling of the
compressor detailed in the 'Maintenance' section at the
beginning of this book. The electric diaphragm twin tone horns
do not require maintenance save the occasional brush clean.
3   The air horns actuating circuit includes the push button on
the steering column, a sealed relay unit mounted on the bodyshell near to the air compressor which is the last item in the
actuating circuit and supplies air to the horns through clear
plastic pipes. (photo).
4   As usual the investigation is to determine which item is
faulty, there is no point in going further and trying to find the
fault within the component. FIAT supply only the complete
components as spare.
5   Check the push button by connecting a jump lead to the No
3 terminal on the relay (black or grey lead) and touching the
nearest part of bare metal on the bodyshell. If the horn sounds,
remove the push button from the centre of the steering wheel
and thoroughly clean.
6   In the instance when the push button check fails to yield a
result, check the voltage on the violet or red wire to the relay.
The voltage should be 12 volts. If that wire is dead, check its
continuity and its connection to the first fuse 'A'.
7   If the horns still do not operate, check the relay by bridging
the No 1 and 4 terminals (violet or red to the blue cable). The
relay should be renewed if the horns sound when these relay
terminals are bridged.
8   The last check is the air compressor itself. Check the voltage
between the blue cable terminals on the compressor motor and
the bodyshell earth when the horn button is pressed. If the
voltmeter records 12 volts when the button is pressed the
compressor motor is suspect.
9   Before you renew the motor/compressor check that its
mounting is clean and electrically conducting, the motor relies
on a return to earth through the mounting for its operation.
Once the motor compressor has been established as faulty, do
not try to effect a repair, individual parts are not available,
renewal of the unit is your only course of action.

10 The twin tone horns. The fault finding procedure to be adopted with the direct electric action twin tone horns is broadly similar to that detailed in paragraphs 3 to 9 for the air horns.

11 It is well to check that power is reaching the horns by disconnecting the feed cables checking the voltage across them and the bodyshell. These horns have the same relay activation as the air horns and the relay and push button is checked as described for the air horn.

12 As with the air compressor for the air horns, the electric diaphragm horns have to be securely fitted to the vehicle structure and properly earthed for correct operation.

22.3 The air horn installation (1) horns, (2) air compressor (3) relay

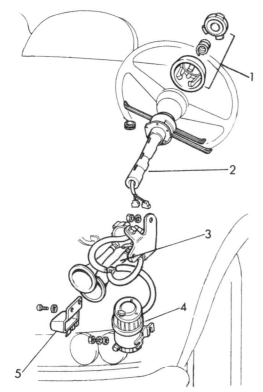

FIG.10.11. THE HORNS AND STEERING COLUMN SWITCH COMPONENTS

1   *Horn push button switch components*
2   *The steering column switch assembly. (main/dip lights; direction indicator; windscreen wiper)*
3   *Horn trumpets*
4   *Air compressor*
5   *Relay employed to switch power to the air compressor*

23 Headlight switch

1   The main/dipped headlight selector switch is mounted in the switch unit on the steering column. Individual switches in this steering column unit cannot be removed and replaced easily and it will be necessary to remove the whole switch unit as described in Section 19 of this Chapter.

2   The supply to the headlight switch comes from the main light switch (for side lights as well) mounted on the dashboard. Depending on the year of car this main switch is either a round switch retained by a chromed threaded ring to the dashboard. Later models have a rectangular switch which is a spring location fit in the dashboard.

3   Either switch is readily removed for checking.

24 Headlights and bulbs - adjustment, removal and replacement

1   The headlight units are mounted on three screws which can be turned in turn to alter the direction in which the headlight beam is directed. All Spider cars have double filament headlamps combining both main and dipped beam capabilities. Early 'AC' body shell Coupe cars has single headlights like the Spider. The later BC and CC Coupe models have twin headlights. The outer lights provide the main beams and dipped beams, being equipped with double filament bulbs. The inner lights are main beam only, and are equipped with a single element bulb.

2   On CC and CS bodyshell (the latest) models of the Coupe and Spiders the headlights are sealed beam units.

3   The headlight alignment procedure is as follows. Begin by removing the radiator trim (Coupes) and the chrome trim around the headlights. The headlights trim is retained by a screw in the base of the bezel which tightens against the headlight support frame.

4   Once all the bezel is removed the three lamps aligning screws are exposed. Check that the tyre pressures are correct and position the car 16 feet away from a suitably darkened vertical opague white screen, make sure the centre line of the car is at right angles to the wall. Bounce the car a little to ensure it has fully settled on the suspension. The headlighs alignment is completed with the vehicle UNLADEN.

5   Draw on the screen two vertical lines 49.37 inches apart and spaced equally each side of the projected centre line of the car. Then draw a horizontal line on the screen a distance equal to the height of the headlight centres above the floor minus 3,15 inches above the base of the screen. Fig. 10.13 illustrates the headlight aiming lines on the screen.

6   Working on the combined dipped/main beam lights adjust the direction of the dipped beam with the lamp alignment screws (Fig. 10.13) until the demarcation area between dark and well lighted areas is on the horizontal line drawn. The centres of the lighted areas should be on the two vertical lines drawn.

7   The position of the double filament bulb in the headlight will ensure that the main beam is correctly aligned once the dipped beam has been used to align the headlight assembly.

8   On later models of the Coupes, fitted with the twin headlights, align the outer main/dipped light assemblies as detailed in paragraphs 5,6 and 7. Then switch onto the main beams of the dipped/main beam lights already aligned, and use the 'targets' provided by the outer lights on the white screen to align the inner main beam lights.

9   Bulb renewal:- Remove the chrome/radiator trim if necessary and then go on to remove the chrome bezal around the headlight to be dismantled. The bezel is held by screws to the light support frame (photo).

10 Do not tamper with the light alignment adjusting screws, they engage the light unit support frame which is not removed when the bulbs are to be renewed.

11 The light unit is retained in its support frame by a ring which slots around the top of the support frame and is secured by a single screw at its base. Once this screw is undone and removed

the lamp retaining ring can be lifted away together with the light unit. Note the light has three loops on its periphery which register in slots in the support frame. (photo).

12 Once the light unit is clear the bulb may be extracted when the retaining spring clip has been lifted aside and the electrical connectors detached. (photo).

13 An additional socket in the reflector of the outer lights provide for a small parking light bulb. The bulb holder and bulb are bayonet fits into the parent lamp unit.

14 Refitting the lights and bulbs follows the reversal of the removal procedure.

15 It will be as well to check the light alignment once the new bulb has been refitted.

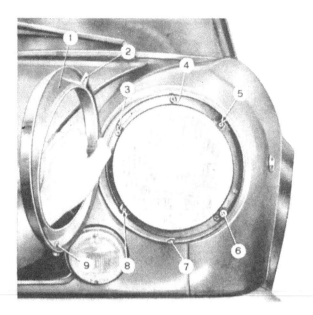

FIG.10.12. THE HEADLIGHT ALIGNMENT ADJUSTING COMPONENTS

1   Bezel
2   Bezel retaining lug
3   Spring clip for clamping headlamp unit
4   Bezel lug seat
5   Beam adjusting screw, horizontal direction
6   Locating pin of headlamp unit
7   Screw hole
8   Beam adjusting screw, vertical direction
9   Bezel fixing screw

24.9 Removal of headlight bezel

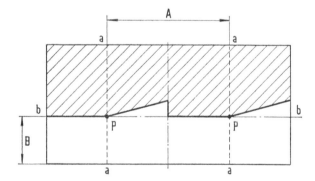

FIG.10.13. THE HEADLIGHT AIMING DIAGRAM

A =   1254 mm (49.37") - B = C - 100 mm (3.94") for a new car - B = C - 80 mm (3.15") for a run-in car - C = height of headlamp centres from floor, measured when aiming.

24.11 Headlight removed showing parking light and main bulb connectors

25 Front side and flasher light assemblies

1   Various patterns of front side and flasher light assemblies have been fitted to the 124 Sport series of cars. The Spiders have always been equipped with a wide assembly incorporating two reflectors and two bulbs, one side light and one flasher light. The Coupes has a single reflector light assembly using a double filament bulb. The differences are minor and involve different contour/colour lenses depending on the country in which it was originally sold.

2   The assemblies comprise a basic shell which incorporates the reflector(s); a gasket, the light lense, bulbs and finally fixing screws.

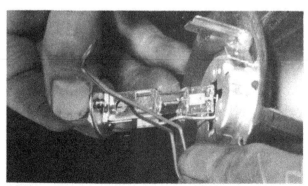

24.12 The bulb can be extracted once the retaining clip has been slid aside

3 To gain access to the bulbs undo and remove the lense screws and lift the lense away with the gasket.

4 The bulbs are a bayonette fit and are easy to remove.

5 As mentioned many times in this manual, cleanliness with electrical systems is essential. Clean the bulb terminals and holder terminals. The lense gasket should always be in near perfect condition, it is essential to prevent foreign matter and water entering the light assembly and corroding the electrical connections.

6 The majority of electrical faults on a car will be caused by the deterioration of electrical connections due to corrosion resulting from the ingress of dirt and water into the electrical system components.

7 Since the bulbs rely on the electrical connection of the light assembly to the bodyshell, it will be necessary to remove the basic shell of the light assembly from the bodyshell and clean the connection areas.

26 Rear side stop and flasher lights

1 Although there have been many shapes and patterns of rear light assembly fitted to the 124 Sport series, they all are of the same design concept. The light lense is retained by screws to the main shell of the light assembly. There is always a rubber gasket between the lense and shell. The shell incorporates the light relfectors and supports the lamp holders. In some instances the shell is screwed onto the outside of the bodyshell and in others the shell is mounted inside, it depends upon the year and model.

2 The light lenses should be removed fairly regularly and cleaned. It is essential that the gasket is kept in good condition because it prevents the ingress of dirt and water into the light.

3 If one of the lights fail, the lense should be removed and the bulb extracted from its socket. They are all bayonette fits.

4 It can easily happen that the lights will fail, but on inspection the bulb is serviceable. In these instances the supply leads to the assembly should be checked and if they are not faulty the earth connection from the light assembly to the bodyshell should be checked.

5 Earth connections from the bulbs to the bodyshell can deteriorate because of corrosion at one of the connection interfaces. Some light installations have a wire or metal strip connecting the bulb sockets to the bodyshell. On others the metallic light assembly base shell itself is used to conduct electricity from the bulb to the bodyshell on which it is mounted.

6 The earth can be proved faulty by connecting a jumper lead from the side of the bulb to the nearest bright metal on the bodyshell. If the lights operate with the jumper lead connected, the earth is at fault and the whole light assembly should be removed and thoroughly cleaned.

7 Clean the mountings on the bodyshell before refitting the light assembly.

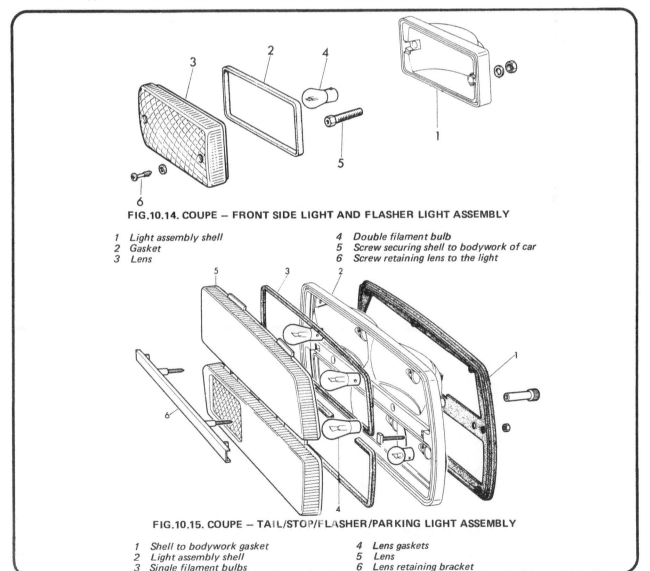

FIG.10.14. COUPE – FRONT SIDE LIGHT AND FLASHER LIGHT ASSEMBLY

1 Light assembly shell
2 Gasket
3 Lens
4 Double filament bulb
5 Screw securing shell to bodywork of car
6 Screw retaining lens to the light

FIG.10.15. COUPE – TAIL/STOP/FLASHER/PARKING LIGHT ASSEMBLY

1 Shell to bodywork gasket
2 Light assembly shell
3 Single filament bulbs
4 Lens gaskets
5 Lens
6 Lens retaining bracket

27 Number plate light

1   Depending upon the model of car the number plate light is either a twin or angle light installation. In both cases the dismantling procedure is the same.

2   Undo and remove the screw securing the light casing to its mounting on the number plate support. Remove the bulb and clean the holder before refitting the bulb.

3   The light mounting is secured to the number plate support with screws which pass from the opposite side of the number plate support. If it is desired to remove the light mounting it is necessary on some models to remove the number plate and its support.

4   If the rear number plate lights fail, check the bulb, then the supply leads and finally the earthing lead.

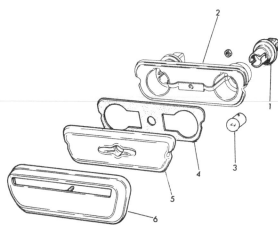

FIG.10.16. COUPE — NUMBER PLATE LIGHT COMPONENTS

1 Light socket	4 Lens gasket
2 Light assembly shell	5 Lens
3 Bulb - 2 off	6 Lens retaining plate, chromed

28 Flasher side repeater lights

1   It is a modern requirement to provide repeater flasher lights on each side of the vehicle.

2   The design of these lights is conventional comprising a lens, light shell, gasket and clamping member. The light shell projects through the bodyshell from the outside, and a clamping member fits onto the inside to pull the shell onto the bodyshell. The lens pushes into the shell. The bulb is as usual a push/twist fit.

3   If the repeater light(s) fail, check the bulb, supply wire and its earth connect in that order.

Fig.10.17. View of the instrument panel removed, showing the block electrical connectors 'A' and the individual instrument light holders 'B', the holders are snap in, and the bulbs are a push and twist fit into the holder

29 Instrument panel lights and switches

1   Each gauge is fitted with an illumination bulb which is mounted in the rear of the instrument. The bulb holders are push fit in the instrument casing, and the bulbs are a push and twist fit in the holder.

2   The intensity of the instrument panel lights is governed by a rheostat which is connected in series with the on/off switch for the panel lights. Both switch and rheostat will need to be checked if the instrument lights fail.

3   The light circuits are straightforward, consisting of a supply wire from the fuse rack, connecting to the on/off switch and then a wire jointing the switch to the rheostat. The lead from the rheostat then supplies the bulbs. The bulbs use the instruments and panel as earth connection.

4   If the lights all fail, the break or fault can be traced by connecting a voltmeter to terminals in the circuit (+ve to the terminal, - ve to the bodyshell), beginning with the rheostat output, to detect the point at which electrical power has been stopped.

30 Instrument panel and instruments

1   The main instrument panel is retained on the dashboard with chromed screws. Electrical connections to the instruments, switches and lights are grouped together into a small number of block connectors.

2   The instruments are clamped in position with brackets retained on the rear of each instrument casing.

3   The on/off switches are either of the toggle pattern which are retained by chrome ring nuts on the facia, or of a rectangular block pattern which is a push/spring fit into the instrument panel.

4   Rheostats are of the radial wiper pattern and are secured in a like manner to the toggle switches.

5   To remove the instrument panel remove the battery connections and undo and remove the chromed screws around its periphery, and lift the complete panel from the dashboard. Marking each block connector to make certain that they can be reconnected correctly, undo the block connectors and lift the panel away from the car.

6   Refitting follows the reversal of the removal procedure.

7   The instruments can be separated from the panel by removing the clamping brackets secured to the rear of the instruments casing. If an instrument is suspect, your only course of action is to take it to a qualified auto-electrian who should have the special equipment to distinguish whether it is the instrument at fault or the transducer (engine temperature, oil pressure, fuel float) that is fitted to the system being monitored.

31 Cigarette lighter

1   The cigarette lighter is supplied by a fairly heavy duty cable, and a similar cable forms the earth connection for the lighter. The earthing cable connects to the earthing terminal on the direction indicator flasher unit behind the dashboard.

2   Should the lighter fail, check the supply and earth cables before removing the unit from the dashboard for inspection.

3   The lighter unit may be checked using a small battery and bulb, after connecting the lighter in series with the bulb and battery the bulb should glow when the unit is actioned to 'light' a cigarette.

4   Once the checks have been completed, the faulty items should be renewed. Individual parts to the lighter unit are not available as spares.

32 Interior lights - interior, courtesey, drop tray, luggage and engine compartment

1   The interior lights are operated by a door jam switch and a separate switch on the light unit. This duplication of switching systems simplifies the task of identifying faulty components when there is a malfunction.

2   Quite simply if the lights fail to shine when both switches are operated then check that power is available on the supply leads. Two of the lights are connected to the cigarette lighter supply. The second rear courtesy light is connected directly to the inspection light recepticle in the engine compartment. (photo).

3   All lights use the bodyshell as the earth return directly.

4   Yet another interior light is fitted to the drop tray. The switch to this light is incorporated in the drop tray unit. Its supply comes from a small black connector behind the dashboard.

5   The lights fitted to illuminate the luggage and engine compartment are operated by push switches mounted in the bodyshell so that they are opened when the respective lids are lowered in position. (photo).

In all cases these push switches are in the earth cable from the light units. Once the lights fail, check the bulb, bulb recepticle, supply lead, earth lead and finally the push switch itself.

32.2 Interior light. Adjacent to rear view mirror

32.5 Luggage compartment light and push switch

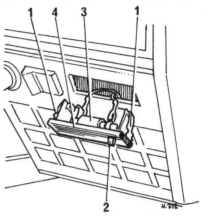

FIG.10.18. THE REAR COURTESY LIGHT
1   *Spring plate, unit mounting*
2   *Bayonet - coupled bulb*
3   *Lamp and lens unit mounting screw*
4   *Switch*
5   *Lens and lamp unit*

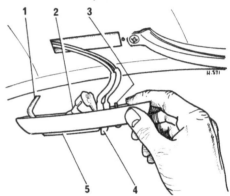

FIG.10.19. THE FRONT INTERIOR LIGHT

1   *Unit mounting, spring plates*   3   *Bulb, push fit*
2   *Switch*   4   *Lens*

33 Ignition switch

1   This switch is mounted on a bracket on the upper steering column support, it is a multi-position switch and incorporates a steering column lock.

2   Unscrew the threaded ring which retains the switch to its support bracket on the steering column.

3   The switch may be extracted from its location by turning the steering wheel slightly whilst rotating the ignition key, the switch will not come clear until the steering column locking pawl has disengaged from the column shaft.

4   Disconnect each lead to the switch in turn, identifying as necessary to ensure correct refitting later.

5   As with so many electrical devices on cars, once it has been identified as faulty; there is no course of action but to renew the switch.

6   Refitting the switch follows the reversal of the removal procedure.

34 Safety systems - brake effectiveness, handbrake and ignition key

1   The FIAT 124 Sport is fitted with several safety systems, mainly according to the country in which it was sold originally.

2   The systems are quite straightforward but do not have any built in test equipment and therefore to some extent you will have to monitor whether the warning lights, light up or do not light up appropriately.

3   The brake effectiveness circuit monitors the pressure in the brake hydraulic system and the pressure transducer completes the warning lights circuits to earth.

4   The brake effectiveness circuit may be checked by bridging the two terminals of the brake stop light pressure switch, the effectiveness light should glow when the brake pedal is released and go out when the pedal is depressed hard. (It may be necessary to start the engine to enable the servo to work during this last check). Do not forget to remove the bridging wire from the stop light switch.

5   Carry out the usual compatability checks on the brake effectiveness circuit as well as inspection of the light and switch.

6   The ignition key "Remove" indicator circuit is supplied from a terminal on the ignition switch and a microswitch in the door jam lights the warning light when the door is opened and the key is still in the switch, the key operates a small switch in the main ignition switch to 'register' its presence. Clearly if this circuit fails to give sensible warnings then the defective component may be identified as follows:

7   The door jam switch may be checked by connecting a bridge lead to the grey/black lead terminal on the light to earth, if the light comes on, the jam switch should be replaced.

8   The bulb and light assembly may be visually checked.

9   The key sensor terminals lead to a small black connector (2 way) near the ignition switch: bridging the two leads will exclude the sensor from the circuit and if the light comes on/off as appropriate the ignition switch should be renewed.

10 Finally there is a single fuse 3 amp situated in the lead to the sensor from the ignition switch, this fuse should be checked but remember if it has blown, it blew for a reason and that reason must be determined immediately.

11 Handbrake on/off warning. This system operates from a push switch positioned on the transmission tunnel immediately beneath the handbrake lever. The switch plunger is depressed when the lever is down in its rest position.

12 If the system is to be checked because of some unappropriate warnings; the centre trim around the handbrake lever, heater controls and transmission tunnel will need to be removed, to gain access to the switch and handbrake lever. See Chapter 12.

13 The switch is protected by a rubber boot and projects through the transmission tunnel, with the leads running inside the tunnel. As usual visual examination and electrical bridging of switches should be all that is nesessary to identify faulty components.

---

35 Radio installation

1   It is not the intention of this manual to cover the complexities of radio fault finding, that is the task for an electrician.

2   For satisfactory operation of the radio suppressors should be added to the following leads if not already there:-

a) A radio suppressor capacitor, between the ground and the terminal on the battery. (A brush of wires, connected to the bodyshell and secured at the rear of the car so that they graze the ground; ensure that static is conducted to the ground.)

b) A radio suppressor capacity between the the terminal on the ignition coil and the bodyshell.

c) A resistive suppressor on the HT lead to the distributor.

d) A resistive suppressor in each HT lead to the spark plugs.

3) If you are intending to fit a radio, then terminal A on the fuse rack should be used to supply the radio set. (Supply as the cigarette lighter and courtesy lights).

---

36 FAULT DIAGNOSIS

Symptom	Reason/s	Remedy
STARTER MOTOR FAILS TO TURN ENGINE		
No electricity at starter motor	Battery discharged	Charge battery.
	Battery defective internally	Fit new battery.
	Battery terminal leads loose or earth lead not securely attached to body	Check and tighten leads.
	Loose or broken connections in starter motor circuit	Check all connections and check any that are loose.
	Starter motor switch or solenoid faulty	Test and replace faulty components with new.
Electricity at starter motor: faulty motor	Starter motor pinion jammed in mesh with ring gear	Disengage pinion by turning squared end of armature shaft.
	Starter brushes badly worn, sticking, or brush wires loose	Examine brushes, replace as necessary, tighten down brush wires.
	Commutator dirty, worn, or burnt	Clean commutator, recut if badly burnt.
	Starter motor armature faulty	Overhaul starter motor, fit new armature.
	Field coils earthed	Overhaul starter motor.
STARTER MOTOR TURNS ENGINE VERY SLOWLY		
Electrical defects	Battery in discharged condition	Charge battery.
	Starter brushes badly worn, sticking, or brush wires loose	Examine brushes, replace as necessary, tighten down brush wires.
	Loose wires in starter motor circuit	Check wiring and tighten as necessary.
STARTER MOTOR OPERATES WITHOUT TURNING ENGINE		
Dirt or oil on drive gear	Starter motor pinion sticking on the screwed sleeve	Remove starter motor, clean starter motor drive.
Mechanical damage	Pinion or ring gear teeth broken or worn	Fit new gear ring, and new pinion to starter motor drive.

Symptom	Reason/s	Remedy
STARTER MOTOR NOISY OR EXCESSIVELY ROUGH ENGAGEMENT		
Lack of attention or mechanical damage	Pinion or ring gear teeth broken or worn	Fit new ring gear, or new pinion to starter motor drive.
	Starter drive main spring broken	Dismantle and fit new main spring.
	Starter motor retaining bolts loose	Tighten starter motor securing bolts. Fit new spring washer if necessary.
BATTERY WILL NOT HOLD CHARGE FOR MORE THAN A FEW DAYS		
Wear or damage	Battery defective internally	Remove and fit new battery.
	Electrolyte level too low or electrolyte too weak due to leakage	Top up electrolyte level to just above plates.
	Plate separators no longer fully effective	Remove and fit new battery.
	Battery plates severely sulphated	Remove and fit new battery.
	Drive belt slipping	Check belt for wear, replace if necessary, and tighten.
	Battery terminal connections loose or corroded	Check terminals for tightness, and remove all corrosion.
	Dynamo not charging properly	Remove and overhaul dynamo.
	Short in lighting circuit causing continual battery drain	Trace and rectify.
	Regular unit not working correctly	Check setting, clean, and replace if defective.
IGNITION LIGHT FAILS TO GO OUT, BATTERY RUNS FLAT IN A FEW DAYS		
Dynamo not charging	Drive belt loose and slipping, or broken	Check, replace, and tighten as necessary.
	Brushes worn, sticking, broken or dirty	Examine, clean, or replace brushes as necessary.
	Brush springs weak or broken	Examine and test, Replace as necessary.
	Commutator dirty, greasy, worn, or burnt	Clean commutator and undercut segment separators.
	Armature badly worn or armature shaft bent	Fit new or reconditioned armature.
	Contacts in light switch faulty	By-pass light switch to ascertain if fault is in switch and fit new switch as appropriate.
WIPERS		
Wiper motor fails to work	Blown fuse	Check and replace fuse if necessary.
	Wire connections loose, disconnected, or broken	Check wiper wiring. Tighten loose connections.
	Brushes badly worn	Remove and fit new brushes.
	Armature worn or faulty	If electricity at wiper motor remove and overhaul and fit replacement armature.
	Field coils faulty	Purchase reconditioned wiper motor.
Wiper motor works very slow and takes excessive current	Commutator dirty, greasy, or burnt	Clean commutator thoroughly.
	Drive to wheelboxes too bent or unlubricated	Examine drive and straighten out severe curvature. Lubricate.
	Wheelbox spindle binding or damaged	Remove, overhaul, or fit replacement.
	Armature bearings dry or unaligned	Replace with new bearings correctly aligned.
	Armature badly worn or faulty	Remove, overhaul, or fit replacement armature.
Wiper motor works slowly and takes little current	Brushes badly worn	Remove and fit new brushes.
	Commutator dirty, greasy, or burnt	Clean commutator thoroughly.
	Armature badly worn or faulty	Remove and overhaul armature or fit replacement.
Wiper motor works but wiper blades remain static	Driving cable rack disengaged or faulty	Examine and if faulty, replace.
	Wheelbox gear and spindle damaged or worn	Examine and if faulty, replace.
	Wiper motor gearbox parts badly worn	Overhaul or fit new gearbox.

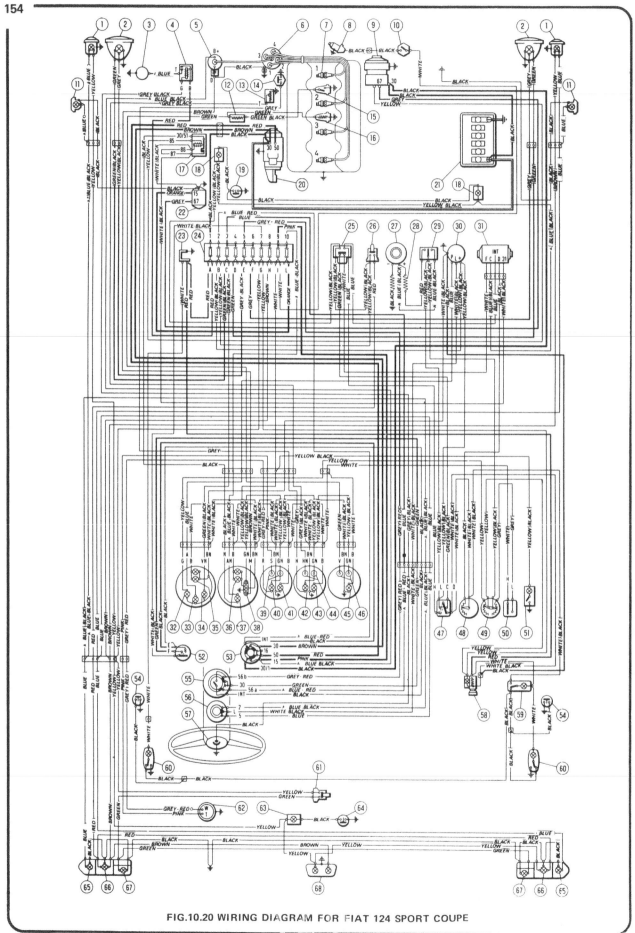

FIG.10.20 WIRING DIAGRAM FOR FIAT 124 SPORT COUPE

FIG.10.20 WIRING DIAGRAM KEY FOR FIAT 124 SPORT COUPE

1	Front parking and direction indicator lamps
2	Headlamps, main beam and anti-dazzle
3	Horn compressor motor
4	Remote control of two-tone horns
5	Ignition coil
6	Ignition distributor
7	Sparking plugs
8	Electro-magnetic fan-clutch brush
9	Alternator
10	Thermostatic switch controlling radiator fan-clutch
11	Side direction lamps
12	Temperature gauge extra resistor
13	Oil gauge sender unit
14	Low oil pressure indicator sender unit
15	Temperature gauge thermal switch: shifts gauge pointer to red end of scale (dangerous water temperature) irrespective of impulses from sender 16
16	Temperature gauge sender unit
17	Relay switch for warning light 37
18	Engine compartment lamps
19	Push switch for engine compartment lights
20	Starter motor
21	Battery
22	Voltage regulator
23	Inspection lamp socket
24	Fuses
25	Pedal switch, windscreen washer and wiper
26	Push switch for stop lights
27	Ventilation fan motor, two-speed
28	Ventilation fan motor extra resistor
29	Three-position ventilation fan selector switch
30	Direction indicator flasher
31	Windscreen wiper motor
32	Side lamps indicator (green)
33	Direction indicator repeater light (green)
34	Main beam repeater light (blue)
35	Speedometer light
36	Engine-speed indicator light
37	Warning light, no-battery charge
38	Engine-speed indicator
39	Fuel gauge
40	Reserve warning lamp
41	Fuel gauge light
42	Low oil pressure warning light (red)
43	Oil gauge light
44	Oil gauge
45	Temperature gauge light
46	Temperature gauge
47	Windscreen wiper switch
48	Screen-wiper motor rheostat
49	Instrument panel lights and side lamps repeater rheostats
50	Instrument panel lights switch
51	Glove compartment lamp with built-in switch
52	Outside lighting switch
53	Key-type switch for ignition, services and starting
54	Push switches for courtesy lights 60
55	Selector switch for outer lighting and anti-dazzle flashes
56	Direction indicator switch
57	Horn button
58	Cigar lighter (with spot lamp)
59	Front interior lamp with built-in switch
60	Rear interior lamps with built-in switch
61	Back-up lamps push switch
62	Fuel gauge sender unit
63	Luggage compartment light
64	Luggage compartment light push switch
65	Rear direction lamps
66	Tail and stop lamps
67	Back-up lamps
68	Number plate lamps

NOTE - The symbol ➞ means that the cable carries a nunbered ring or sleeve

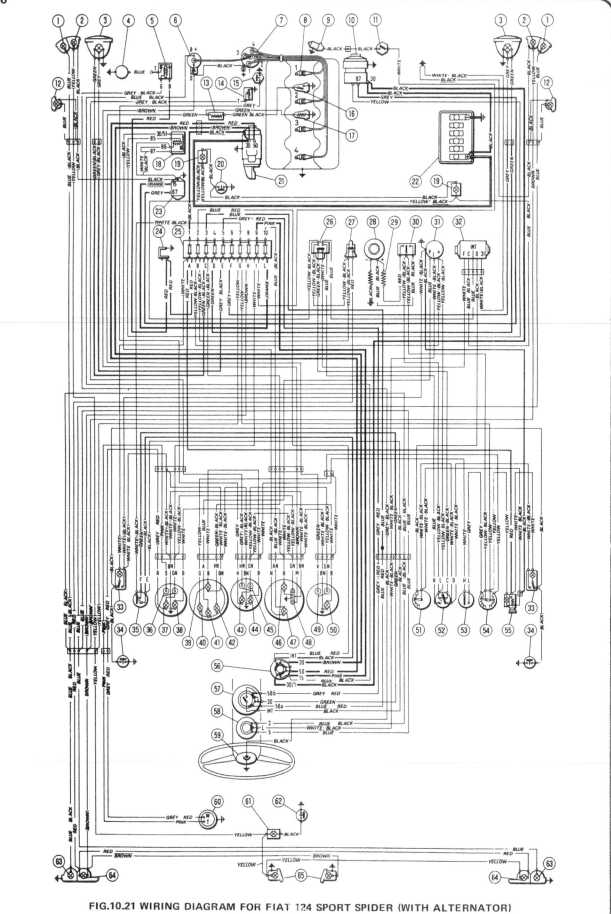

FIG.10.21 WIRING DIAGRAM FOR FIAT 124 SPORT SPIDER (WITH ALTERNATOR)

FIG.10.21 WIRING DIAGRAM KEY FOR FIAT 124 SPORT SPIDER (WITH ALTERNATOR)

1 Front direction lamps
2 Front lamps
3 Headlamps, main beam and anti-dazzle
4 Horn compressor motor
5 Remote control of two-tone horns
6 Ignition coil
7 Ignition distributor
8 Sparking plugs
9 Electro-magnetic fan-clutch brush
10 Alternator
11 Thermostatic switch controlling radiator fan-clutch
12 Side direction lamps
13 Temperature gauge extra resistor
14 Oil gauge sender
15 Low oil pressure warning sender
16 Temperature gauge thermal switch: shifts gauge pointer to red end of scale (dangerous water temperature) irrespective of impulses from sender 17
17 Temperature gauge sender
18 Relay switch for warning light 47
19 Engine compartment lamps
20 Push switch for engine compartment lights
21 Starter motor
22 Battery
23 Voltage regulator
24 Single-pole socket for inspection lamp
25 Fuses
26 Pedal switch, windscreen washer and wiper
27 Push switch for stop lights
28 Ventilation fan motor, two speed
29 Ventilation fan motor extra resistor
30 Three-position ventilation fan selector switch
31 Direction indicator flasher
32 Screen wiper motor
33 Dashboard lamps with built-in switches
34 Push switches for courtesy lights
35 Outside lighting switch
36 Fuel gauge
37 Reserve warning lamp
38 Fuel gauge light
39 Speedometer light
40 Side lamps indicator (green)
41 Direction indicator repeater light (green)
42 Main beam repeater light (blue)
43 Low oil pressure indicator light (red)
44 Oil gauge lamp
45 Oil pressure gauge
46 Engine speed indicator lamp
47 Warning light, no battery charge
48 Engine speed indicator
49 Temperature gauge
50 Temperature gauge indicator lamp
51 Screen-wiper motor rheostat
52 Screen-wiper motor on-off switch
53 Instrument panel lights switch
54 Instrument panel lights and side lamps repeater rheostat
55 Cigar lighter (with stop lamp)
56 Ignition, services and starting switch
57 Selector switch for outer lighting and anti-dazzle flashes
58 Direction indicator switch
59 Horn button
60 Fuel gauge sender
61 Luggage compartment light
62 Luggage compartment light push switch
63 Rear direction lamps
64 Tail and stop lamps
65 Number plate lamps

NOTE - The symbol ──means that the cable carries a numbered ring or sleeve

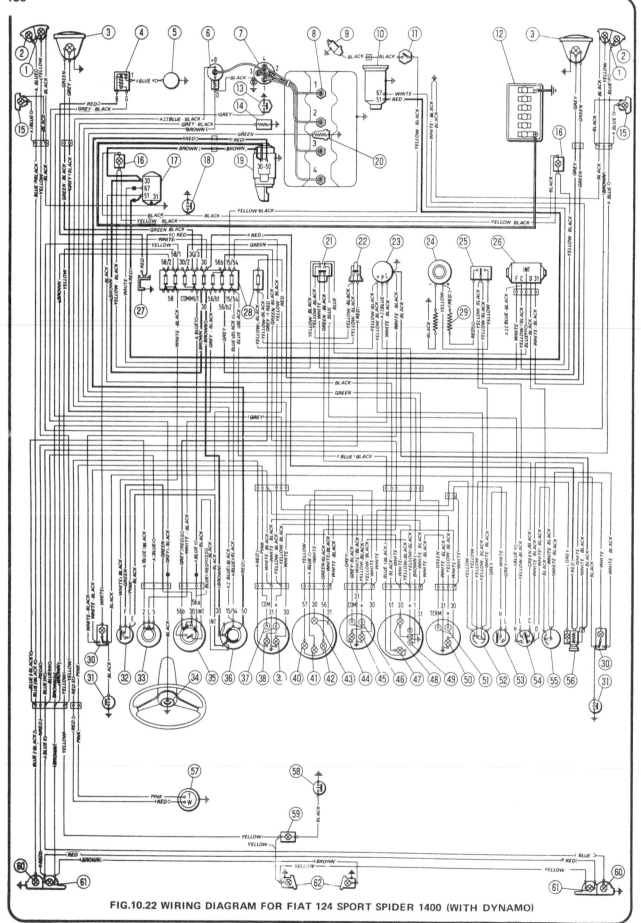

FIG.10.22 WIRING DIAGRAM FOR FIAT 124 SPORT SPIDER 1400 (WITH DYNAMO)

FIG.10.22 WIRING DIAGRAM KEY FOR FIAT 124 SPORT SPIDER 1400 (WITH DYNAMO)

1	Front lamps
2	Front direction lamps
3	Headlamps, main beam and anti-dazzle
4	Remote control of two-tone horns
5	Horn compressor motor
6	Ignition coil
7	Distributor
8	Sparking plugs
9	Electro-magnetic fan-clutch brush
10	Generator
11	Thermostatic switch controlling radiator fan-clutch
12	Battery
13	Low-oil-pressure warning sender
14	Oil gauge sender
15	Side direction lamps
16	Engine compartment lamps
17	Regulator unit
18	Push siwtch for engine compartment lights
19	Starter motor
20	Cooling water temperature gauge sender
21	Pedal switch, windscreen-washer and wiper
22	Push switch for stop lights
23	Direction indicator flasher
24	Ventilation fan
25	Three-position ventilation fan selector switch
26	Screen-wiper motor
27	Single-pole socket for inspection lamp
28	Fuses
29	Ventilation fan extra resistor
30	Dashboard lamps with built-in switches
31	Push switches for courtesy lights (30)
32	Outside lighting switch
33	Direction indicator lever
34	Horn button
35	Selector switch for outer lighting and anti-dazzle flashes
36	Ignition, services, starting and anti-theft switch
37	Fuel gauge
38	Reserve warning lamp (red)
39	Fuel gauge light
40	Side lamps indicator (green)
41	Direction indicator repeater light (green)
42	Main beam repeater light (blue)
43	Speedometer lamp
44	Low oil pressure indicator light (red)
45	Oil gauge lamp
46	Oil pressure gauge
47	Warning light, generator not charging (red)
48	Engine-speed indicator lamp
49	Engine-speed indicator
50	Cooling water temperature indicator lamp
51	Cooling water temperature indicator
52	Instrument panel lights and side lamps repeater rheostat
53	Instrument panel lights switch
54	Screen-wiper motor on-off switch
55	Screen-wiper motor rheostat
56	Cigar lighter and lamp
57	Fuel gauge sender
58	Luggage compartment light push switch
59	Luggage compartment bulb
60	Rear direction lamps
61	Tail and stop lamps
62	Number plate lamps

NOTE - The symbol ── means that the cable carries a numbered ring or sleeve

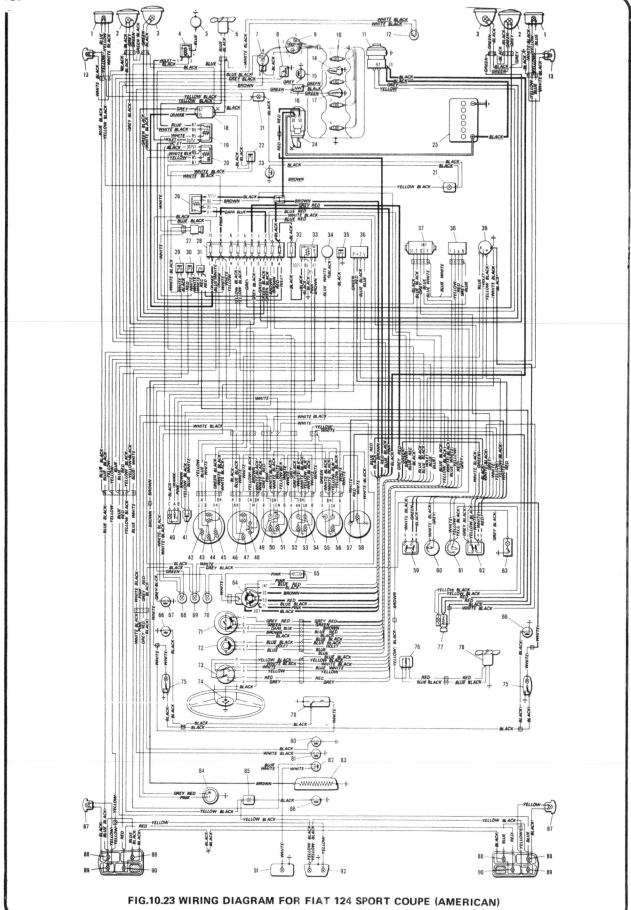

FIG.10.23 WIRING DIAGRAM FOR FIAT 124 SPORT COUPE (AMERICAN)

FIG.10.23 WIRING DIAGRAM KEY FOR FIAT 124 SPORT COUPE (AMERICAN)

1 Front parking lights and direction indicators
2 Outboard headlamps (high and low beams)
3 Inboard headlamp (high beams)
4 Electro-pneumatic horn relay switch
5 Compressor for electro-pneumatic horns
6 Engine cooling electro-magnetic fan motor
7 Ignition coil
8 Oil pressure gauge sending unit
9 Ignition distributor
10 Spark plugs
11 Alternator
12 Thermostatic switch for motor 6
13 Front side marker lamps
14 Engine water heat gauge thermostatic switch - moves gauge
   pointer to scale end (dangerous temperature) regardless
   of sending unit 17
15 Insufficient oil pressure indicator sending unit
16 Additional resistor for heat gauge
17 Water heat gauge sending unit
18 Voltage regulator
19 Relay switch for motor 6 control
20 Relay switch for indicator 47
21 Engine compartment lamps
22 Press switch, for electro-valve energizing during fast idle
   rate adjustment
23 Engine compartment lamp jam switch
24 Starter
25 Battery
26 Relay switch for headlamp high beams
27 Electro-valve, exhaust emission control device
28 Fuses
29 Switch, on clutch pedal, for exhaust emission control
   electro-pneumatic device
30 Pressure operated stop light switch
31 Inspection lamp socket
32 Fuse for demister 83 (optional equipment)
33 Relay switch for demister 83 (optional equipment)
34 Windshield washer electric pump
35 Switch, on the hydraulic circuit, for indicator 69
36 Flasher, vehicular hazard warning signal
37 Windshield wiper motor
38 Windshield wiper intermittent operation cycling switch
   unit
39 Direction indicator flasher
40 Switch, with indicator, for demister 83 (optional equipment)
41 Windshield washer motor button
42 Parking light indicator (green)
43 Speedometer light
44 Directional signal tell tale (green)
45 Headlamp high beam indicator (blue)
46 Engine tachometer light
47 Battery charge indicator (red)
48 Engine tachometer

49 Fuel gauge
50 Fuel reserve indicator
51 Fuel gauge light
52 Insufficient oil pressure indicator (red)
53 Oil pressure gauge light
54 Oil pressure gauge
55 Engine water heat gauge
56 Engine water heat gauge light
57 Electric clock
58 Electric clock light
59 Outer lighting switch
60 Windshield wiper sweep rate rheostat
61 Panel light rheostat switch
62 Vehicular hazard warning signal switch
63 Drop tray light w/incorporated switch
64 Lock switch
65 Separate 3 amp fuse for indicator 70 and its switch
66 Jam switches, between doors and pillars, for lights 75
67 Jam switch, on left door pillar, for indicator 70
68 Vehicular hazard warning signal indicator
69 Brake system effectiveness and brake on indicator (red)
70 Remove key indicator
71 Headlamp high/low beam change-over and flashes switch
72 Director indicator switch
73 Windshield wiper three-position switch
74 Horn button
75 Rear interior lights (courtesy) with incorporated switch
76 Electro-fan three-position switch
77 Electric cigarette lighter (with housing indicator)
78 Electro-fan motor, two-speed
79 Front interior light with incorporated switch
80 Switch, on handbrake lever, for indicator 69
81 Switch, on transmission (3rd and 4th gears) for exhaust emission
   control electro-pneumatic device
82 Back-up light pressure switch
83 Electric back window demister (optional equipment)
84 Fuel gauge sending unit
85 Luggage compartment lamp
86 Luggage compartment lamp jam switch
87 Rear side marker lamps
88 Rear direction indicators
89 Rear parking lights
90 Rear stop lights
91 Back-up light
92 Number plate lights

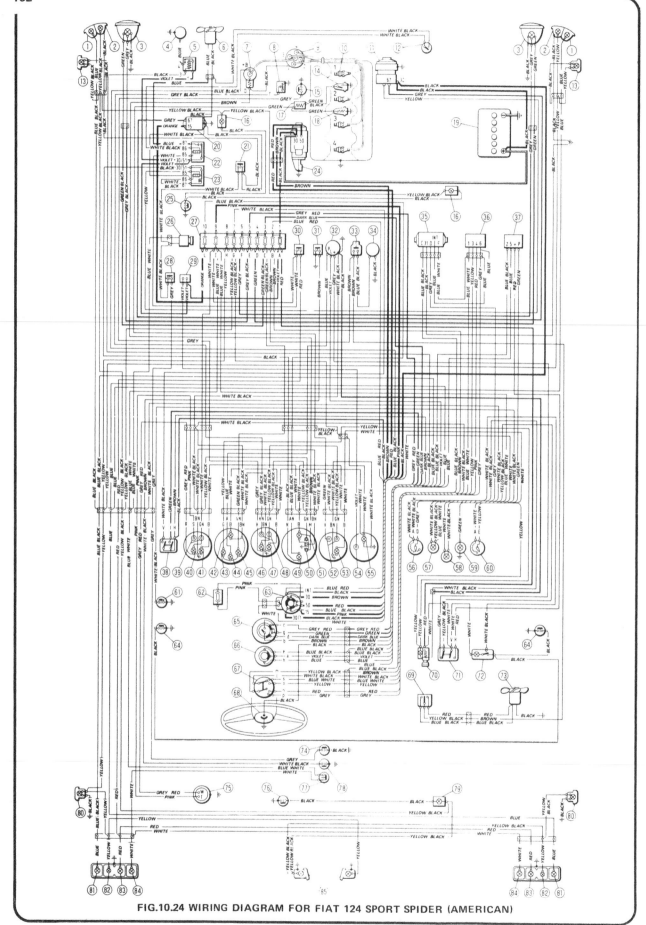

FIG.10.24 WIRING DIAGRAM FOR FIAT 124 SPORT SPIDER (AMERICAN)

FIG.10.24 WIRING DIAGRAM KEY FOR FIAT 124 SPORT SPIDER (AMERICAN)

1 Front direction indicators
2 Front parking lights
3 Headlamps (high and low beams)
4 Motorcompressor for electropneumatic horns
5 Horn control relay switch
6 Engine cooling fan motor
7 Ignition coil
8 Oil pressure gauge sending unit
9 Ignition distributor
10 Sprk plugs
11 Alternator
12 Engine fan motor thermostatic switch
13 Front side marker lamps
14 Heat gauge thermostatic switch: sends heat gauge pointer
   to scale end (excessive water temperature) independently
   of sending unit 18
15 Insufficient oil pressure indicator sending unit
16 Engine compartment lamps
17 Heat gauge additional resistor
18 Heat gauge sending unit
19 Battery
20 Voltage regulator
21 Press switch for electro-valve energizing during fast idle
   rate adjustments
22 Engine fan motor relay switch
23 Battery charge indicator relay
24 Starter motor
25 Engine compartment lamps jam switch
26 Electro-valve, exhaust emission control device
27 Fuses
28 Switch on clutch pedal for exhaust emission control
   electro-pneumatic device
29 Inspection lamp socket
30 Stop lights switch
31 Switch, on the hydraulic circuit, for indicator 57
32 Flasher, direction indicator
33 Windshield washer motor button switch
34 Electric pump windshield washer
35 Windshield wiper motor
36 Windshield wiper intermittent operation cycling switch
   unit
37 Flasher, indicator 57
38 Outer lighting 2-position switch
39 Fuel gauge
40 Fuel reserve indicator
41 Fuel gauge light
42 Speedometer light
43 Parking lights indicator (green)
44 Directional signal arrow tell-tale (green)
45 Headlamp high beam indicator (blue)
46 Insufficient oil pressure indicator (red)

47 Oil pressure gauge light
48 Oil pressure gauge
49 Battery charge indicator (red)
50 Engine tachometer
51 Engine tachometer light
52 Engine water heat gauge
53 Engine water heat gauge light
54 Clock
55 Clock light
56 Windshield wiper sweep rate rheostat
57 Brake system effectiveness and handbrake ON indicator
   (red)
58 Remove key indicator
59 Vehicular hazard warning signal indicator
60 Panel light rheostat switch
61 Jam switch, on left door pillar, for indicator 58
62 Separate 3 amp fuse for indicator 58 and its switch
63 Lock switch
64 Jam switches, between doors and pillars, for courtesy
   lights
65 Headlamp high/low beam change-over and low beam flashes
   switch
66 Direction indicators switch
67 Windshield wiper control three-position switch
68 Electropneumatic horn control button
69 Electro-fan 3-position switch
70 Electric cigarette lighter (with housing indicator)
71 Vehicular hazard warning signal switch
72 Courtesy light, with incorporated switch
73 Electro-fan motor, two-speed
74 Switch, on transmission (3rd and 4th gears) for exhaust emission
   control electro-pneumatic device
75 Fuel gauge sending unit
76 Luggage boot lamp jam switch
77 Back-up light switch
78 Switch, on handbrake lever, for indicator 57
79 Luggage boot lamp
80 Rear side marker lamps
81 Rear direction indicators
82 Rear parking lights
83 Rear stop lights
84 Back-up lights
85 Number plate lights

Chapter 11 Suspension and steering

Contents

Specifications

Front suspension:

Type	Independent, with hydraulic telescopic shock absorber coil springs and anti-roll bar
Anti-roll bar	Cross bar mounted in rubber blocks
Upper control arms:	
Attachment to bodyshell	Pin in rubber bushings
Attachment to kingpin	Ball joint
Lower control arms:	
Attachment to crossmember	Pin in rubber bushings
Attachment to kingpin	Ball joint
King pin inclination	6°
Caster (car loaded)*	3° 30' + 30' − 10'
*Coupe 3 persons plus 66 lb luggage	
*Spider 2 persons plus 44 lbs luggage	
Method of adjusting caster angle	Shims
Front wheel camber (car loaded)*	0° 30' ± 20'
camber adjustment	Shims
Toe in (car loaded)*	0.118" ± 0.039"
Toe in adjustment	Threaded sleeve on track rod
Shock absorber	Hydraulic, telescopic double acting
Coil springs - No. of turns	7½
Wire diameter	0.492 in 1400 Coupe
	0.512 in all other models
Free length of spring	13.54 in
Front wheel track	53.15 in

Rear suspension:

AC and AS bodyshells	Solid axle attached to bodyshell by two major trailing arms, one panhard rod (transverse), a tubular extension to the axle, coil springs and hydraulic telescopic shock absorbers.
BC, BS and current models	Solid axle attached to bodyshell by four trailing arms, one panhard rod (transverse), coil springs and telescopic shock absorbers

Coil springs:
Wire diameter	0.465 in
Number of turns	7
Free length of spring	16.33 in

Steering system:

Steering gear ratio	16.4 : 1
Type of steering gear	Worm and roller
Worm shaft bearings	Two ball bearing races
Roller shafts supports	2 roller bearings and one bronze bushing
Bearing adjustment	Upper rings and plates at lower ends
Worm and roller adjustment	With screw and plate on roller shaft
Roller shaft bushing I.D.	1.1298/1.1306 in
Roller shaft diameter	1.1295/1.1286 in

Worm shaft turning torque:
Complete with roller shaft	30º	0.643 to 1.214 ft lbs
	30º	0.500 ft lbs

Turning circle	- Coupes	36' 1"
	- Spiders	34' 1"

Lubrication for steering box	SAE 90EP oil

Torque wrench settings:

Front suspension

	ft lbs
Crossmember to chassis nuts and bolts	25
Upper control arm spindle to bodyshell nut	72
Lower control arm spindle to crossmember nut	44
Lower control arm spindle end nut	72
Upper and lower kingpin ball joint to control arm	15
Kingpin ball joint spigot nut	45
Brake caliper assembly to kingpin member nut	44
Steering knuckle arm to kingpin member nut	44
Brake caliper bracket to plate bolt	25
Top shock absorber nuts	11
Lower shock absorber nut	36
Anti-roll bar location on bodyshell and lower control arm nuts	11
Wheel to hub stud	51

Rear suspension

Trailing and panhard (transverse) rods to bodyshell and axle casing nuts	72
Shock absorber to bodyshell nut	11
Shock absorber to axle casing nut	11
Anti-roll bar to bodyshell nut	11
Anti-roll bar to tie rod nut	11
Tie rod to axle casing nut	25

Steering

Steering wheel to column nut	36
Steering column support bracket bolt	7¼
Steering box to bodyshell nut	29
Idler lever pivot box to bodyshell nut	29
Steering arm to roller shaft nut	173
Steering ball joint pin nuts	25
Upper and lower steering columns to intermediate shaft bolts	18

1 General description

In the independent front suspension system fitted, the front wheels, hubs and brakes are all mounted on a kingpin member. On the upper and lower ends of the kingpin member there are ball joints which connect it to the upper and lower suspension control arm. The upper arm pivots on rubber bushes on a spindle which is attached to the bodyshell. The lower arm is similarly attached to the bodyshell and to a box crossmember which stiffens the body structure locally around the lower control arm attachment.

The coil spring, telescopic shock absorbers and anti-roll bar, all act on the lower control arm.

The steering system comprises an articulated steering column, a worm and roller arrangement steering box and track and link rods to the steering knuckle arms mounted on the kingpin member. There is a relay lever and pivot/damper box mounted on the opposite side of the vehicle to the steering box to ensure symmetry of the track rod geometry.

Finally the rear suspension, this is quite a strightforward system comprising a solid axle located on the bodyshell by two major trailing arms, a panhard rod (transverse), coil springs and telescopic dampers. Methods of taking the wheel torque reaction have changed. On early models of the 124 Coupe or Spider the final drive housing had a tubular extension bolted to it. This extension fed the torque from the axle to a rubber pillow block situated near to mid-way between the gearbox and the rear axle. Later and current models have an additional two trailing arms which together with the major arms mentioned earlier accept the wheel torque from the axle.

2 Front suspension shock absorber - removal and refitting

1 The shock absorbers may be checked without removing them from the car. Quite simply if the ride in your car is rough compared with its contempories, or there is tendency for the car to pitch excessively when on an undulating road, the shock absorbers are worn and should be renewed.

2 Note that shock absorbers should only be renewed in pairs. The vehicle will not handle properly even to the extent of being dangerous, if shock absorbers at different states of wear and/or rate are used on the front suspension.

3 There is no need to remove any major suspension components before the shock absorbers may be extracted from the front suspension.

4 Raise the front of the car onto car ramps or chassis stands. Chock the rear wheels and ensure that the vehicle is safely

supported.

5 From inside the engine compartment, remove the nuts which secure the top of the shock absorbers to the bodyshell.

6 Retrieve the washers and rubber bushes from the shock absorber spindle.

7 Then from underneath the car, remove the nut and long bolt which locates the bottom of the shock absorber in a bracket in the lower control arm.

8 The shock absorber may then be lowered through the aperture in the lower control arm.

9 Refitting is the exact reversal of removal, but remember to tighten the mounting nuts and bolts to the specified torques.

3 Front suspension spring - removal and refitting

1 Like the shock absorbers, springs should only be renewed in pairs. It is worth noting however that coil suspension springs rarely give trouble and if the vehicle ground clearance is less than it should be, the shock absorbers will probably be suspect and not the spring. This may appear a little strange, but none the less it is true that if the shock absorber is worn or faulty, it will effect the steady height of the car on the suspension.

2 Section 2 of this Chapter details a few tests that can be carried out with the shock absorbers on the car to determine their condition.

3 Being satisfied that the springs do merit removal and closer inspection, begin by raising the front of the car onto chassis stands placed underneath the ends of the front suspension crossmember. Chock the rear wheels and ensure that the whole vehicle is safely supported.

4 Remove the shock absorbers as directed in Section 2.

5 The next task can only safely be completed with the aid of FIAT tool No A74174 which is used to compress the spring in situ so that it can be extracted from the suspension linkage.

6 Assemble the tool into the spring and compress the spring to the point when there is no force on the lower control arm of the suspension.

7 Remove the two nuts which secure the anti-roll bar locating bush and cap to the lower control arm. Lift the anti-roll bar and associated components from the lower suspension arm.

8 Next undo the nut which secures the lower kingpin ball joint pin in the kingpin member and with a soft faced hammer tap the pin out of the tapered hole in the kingpin. FIAT tool No A47042 is especially designed to press the ball joint pin from the kingpin. It is essential to support the wheel, brake and kingpin assembly adequately for the duration of the remaining tasks. Brake hoses and steering joints could be severly strained if the

assembly is not supported when the lower control arm is disconnected from the kingpin/wheel assembly.

9 Finally the two nuts which secure the lower suspension arm
pivot spindle to the crossmember may be removed to allow the
spring, compressing tool and lower arm to be lifted away from
the vehicle together.

10 Retrieve the spacer shims that fit on the bolts between the
lower arm pivot spindle and the crossmember. Store them so
that they are replaced in the exact positions from which they are
taken. The shims are the means by which the camber and castor
angle of the front wheels are adjusted.

11 Recover the spring seatings from both upper and lower
suspension arms.

12 The spring compressor is now relaxed and removed from the
spring.

13 Clean all the components for inspection.

14 Measure the free length of the spring and compare the result
with the figure in the specification at the beginning of this
Chapter. (13.5 in) If either is too short, renew both springs.

15 It should be noted that two rates of springs are fitted to the
suspension of the FIAT 124 Sport. They are identified by a
green or yellow strip on the coils. The yellow marked spring will
have a height greater than 8½ inches when loaded to 900 lbf and
the green marked spring will have a height less than 8½ inches
with the same load.

16 Naturally both front springs must be of the same rate class,
but also the rear springs must be matched as follows:-

17 The rear springs are marked as the front springs into two rate
classes. (Note the springs are quite different, only the rates are
classified.)

18 Yellow striped front springs may be fitted with either yellow
or green striped rear springs. Green striped front springs may
only be fitted with green striped rear springs, never with yellow
striped rear springs.

19 Inspect the rubber spring seatings and if cracked or worn
replace them.

20 Having satisfied yourself as to the condition (used or new) of
the springs, their rate class (yellow or green) and of their
compatibility with the rear springs, then the springs may be
refitted to the car.

21 Refitting the front springs is quite straightforward being the
reversal of the removal procedure. However ensure that correct
spacer shims are fitted on the appropriate bolts between the
lower arm pivot spindle and the bodyshell/crossmember. Make
certain that all securing nuts have been tightened to their
specified torques.

4 Front wheel hub - removal, dismantling, renovation and refitting

1 Place a jack beneath the lower control arm of the suspension,
as near to the road wheel as practicable and raise the wheel off
the ground.

2 Undo the four bolts which secure the wheel to the hub and
remove the wheel.

3 Undo the two bolts which secure the brake caliper block to
the kingpin member. Lift the caliper block away from the brake
disc and carefully support it nearby, tying it to the upper central
arm and or spring will be acceptable.

4 In any event do not strain the flexible brake hose which
connects the caliper to the braking system and wedge the brake
pads apart whilst the caliper assembly is separated from the hub
assembly.

5 Undo the two spigot bolts which retain the brake disc on the
hub and remove the disc.

6 Prise the hub cap off the end of the hub. FIAT supply a
special percussion puller for this cap, its number is A47014.

7 Undo the nut that holds the hub and bearings onto the stub
axle. Note that the stub axle is integral with the kingpin
member. Note also that the RH stub axle nut has a LH thread.

8 The hub may then be drawn off the stub axle with a sharp
tug, or if you are unlucky a hub puller will be necessary. Again

FIAT supply a purpose made pulley, tool number A47015.

9 The complete outer roller bearing, and inner bearing outer
ring and roller assembly will have come away with the hub. The
oil seal will be inplace too. Prise the oil seal out of the hub casing
with a screwdriver and lift the outer bearing inner race and both
outer and inner bearing roller assemblies from the hub.

10 The two outer rings of the bearings left in the hub may be
gently tapped out with a soft metal drift.

11 The inner race of the inner bearing which at this stage will be
still on the stub axle will require the special tool FIAT number
A47001 to remove it from the stub axle. A conventional
sprocket puller may perform the task, but there is only a small
shoulder on the inner ring on which the puller may purchase to
move the ring off the stub axle.

12 With all hub components dismantled they may be cleaned in
solvent such as 'gunk' ready for inspection.

13 Examine the bearing roller and ring surfaces fro pitting,
scoring and general deterioration. If wear or damage is found,
renew both inner and outer bearings complete.

14 Examine the oil seal, if damaged or not in a virtually new
condition renew the seal.

15 Once all the hub components have been inspected and
renewed or accepted for continued service, assembly of the hub
commences as follows:-

16 Tap the inner ring of the inner bearing onto its seating on the
stub axle with a tubular drift. Ensure that it is fitted the correct
way round. (Fig. 11.3).

17 Next insert the two outer rings of the bearings into their
appropriate seatings in the hub casing. Use a soft metal drift and
be careful to tap them evenly around to avoid the ring being
forced askew in the casing. Again refer to Fig. 11.3 to ensure that
the rings are being fitted the correct way round.

18 Coat the roller assemblies liberally with medium lithum base
grease such as Castrol LM Grease and place the roller assembly
on their respective inner rings.

19 Pack the inside of the hub casing with the same grease and
then fit the ring spacer and seal into position in the inside end of
the casing.

20 Mount the hub onto the stub axle which has the inner
bearing inner ring and roller assembly already in place at its root.

21 Slide the outer bearing inner race and roller assembly onto
the stub axle to locate in the outer end of the hub casing.

22 Slip the tag penny washer into position next to the outer
bearing and then screw the nut into position. Torque the nut to
15 ft lb to seat the bearings completely, loosen the nut and
retighten to 5 ft lb. Then slacken the nut 30° (one half flat) and
crimp the nut end onto the stub axle to lock in position.

23 Pack and refit the hub cap, refit the brake disc, then the
caliper block and finally the wheel. Make certain that the
securing bolts are tightened to their appropriate torques.

5 Front wheel hub bearing adjustment

1 The front hub bearings may be adjusted without removing
the road wheel or any other major item.

2 If the play on the front roller bearings is considered
excessive, proceed as follows:-

3 Place a jack under the front suspension and raise the road
wheel from the ground. Remove the wheel trim and then the cap
on the end of the hub.

4 Wipe the exposed stub axle nut free from grease and loosen it
completely. Ensure that the previous crimping does not interfere
with the nuts movement.

5 Tighten the nut back against the bearings to a torque of 15 ft
lbs. Loosen the nut and retighten to 5 ft lbs.

6 Mark the washer behind the nut midway along one of the
flats on the nuts and then slacken the nut by half a face (30°).
With either a cold chisel or FIAT tool No A74126 crimp the nut
end onto the stub axle to lock it into its new position.

7 Repack the hub cap with medium grease and refit the cap.

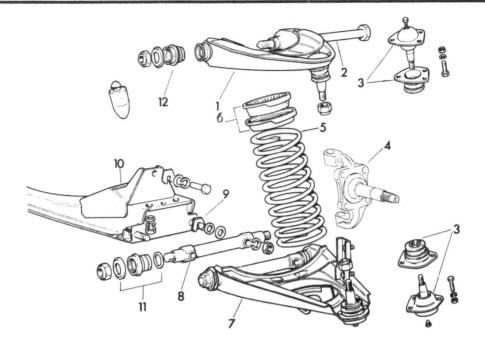

FIG. 11.1. EXPLODED VIEW OF FRONT SUSPENSION COMPONENTS

1 Upper suspension control arm
2 Upper arm pivot bolt
3 Kingpin ball joint replacement assembly
4 Kingpin/stub axle member
5 Helical coil spring
6 Spring seatings

7 Lower control arm assembly
8 Lower arm pivot spindle
9 Wheel alignment shims
10 Suspension cross member
11 Pivot spindle/lower arm bushing
12 Upper arm bushes

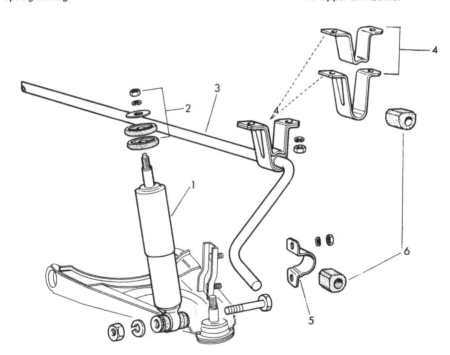

**FIG. 11.2. FRONT SUSPENSION SHOCK ABSORBER AND ANTI-ROLL BAR
COMPONENTS**

1 Shock absorber
2 Rubber bushes, penny washer and retaining nut
3 Anti-roll bar

4 Bodyshell attachment bracket
5 Attachment cap to lower control arm
6 Anti-roll bar bushes

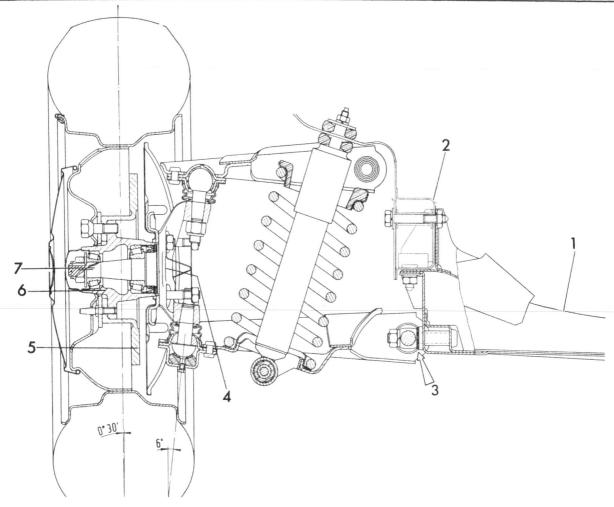

FIG. 11.3. SECTION OF L.H. FRONT WHEEL AND SUSPENSION

1 Cross member	3 Lower arm pivot spindle	4 Kingpin member	6 Wheel hub
2 Bodyshell	shims - wheel alignment	5 Brake disc	7 Stub axle

6 Front suspension kingpin ball joints - inspection, removal and refitting

1   The condition of the ball joints on the top and bottom of the kingpin member can be checked by placing a jack under the lower suspension arm, acting on the damper union and raising the road wheel off the ground.

2   Grasp the wheel and try to move it up and down, side to side and finally in and out. If relative movement of the kingpin and lower suspension member is observed, the ball joint must be worn and merits replacement.

3   The upper ball joint may be checked by grasping the upper suspension arm and kingpin member and forcing them in opposition. If relative movement across the ball joints is observed then it too is worn and requires replacement.

4   A point to be mentioned before the removal task is commenced, is that when assembled in the FIAT factory, the ball joint housing was rivetted to the appropriate suspension arm and therefore if it is the original joint to be replaced, the rivets must be drilled out, and the proper nuts, bolts and washers purchased with the new ball joint.

5   Proceed to remove the ball joints as follows:- raise the suspension and road wheel as directed in paragraph 2 of this section. Remove the road wheel.

6   Undo and remove the nut on the ball pin of the joint to be

removed. With either the turn buckle tool FIAT No A47042 or a universal ball pin extractor, separate the kingpin member from the ball joint pin.

7   Support the kingpin/wheel hub/brake assembly adequately to prevent the flexible brake hose and/or the steering linkage being strained.

8   Remove the three nuts and bolts or centre punch and drill out the three rivets which secure the ball joint housing onto the suspension arm.

9   Lift the old ball joint away together with the pin boot and its base. Discard the old ball joint and thoroughly inspect the old pin boot. It is false economy not to renew both ball joint and boot, because a new joint will be quickly ruined by water and dirt admitted by a crack in the old boot.

10 Place the new ball joint and boot in position under the suspension arm with the ball pin projecting in the appropriate direction. Bolt the joint and cover in position and tighten the nuts.

11 Insert the ball pin into its mating hole in the kingpin member. Slip the pin washer and a new nut into position and tighten the pin nut to the specified torque.

12 Refit the road wheel and check the wheel alignment. It is improbable that the replacement ball joint will significantly effect wheel alignment, but if it had been necessary to drill out old rivets and fit the replacement joint with nuts and bolts, it is a reasonable precaution to check the wheel alignment as directed in Section 24 of this Chapter.

7 Front suspension - upper control arm - removal and refitting

1 The upper suspension control arm is attached to the body-shell by a long bolt which passes through both bushed ends of the inner edge of the triangular suspension arm. The outer end of the arm holds the top kingpin member ball joint.

2 The ball joint, as already described in Section 6, is replaceable individually, and the rubber bushes which fit in the ends of the inner edge of the arm are also replaceable, though as with so many components on cars they should both be renewed even if only one bush is suspect.

3 Commence to remove the upper suspension arm as follows:-
Place a jack underneath the lower suspension arm and raise the road wheel off the ground.

4 Remove the road wheel and then undo and remove the nut which retains the pin of the upper kingpin ball joint in the kingpin member.

5 With a universal ball joint pin extractor, or the special FIAT tool No A47042 push the ball joint pin from the kingpin member.

6 Once the pin has been freed, support the kingpin/brake assembly to prevent it from moving outwards and straining the brake hose and steering linkage.

7 Next undo the self-locking nut on the long bolt which retains the arm to the bodyshell. Then extract the long bolt. The upper suspension arm is now free and can be taken to the work-bench for inspection and renovation.

8 Refitting the suspension arm follows the reversal of the removal procedure. Remember to use new self-locking nuts on the ball joint pin, and long bolt associated with the arm. Those nuts should also be tightened to the torques specified at the beginning of this Chapter.

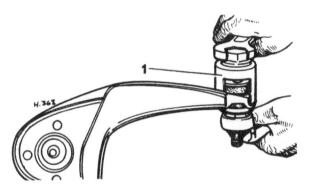

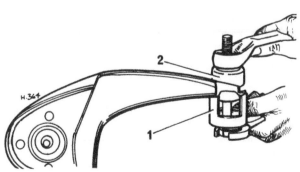

FIG. 11.4. FIAT TOOL A47046 (1)
being used to remove the bush in the upper control arm. Top illustration.

The same tool with a collar (2) being used to insert a new bush into the suspension arm.

8 Front suspension - upper control arm - renovation

1 The ball joint can be removed and renewed as described in Section 6 of this Chapter.

2 The rubber bushes in the ends of the inside edge of the triangular arm need to be pushed out of the arm with a drift or mandrel acting from the middle of the arm. As usual there is a special FIAT tool No. A47046, which is illustrated in Fig. 11.4 to remove and refit these bushes. The tool can easily be improvised with a piece of tube, some penny washers and a long bolt.

3 As mentioned in Section 7 both bushes in the arm should be renewed, even if only one is suspect.

4 While the suspension arm is on the work-bench it is wise to take the opportunity of thoroughly cleaning it, inspecting for rust or cracks and finally rust proofing and painting. Obviously this arm is a vital component on the car and its condition must not be allowed to deteriorate.

9 Front suspension - lower control arm - removal and refitting

1 The removal procedure for the lower suspension arm is exactly the same as the procedure for the removal of the front spring. The spring acts between seatings on the bodyshell and similar seatings mounted on the suspension arm.

2 Section 3 specifies the use of a spring compression tool, but it is possible to avoid the use of such a tool when only the suspension arm is to be removed.

3 At the stage in Section 3 when the spring compression tool would have been fitted, the spring coils should be tied together with quite a few loops of steel cable. This measure will hold the spring at its loaded length, enabling the removal and refitting of the suspension arm without resistance from the spring. It is worth adding that if the spring is tied together, eliminating the need for a compression tool, it will also be unnecessary to detach the upper end of the shock absorber from the bodyshell.

4 The refitting of the lower suspension arm follows the reversal of the removal procedure.

5 Remember to use new self-locking nuts on the lower ball joint pin and on the ends of the pivot spindle. Tighten all nuts to the torques specified at the beginning of this Chapter.

6 Finally the alignment of the front wheels should be checked, because renewed bushes or ball joint may have altered the correct position sufficiently to merit different shims between the pivot spindle and bodyshell.

10 Front suspension - lower central arm - renovation

1 The removal and refitting of the ball joint has been described in Section 6.

2 Again the bushes in the ends of the inside edge of the triangular arm require tooling for their extraction and insertion. On this occasion the spindle itself is used as part of the tooling.

3 FIAT tool No A47045 is tubular and supports the bush housing whilst the spindle is employed to push the bush from the arm. The tool may easily be improvised from a piece of steel tube, the bore of which would allow the bush to slip through. Fig. 11.5 illustrates the use of tool No. A47045.

4 When it comes to refitting bushes into the suspension arm FIAT provide a support tool No A47177/2 and a collar No A47177/1 to ensure that the bushes and arm are not damaged during the refitting Fig. 11.6 illustrates the use of these tools.

5 Fortunately again these tools may be improvised from suitable steel tubing and penny washers.

6 The opportunity should be taken to thoroughly clean the lower suspension arm when it is on the bench and to inspect it for rust and cracks. The arm can then be rust proofed and painted. If cracks or joint separation is found the arm should be discarded and a new one refitted. The suspension arm is a vital component on the car and it must be kept in virtually as new condition.

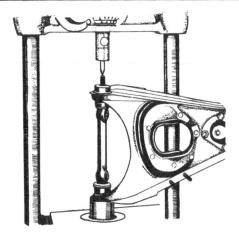

FIG. 11.5. FIAT TOOL A47045 (3)

being used together with the lower arm pivot spindle (1) to extract an old bush (2) from the arm.

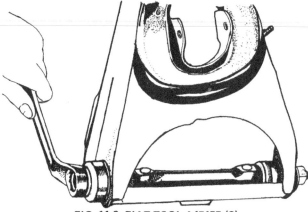

FIG. 11.6. FIAT TOOL A47177 (2)

(a collar (3)) being used with A47177/2 (a spacer (2) and the pivot spindle (1) to insert a new bush into the suspension arm.

11 Front suspension - crossmember - removal and refitting

1   This is a most complex task and fortunately it is very rarely necessary since there is little to go wrong. It is a box beam which supports the lower suspension arms at its ends and the engine at mountings near the ends. The beam is secured to a strengthened area of bodyshell between the wheel arch and engine compartment.

2   Raise the front of the car onto chassis stands located just forward or rear of the crossmember location on the bodyshell. Remove the road wheels.

3   Continue by removing both lower suspension arms as described in Sections 3 and 9 of this Chapter.

4   The engine should now be supported with a jack acting on the sump, or a sling around the front of engine connected to a hoist, so that the rubber engine suspension blocks may be detached from the cross beam.

5   FIAT tool No A70526 is a beam which spans the top of the engine compartment and is used to support the engine whilst the crossmember is detached.

6   The two nuts and single bolt which attach each end of the cross beam to the bodyshell can now be removed and the beam itself lifted clear. Retrieve the spacers and washers on the bolt and two nuts and store them so that they may be replaced in the exact position from which they were removed.

7   Refitting is the exact reversal of removal, but remember to use new spring locking washers on the crossmember attachments and new self-locking nuts the lower kingpin ball joint pins.

8   Torque all nuts and bolts to their specified torques at the beginning of this Chapter.

12 Front suspension - anti-roll bar - removal and refitting

1   This task is quite straightforward, the bar is located at four points. Its ends are attached to the lower suspension arms and two caps and split bushes hold the centre length of the bar to strengthened areas of the bodyshell beneath the forward end of the engine compartment.

2   Begin removal of the bar by raising the front of the car onto car ramps, or chassis stands placed beneath the ends of the suspension crossmember.

3   Continue by removing the nuts which secure the caps and bushes at the ends of the bar to brackets on the lower suspension arm.

4   Finally remove the nuts which secure the main bar support brackets to the bodyshell and lift away the bar together with all the bushes, caps and centre brackets.

5   The centre brackets come in two parts which are available individually as spares. The bar end bush caps are also available individually.

6   The bushes can be slid off the bar and renewed if necessary.

7   Refitting is the exact reversal of removal, but remember to use new spring washers behind the retaining nuts and tighten all nuts to the torques specified at the beginning of this Chapter.

13 Rear suspension - shock absorber removal and refitting

1   The rear suspension employs two helical coil springs with telescopic shock absorbers mounted coaxially. It will be necessary to remove both spring and shock absorber whenever either component is suspect and being replaced. The procedure below therefore, includes spring removal and refitting.

2   The condition of the shock absorbers is indicated by the behaviour of the vehicle on the road. If there is an excessive pitching motion when motoring on an undulating road and/or and excessively bumpy ride compared with its contempories, the shock absorbers merit renewal. Another indication of shock absorber malfunction is when the vehicle is pressed and bounced at each corner, and it does not immediately return to its usual rest position. If either the car is slow to return, or over returns and oscillates before recovering, the shock absorbers merit replacement.

3   Raise the rear of the car onto chassis stands located beneath suspension anchorages. Chock the front wheels and ensure that the vehicle is safely supported.

4   Remove the road wheels, release the handbrake and disconnect the handbrake cables from the caliper assemblies. Then unclip the cable and sleeve from the major trailing arm location.

5   Undo the nuts and bolts which secure the brake regulator mechanism links to the rear axle casing. Lift the links clear.

6   Next remove the caps from the brake fluid reservoirs in the engine compartment. Stretch a sheet of thin polythene over the top of the reservoirs and replace the caps. This measure will prevent excessive loss of fluid from the brake system when pipes and hoses are subsequently disconnected.

7   Disconnect the flexible brake hose which runs from the fixed metal brake pipe on the rear axle casing. Tape over the ends to prevent the ingress of dirt.

8   Place two jacks underneath the axle casing each side of the final drive, preferably acting on the major trailing arm anchorages to the casing.

9   Raise the axle slightly to relieve the shock absorbers which were at their most extended position.

10 Front within the luggage compartment, remove the nuts that hold the top of the shock absorbers in place and the large washer and rubber spacer.

11 The jacks supporting the axle can now be used to lower the axle past its usual limit so that the top shock absorber locating stud pulls out of the bodyshell. The axle will need to be lowered several inches before the spring is completely unloaded.

12 The two nuts which secure the shock absorber locating bracket to the rear axle casing can now be undone and both

spring and shock absorber lifted away. Retrieve the spring seatings from the axle casing and bodyshell.

13 Finally the single bolt and nut which hold the shock absorber in its lower location bracket are removed.

14 The shock absorber is supplied without the top rubber spacers and it is as well to renew these whenever the shock absorbers are renewed.

15 As with the front wheel shock absorbers the rear shock absorbers should be renewed as a pair. The handling of the car will be seriously, even dangerously effected if the rear shock absorbers are at a different state of wear (and therefore a different rate).

16 The refitting procedure for the rear shock absorbers (and springs) is simply the reversal of the removal procedure. Remem-

ber to position a rubber spacer on the top of the shock absorber before it is assembled onto the axle and bodyshell. It should also be appreciated that when the springs and shock absorber are initially mounted on the axle casing, the jacks will have to raise the axle and compress the springs by several inches before the top end of the shock absorber manages to project through the bodyshell into the forward end of the luggage compartment.

17 Use new spring locking washers when refitting spring and damper components to the car. Tighten all nuts and bolts to the torques specified at the beginning of this Chapter.

18 Once the brake pipes and hoses have been reconnected the system must be bled to thoroughly clear it of air bubbles.

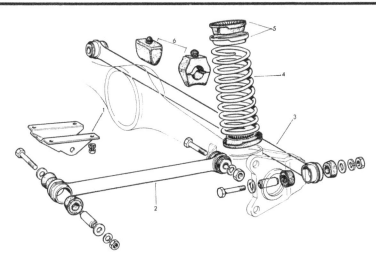

FIG. 11.7. EXPLODED VIEW OF REAR SUSPENSION COMPONENTS

1 Major trailing arm anchorage on bodyshell
2 Major trailing arm assembly
3 Transverse, panhard rod assembly
4 Spring
5 Spring seatings
6 Axle travel buffers

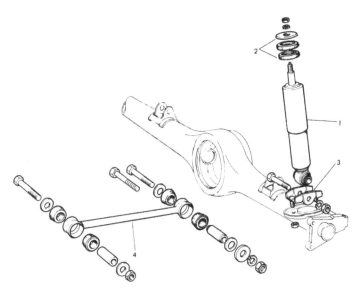

FIG. 11.8. REAR SUSPENSION SHOCK ABSORBER LOCATION AND MINOR TRAILING ARM

1 Shock absorber
2 Rubber seatings for top of damper
3 Damper/axle casing locating bracket
4 Minor trailing arm assembly. Not fitted to early pattern of rear axle which incorporated a tubular extension covering the propeller shaft.

14 Rear suspension - spring removal and refitting

1 The spring removal and refitting procedures are exactly as those for the shock absorbers, details in Section 13.
2 Once the springs are clear of the car, they can be brushed clean and inspected carefully for rust and cracks. The free length of the springs should also be checked, details are included in the specification at the beginning of this Chapter. The springs should be renewed if their condition is less than perfect, they are stressed and vital parts on the car and failure could be catastrophic.
3 Attention should be paid to the colour markings on the spring. There will be either a yellow or green stripe on the spring coils; these indicate the class in which the rate of the spring is in.
4 Both rear springs should be the same class (colour) and the rules set down in Section 3 of this Chapter should be observed when the springs are being renewed and the matching of the front and rear springs needs consideration.
5 The class rates are given as follows:- if when the spring is subjected to a load of 630 lbf the height of the spring exceeds 11¾ inches, it is a class A, yellow stripe spring. If on the other hand its height is less than 11¾ inches the spring is class B and a green stripe will be found.

15 Rear suspension - axle location arms - removal, inspection, renovation and refitting

1 There are several groups of location arms acting on the rear axle. To begin with there are the main pair of rods attached to anchorages bolted onto the bodyshell and to hardpoints welded to the axle casing. Then there is the transverse stabilising rod, known as a 'panhard' road and finally on the latest pattern of axle (without a forward extension tube covering the propeller shaft) there are a pair of rods above the axle which accept the torque from the road wheels.
2 All rods have a similar design, being basically a thick walled tube with bush housings welded to each end. Rubber bushes press into each housing around a tube spacer.
3 Never attempt to remove any more than one rod at any time, so if the task is to renew all the rubber bushes in the ends of the locating arms remove one, renew the bushes, refit it and then proceed to the next arm.
4 The procedure for their removal is as follows:
   Raise the rear of the car onto chassis stands located beneath hardpoint on the bodyshell which will not result in interference with later tasks. Do not put the chassis stands underneath the axle, it is necessary for the suspension linkage to be relaxed.
5 Place a pair of jacks beneath the rear axle casing and raise to take the weight of the axle assembly. The suspension rods should now be relaxed and ready for removal.
6 Undo the nuts on the long bolts which pass through each end of the location rods, extract the long bolts and pull the rod out of the anchorage.
7 The rod may now be brushed clean and inspected. It should be perfectly straight and the ends round. Check the welded ends carefully for cracks and corrosion. As with springs the rods must be in perfect condition, any fault or surface deterioration could cause catastrophic failure. Replace suspect or bent rods immediately.
8 With the rod on a bench, press the centre tube spacer inside the rubber brushes out with a suitable drift and prise the old bushes out with a screwdriver. New bushes may be pressed into position followed by the central spacer tube.
9 The refitting procedure is simply the reversal of that for removal. Always use new self locking nuts or spring lock washers as appropriate and tighten these nuts and bolts to the torques specified at the beginning of this chapter.

16 Rear suspension - anti roll bar - removal and refitting

1 On some models a rear anti roll bar is fitted. It was fitted on all models fitted with the early pattern of rear axle with a tubular extension covering the propeller shaft. Commence removal by raising the rear of the car onto car ramps.
2 The removal of the bar is then straightforward. It is retained to the bodyshell by two caps around two rubber bushings. Remove the two nuts per cap, remove the caps and lift away the the bar and bush from the bodyshell.
3 The bar is located on the axle by two forks and links. Undo the nuts on the ends of the bolts through the link joints, extract the bolts and lift away the bar. The links may be removed from the axle after the bolts through their anchorages have been extracted.
4 Inspect the bar closely for cracks and surface deterioration. Renew all rubber bushes as their condition merits.
5 Refitting the bar and linkage is simply the reversal of removal. Remember to use new spring lock washers and self locking nuts as appropriate.

17 Steering - steering wheel removal and refitting

1 The steering wheel is retained on the top end of the column by a single hexagon nut.
2 As usual it is necessary to clear the centre of the wheel from trim and electrical equipment so that a spanner can reach that single nut and undo it.
3 Proceed as follows: Begin by disconnecting the battery terminals and stow them safely by, so that there is not any chance of sparks being generated when other electrical leads are subsequently disconnected.
4 Then press and turn the motif/horn button assembly in the centre of the wheel. The turn should release the bayonette location of this assembly. Retrieve the button return spring.
5 The single nut which retains the wheel is now exposed and may be undone with a suitable socket spanner or FIAT tool number A57005.
6 Pull the wheel off the end of the column.
7 Refitting is straightforward, being the reversal of the removal procedure.

18 Steering - steering column assembly - removal, renovation and refitting

1 FIAT manufacture three patterns of steering column, two are articulated and the third is a straight bar running from the steering wheel directly to the steering box. The third pattern described, only fitted to early Coupes, does not have a collapsing capability and therefore is only distributed to a small number of countries. The two articulated columns are slightly different and one is fitted to all Coupes and the other pattern is fitted to all Spiders.
2 The upper section of the steering column is held in a mounting retained to the underside of the dashboard by four bolts.
   The lower end of the column fits onto the splined input shaft of the steering box.
3 Begin removal of the steering column by removing the steering wheel as directed in Section 7 of this Chapter.
4 Undo the screws retaining the half covers on the direction indicator/light/windscreen wiper switch complex on the column. Then remove the covers.
5 The collar which holds the switch complex to the column support bracket can now be slackened and the plug connectors on the electrical leads from the switches disconnected. The switch complex may now be slid off the mounting bracket and steering column.
6 Undo the threaded ring which retains the ignition switch in its mounting on the steering column support bracket.
7 Working in the bottom of the engine compartment and/or in the drivers foot well, undo the nuts on the end of the bolts which secure the columns universal joints yokes to the top column and steering box column.
8 Finally the four bolts which retain the steering column

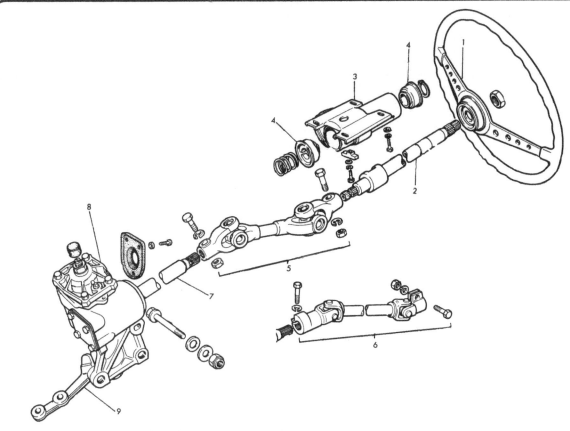

FIG. 11.9. AN EXPLODED VIEW OF THE STEERING COLUMN COMPONENTS

1 Steering wheel
2 Upper steering column
3 Upper column support bracket
4 Upper column bearings
5 Centre section of steering column SPIDERS

6 Centre section of steering column COUPES - later models
7 Lower column/steering box input shaft
8 Steering box
9 Pitman arm

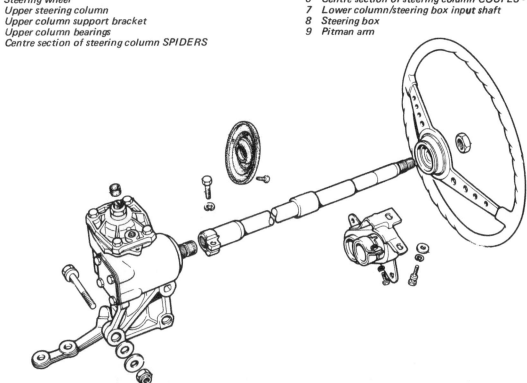

FIG. 11.10. STEERING COMPONENTS FITTED TO THE COUPE EARLY MODELS BEFORE BODYSHELL NO. 164602

The removal, dismantling, and reassembly of this design follows the same lines as described in detail for the later articulated steering column

support bracket to the underside of the dashboard can be removed and the column assembly lifted from the car.

9 REFITTING the column is straightforward and follows the reversal of the removal procedure. As usual always use new spring lock washers when reassembling the column assembly and tighten the nuts and bolts to their correct torques as specified at the beginning of this Chapter.

RENOVATION:

10 **Upper column support bearings:** These bearings are a push fit in the column support bracket. Begin their removal by separating the centre column section from the top column. The bolt which secures the universal joint yoke should have already been removed. Remove the circlip holding the inner ring of the upper most bearing in place on the column. Then, with a screwdriver prise out the bearing outer rings and the rest of the bearing, from each end of the support bracket. The steering column can now be extracted from the support bracket and all the components cleaned and inspected. Renew any component which is worn or damaged. The reassembly of the top column and its bearings into the support bracket follows the reverse procedure to removal. Note: The bearing retaining spring, next to the lower bearing in the bracket, must be in good order. Its function is to take up the play between the balls and races of both bearings on the upper steering column.

11 **Centre section of steering column - universal joints**

The centre section comprises a single rod with universal joints at each end which connect onto the top steering column and the steering box input shaft. The universal joint spider and bearing assemblies are quite conventional and come apart in much the same manner as their more robust types used on the propeller shaft in the vehicles transmission.

The universal joint spider bearing cups are either a push fit into the yokes, or are retained by star springs. The steering column assembly fitted to Sport Spider cars uses star springs to retain the bearing cups on the U.J. Spider.

The joints are dismantled by removing the bearing cup retainers as necessary, then tapping each yoke in turn to push the bearing cups out of the mating yoke. Once the cups protrude

from the yoke, they may be extracted with a pair of pliers. Once all the cups have been removed the spider may be removed from the joint yokes.

As with the transmission type joints a small drift may be used to push the bearing cups out of the yokes directly from the centre of the joint if the technique described earlier is unsuccessful. Now that the centre column components have been dismantled, they should be cleaned and thoroughly inspected. If worn splines are found, or cracks and other deterioration, renew the appropriate parts immediately. Remember to check the mating splined members too, the splines in a new yoke will soon be ruined if they are mated with worn splines on the upper column or steering box input shaft.

The universal joints reassembly follows the reverse procedure to removal. Remember to use bearing seals and retaining clips, and ensure that the bearing cups are pressed into the same position in each yoke, so that the spider is correctly aligned relative to the yokes.

19 Steering box - removal and refitting

1 Begin by raising the front of the car onto car ramps or chassis stands located beneath the ends of the suspension cross member.
2 Undo the nuts which retain the steering linkage ball joints pins into the pitman arm on the steering box. The pins may be extracted from the pitman arm with either a universal pin extractor or a sprocket puller used appropriately.
3 From inside the car, remove the nut and bolt which secure the lower yoke on the steering column to the steering box input shaft.
4 Finally undo the nuts and bolts which retain the steering box to the bodyshell and lift the box away from the car, taking it to a clean bench for dismantling and repair.
5 Refitting the steering box follows the reversal of the removal procedure. Use new spring lock washers and self locking nuts as appropriate and tighten all nuts and bolts to the torques specified at the beginning of this Chapter.

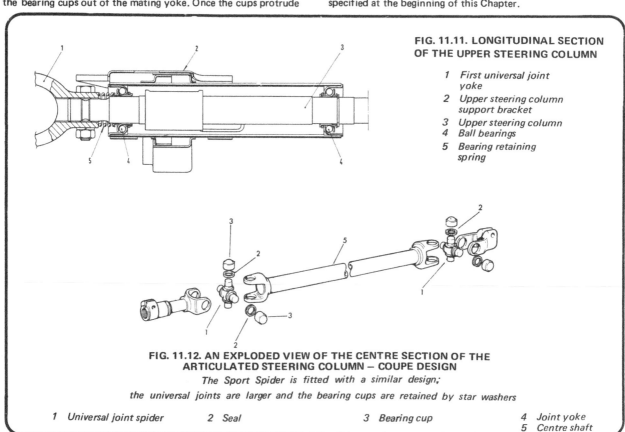

FIG. 11.11. LONGITUDINAL SECTION OF THE UPPER STEERING COLUMN

1 First universal joint yoke
2 Upper steering column support bracket
3 Upper steering column
4 Ball bearings
5 Bearing retaining spring

FIG. 11.12. AN EXPLODED VIEW OF THE CENTRE SECTION OF THE ARTICULATED STEERING COLUMN – COUPE DESIGN

The Sport Spider is fitted with a similar design;
the universal joints are larger and the bearing cups are retained by star washers

1 Universal joint spider   2 Seal   3 Bearing cup   4 Joint yoke
                                                              5 Centre shaft

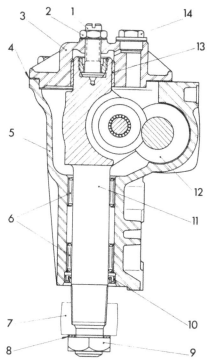

FIG. 11.13. SECTION OF STEERING BOX THROUGH OUTPUT ROLLER SHAFT

1   Gear mesh adjusting screw
2   Lock nut
3   Steering box major cover
4   Gasket
5   Steering box casing
6   Needle roller bearings
7   Pitman arm
8   Locking tab washer
9   Nut
10 Oil seal
11 Roller shaft
12 Input shaft worm gear
13 Bushing
14 Oil filter plug

20 Steering box - dismantling, examination and reassembly

1   Wash down the exterior of the box and wipe dry with a non-fluffy rag.
2   Remove the oil drain/filler plug and allow the oil to drain out.
3   Mount the box firmly in a vice and unscrew the nut which holds the pitman arm on the box output shaft.
4   Using a universal puller draw the pitman arm from the spline on the output shaft. Retrieve the tag washer. There is a master spline on these two parts so that they may only be fitted together in one way.
5   Undo and remove the four bolts which secure the major cover onto the box casing. Lift away the output shaft, cover and roller sector, from the steering box. Retrieve the gasket fitted between the major cover and casing.
6   Duly remove the needle roller bearings in which the box output shaft ran, if absolutely necessary, they may be pushed out of the bore in the steering box casing using a soft drift. The oil seal in the base of the bore should be prised out with a screwdriver.
7   The main cover can be separated from the output shaft once the locknut on the gear meshing adjusting screw in that cover has been undone and adjusting screw turned sufficiently to work right through and disengage the cover.
8   The screw is retained in the output shaft by a threaded insert. It will not normally be necessary to remove this insert to extract the gear mesh adjusting screw.
9   If play is found between the roller gear on the output and the spindle which retains it to the shaft, it will be necessary to drift the spindle out to free the roller gear. The needle roller bearings in the roller, and the surface of the spindle can then be inspected for wear.
10 The steering box input shaft and worm gear can be extracted from the box once the four bolts which retain the minor cover in

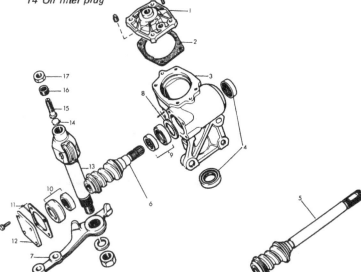

FIG. 11.14. AN EXPLODED VIEW OF THE STEERING BOX COMPONENTS

1   Major cover
2   Gasket
3   Box casing
4   Oil seals
5   Pattern of input shaft fitted to cars with the articulated steering column
6   Early pattern input shaft used with solid continuous steering column - COUPES ONLY
7   Pitman arm
8   Worm gear positioning shims (0.004 and 0.006 inch)
9   Spacer and bearing - top end
10 Spacer and bearing - bottom end
11 Shims governing running torque of input shaft (0.004, 0.005, 0.008 inch)
12 Minor cover
13 Output shaft and roller assembly
14 Pad
15 Gear mesh adjusting screw
16 Insert which retains screw (15) in output shaft
17 Lock nut

place have been removed. Take care to collect the shims which fit between the cover and the box casing, they are the means by which the end play on the input shaft bearings is adjusted.

11 Once the cover is off the box, the input shaft can be pushed through and out of the box, together with a spacer and bearing.

12 Prise the input shaft oil seal out of its seating in the box casing with a screwdriver, and then tap the outer ring of the input shaft bearing out of its seating in the box. Retrieve the spacer and worm gear positioning shims.

13 All the steering box components are now completely dismantled and should be washed in a solvent such as 'Gunk' in preparation for inspection.

14 The splined ends of the input and output shaft should be checked for wear. Both input and output shaft have surfaces which serve as the inner ring for their respective bearing assemblies. The output shaft shank surface should be closely inspected for scoring, pitting and other surface deterioration. The diameter of the shank should be between 1.1295 and 1.1286 inches in the bearing region. Renew the shaft and bearings if found to be worn.

15 Inspect the ball grooves on each side of the worm gear on the input shaft for signs of pitting and wear. The gear surfaces should be inspected similarly.

16 Finally the bush in the major cover which accepts the top end of the output shaft should be examined. It internal diameter should be between 1.1298 and 1.1306 inches: if worn, prise it out of the cover and replace with a new bush. Remember to check the diameter of the top end of the output shaft which runs in the bush; it should be between 1.1259 and 1.1286 inches.

17 Steering box reassembly basically the procedure is the reversal of dismantling. The following paragraphs indicate how the two major shafts on the steering box are set correctly in their bearings.

18 Beginning with the steering box input shaft; always use a new oil seal in the box seating and retain the bearing balls in position on their respective races with medium grease whilst the shaft is inserted into the box casing. Once the shaft and associated components have been assembled, the shims fitting between the minor cover and casing which adjust the running of the shaft bearings can be selected. Reduce the number of shims to increase running torque or increase to reduce the torque which should be between 2 and 6 lb in.

19 Having achieved the correct running torque for the input shaft, and completed the assembly of the input shaft components; the output shaft may be refitted into the steering box.

20 Lightly tap the needle race assemblies into position in the bore in the steering box, and fit a new oil seal in the seating in the base of the bore. Insert the output shaft complete with the roller gear fitted.

21 Then lower the major cover and its gasket over the box and turn the gear mesh adjusting screw to draw the cover onto the output shaft and steering box casing.

22 Lift the cover and output shaft temporarily to disengage the roller from the worm. Turn the worm to bring a profile so that when the roller is lowered into mesh, it lies exactly midway along the worm gear.

23 Screw the mesh adjusting screw so that the major cover can seat fully on the box casing. Bolt the cover into position and then without turning either shaft, push the pitman arm onto the splined end of the output shaft so that it lies parallel with the mounting base of the box. Fit a new spring lock washer and screw the retaining nut into position.

24 Turn the mesh adjusting screw to eliminate any backlash of the roller and worm gear, and then turn the input shaft firstly to check that the output shaft components and pitman arm have been correctly assembled into the steering box and secondly to check that the roller moves evenly along the worm gear.

25 The range of pitman arm movement required is shown in Fig. 11.16. If it is found that as the input shaft is turned and the roller moves along the worm gear, that the roller becomes progressively tight towards one end and slack towards the other:

The worm gear/input shaft is not centred properly. This can be corrected by changing the thickness of the shims behind the upper input shaft bearing outer ring. Remember that a corresponding change must be made to the shims behind the minor cover which govern the running torque of the input shaft.

26 Finally having satisfied yourself that the steering box components are properly adjusted and set; fit the locking nut onto the mesh adjusting screw, tighten and retain in position by bending a tab on the washer onto one of the nut faces. Tighten the pitman arm retaining nut to the specified torque, and check the tightness of the major and minor cover bolts. Refill the steering box with an SAE 90EP oil. The box is now ready to be refitted to the car.

21 Steering box gears - adjustment

1 There is no real advantage in trying to tackle this task with the box in place on the car. The steering rods need to be detached from the pitman arm on the steering box, and the steering column should also be detached from the box when the running of the gears is to be adjusted, therefore, it will be easier and the job will be completed more accurately if the box is removed as described in Section 19.

2 The running of the steering box gears can now be adjusted as described in paragraphs 18,24 and 25 of Section 20 in the Chapter.

3 If there is still some slackness, sloppiness or alternatively stiffness in the steering box gear system; then the box should be dismantled and inspected as detailed in Section 20 of this Chapter.

22 Idler arm, pivot - removal and refitting

1 The idler arm pivot incorporates a damping device which accepts shocks and road vibration coming from the wheels and thereby serves to protect the steering system. From the drivers point of view, the steering will feel light and smooth with very little road vibration being fed back to the steering wheel.

2 The idler arm and pivot are mounted in the opposite position to the steering box on the body shell. The idler arm and pitman arm on the steering box are linked by a fixed length steering rod.

Adjustable length track rods connect the idler arm and pitman arm to the steering knuckles on the kingpin members.

3 Removal of the idler arm and pivot is quite straightforward; begin by raising the front of the car onto a pair of car ramps or chassis stands. Chock the rear wheels and check that the car is safely supported; you will be working underneath it!

4 Continue by detaching the track rod and link rod from the idler arm; using a universal ball pin extractor to press the ball joint pins from the idler arm once the retaining nuts have been undone.

5 The idler arm and pivot may then be unbolted from the bodyshell.

6 Do not bother to dismantle the idler arm/pivot assembly, FIAT do not supply individual parts and its design and function are such that repair should not be attempted.

7 If on inspection the pivot is sloppy, and obviously worn, and if when the car was in use, the damper in the pivot assembly felt ineffective: the whole pivot and idler arm assembly will have to be renewed.

8 Refitting the idler arm/pivot assembly follows the reversal of the removal procedure. Remember to use new self locking nuts on the balljoint pins and bolts which secure the pivot housing to the bodyshell. Tighten nuts and bolts to the torques specified at the beginning of this Chapter.

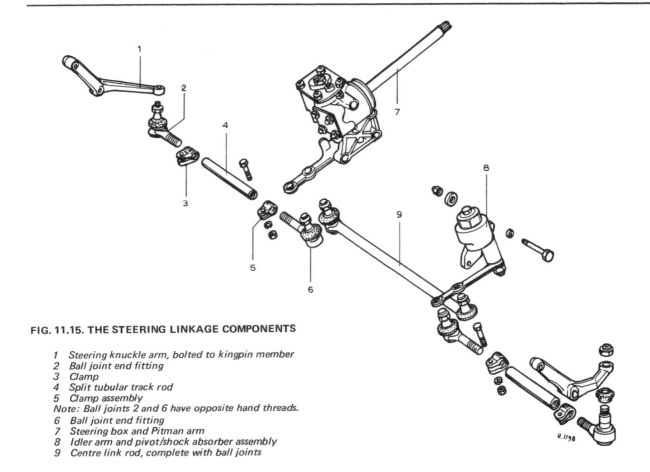

FIG. 11.15. THE STEERING LINKAGE COMPONENTS

1   *Steering knuckle arm, bolted to kingpin member*
2   *Ball joint end fitting*
3   *Clamp*
4   *Split tubular track rod*
5   *Clamp assembly*
Note: Ball joints 2 and 6 have opposite hand threads.
6   *Ball joint end fitting*
7   *Steering box and Pitman arm*
8   *Idler arm and pivot/shock absorber assembly*
9   *Centre link rod, complete with ball joints*

23 Steering rods and balljoints - general

1   The pitman arm is connected to the idler arm by a fixed length link rod, the two track rods are connected to the pitman and idler arms on the one end and on the other end to the steering knuckle arms.

2   To allow for front wheel 'toe-in' adjustment each end of the track rods has an internal thread into which the ball joint member is screwed. A clamp around each end tightens the split track rod onto the ball joint member and locks the two together.

3   Regular inspections should be made of the ball joints on each end of the track rods and link rod. If the joints are found to be sloopy and worn they must be renewed.

4   Removal of the ball joint assemblies from the track rods is straight forward: with the rod assembly on a bench, measure the distance between the centres of the ball joints at each end of the rod. Slacken the end clamps and unscrew the joint from the rod, counting the number of turns taken. Screw in a new joint into the rod to the same position as the old one had occupied. Check the ball joint centre distance and tighten the rod clamps.

5   Refit the track rods to the vehicle and check the front wheel toe-in as detailed in Section 24.

6   The ball joints on the ends of the centre link rod cannot be removed since they are integral with the rod. If one or both joints are worn it will be necessary to purchase a new link rod asssembly.

7   As usual new self locking nuts should be used when refitting rod assemblies back onto the car, and all nuts should be tightened to the torques specified at the beginning of this Chapter.

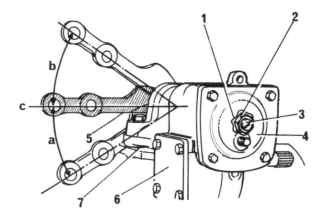

FIG. 11.16. VIEW OF STEERING BOX SHOWING:-

(i)   Pitman arm central position with range of movement
(ii)   Gear mesh adjusting screw and locknut
Angle 'a' =   angle 'b' =   30° 40' ± 1° 40'
'c' centre position   -   straight ahead

1   *Lock washer*
2   *Lock nut*
3   *Gear mesh adjusting screw*
4   *Major cover*
5   *Pitman arm*
6   *Minor cover*
7   *Steering box mounted on horizontal surface for pitman arm travel, and gear meshing checks to be made*

24 Front wheel alignment - adjustment and symptoms

1   Provided that any repair work on the steering and suspension system involved only the renewal of joints and/or bushes and not disturbance of the lengths of the track rods; then you should be able to drive carefully to your nearest FIAT dealer, where the wheels may be aligned with the specialised equipment which is essential for this task.

2   Of all the settings to be considered: castor angle, camber angle and toe-in, it will only be the latter which is likely to be seriously effected during repair work on the car. As long as the same shims are used on reassembly, as were found on the front suspension lower control arm pivot spindle, then the camber and castor angles should be good enough for the drive to the local FIAT dealer.

3   To check the toe-in, the car should be positioned on a flat and level surface and loaded with two persons and 44 lbs of luggage in the case of a Spider or three persons and 66 lbs of luggage in the case of the Coupe. The tyres should be at their correct pressures.

4   The front wheel toe-in is the difference in the distance between the middle of the tyre treads when measured at hub level on the most forward and the most rearward top of the tyres.

5   Check the toe-in against the figure given in the specification at the beginning of this Chapter.

6   Adjust the toe-in by loosening the clamps on each end of the track rods, then rotate each rod equally to move the ball joints out or in as appropriate. The track road assembly behaves as a turn buckle because the threads in which the end ball joints seat are opposite hands.

7   Once the correct toe-in has been achieved, tighten the track rod end clamps and as detailed earlier drive carefully around to your local FIAT garage where an accurate job of the wheel alignment can be made.

8   On many occasions the alignment of the front wheels may be unwittingly altered when the wheels accidently hit a kerb or similar. Always keep a watch on tyre wear because it will indicate whether the wheel is correctly aligned or not.

Excessive wear on the inner or outer edges of both front tyres indicates that the toe-in setting is incorrect.

Scrubbing of the tyre, accompanied by feathering on the edges of the tread and uneven wear across the tyre, indicates an error in camber angle. A tendency for the car to pull either to the left or right can be a brake fault, an error in castor angle, or one tyre tread being significantly deeper than the other.

9   It must be appreciated that in all probability it will not be just one setting in error, but a combination of errors; and this is why it is advisable to entrust the wheel alignment task to your local FIAT garage.

Fault finding chart

Symptom	Reason/s	Remedy
STEERING FEELS VAGUE, CAR WANDERS AND FLOATS AT SPEED		
General wear or damage	Tyre pressure uneven	Check pressures and adjust as necessary.
	Dampers worn	Test, and replace if worn.
	Steering gear ball joints badly worn	Fit new ball joints.
	Suspension geometry incorrect	Check and rectify.
	Steering mechanism free play excessive	Adjust or overhaul steering mechanism.
	Front suspension and rear suspension pick-up points out of alignment	Normally caused by poor repair work after a serious accident. Extensive rebuilding necessary.
STIFF AND HEAVY STEERING		
Lack of maintenance or accident damage	Tyre pressure too low	Check pressures and inflate tyres.
	No oil in steering gear	Top up steering gear.
	No grease in steering and suspension ball joints	Clean nipples and grease thoroughly.
	Front wheel toe-in incorrect	Check and reset toe-in.
	Suspension geometry incorrect	Check and rectify.
	Steering gear incorrectly adjusted too tightly	Check and re-adjust steering gear.
	Steering column badly misaligned	Determine cause and rectify (usually due to bad repair after severe accident damage and difficult to correct).
WHEEL WOBBLE AND VIBRATION		
General wear or damage	Wheel nuts loose	Check and tighten as necessary.
	Front wheels and tyres out of balance	Balance wheels and tyres and add weights as necessary.
	Steering ball joints badly worn	Replace steering gear ball joints.
	Hub bearings badly worn	Remove and fit new hub bearings.
	Steering gear free play excessive	Adjust and overhaul steering gear.
	Front springs weak or broken	Inspect and renew as necessary.

Chapter 12 Bodywork and heater

Contents

1 General description

The vehicle structure for both Coupe and Spider is a welded fabrication of many individually shaped parcels to form a 'monocoque' bodyshell. Certain areas are strengthened locally to provide for suspension system, steering system, engine support anchorages and the like. The resultant structure is very strong and rigid.

It is as well to remember that monocoque structures have no discreet load paths and all metal is stressed to an extent. It is essential therefore to maintain the whole bodyshell both top and underside, inside and outside, clean and corrosion free. Every effort should be made to keep the underside of the car as clear of mud and dirt accumulations as possible. If you were fortunate enough to acquire a new car then it is advisable to have it rust proofed and undersealed at one of the specialist workshops who guarantee their work.

This chapter describes the 'everyday' measures that can be taken to ensure that your car will look good and be structually safe to ride in for many years. It does not attempt to describe the methods of structural repair, such tasks have always and will remain outside the scope of the owner — 'mechanic'.

2 Maintenance - exterior

1 The general condition of a car's bodywork is the one thing that significantly affects its value. Maintenance is easy but needs to be regular and particular. Neglect - particularly after minor damage - can quickly lead to further deterioration and costly repair bills. It is important to keep watch on those parts of the bodywork not immediately visible, for example the underside, inside all the wheel arches and the lower part of the engine compartment.
2 The basic maintenance routine for the bodywork is washing - preferably with a lot of water from a hose. This will remove all the loose solids which may have stuck to the car. It is important to flush these off in such a way as to prevent grit from scratching the finish. The wheel arches and underbody need washing in the same way, to remove any acccumulated mud which will retain moisture and tend to encourage rust. Paradoxically enough, the best time to clean the underbody and wheel arches is in the wet weather when the mud is thoroughly wet and soft. In very wet weather the underbody is usually cleaned of large accumulations automatically and this is a good time for inspection.
3 Periodically, it is a good idea to have the whole of the under-side of the car steam cleaned, engine compartment included, so that a thorough inspection can be carried out to see what minor repairs and renovations are necessary. Steam cleaning is available at some garages and is necessary for removal of accumulation of oily grime which sometimes collects thickly in areas near the engine and gearbox. If steam facilities are not available there are one or two excellent grease solvents available which can be brush applied. The dirt can then be simply hosed off. Any signs of rust on the underside panels and chassis members must be attended to immediately. Thorough wire brushing followed by treatment with an anti-rust compound, primer and underbody sealer will prevent continued deterioration. If not dealt with the car ccould eventually become structurally unsound and therefore unsafe.
4 After washing the paintwork wipe it off with a chamois leather to give a clear unspotted finish. A coat of clear wax polish will give added protection against chemical pollutants in the air and will survive several subsequent washings. If the paintwork sheen has dulled or oxidised use a cleaner/polisher combination to restore the brilliance of the shine. This requires a little more effort but it usually because regular washing has been neglected! Always check that door and drain holes and pipes are completely clear so that water can drain out. Brightwork should be treated the same way as paintwork. Windscreens and windows can be kept clear of smeary film which often appears if a little ammonia is added to the water. If glass work is scratched, a good rub with a proprietary metal polish will often clean it. Never use any form of wax or other paint/chromium polish on glass.

3 Maintenance - interior

The flooring cover (usually carpet) should be brushed or vacuum cleaned regularly to keep it free from grit. If badly stained, remove it from the car for scrubbing and sponging and make quite sure that it is dry before replacement. Seat and interior trim panels can be kept clean with a wipe over with a

These photos illustrate a method of repairing simple dents. They are intended to supplement *Body repair - minor damage* in this Chapter and should not be used as the sole instructions for body repair on these vehicles.

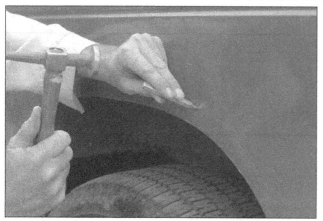

1   If you can't access the backside of the body panel to hammer out the dent, pull it out with a slide-hammer-type dent puller. In the deepest portion of the dent or along the crease line, drill or punch hole(s) at least one inch apart . . .

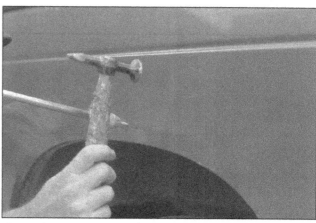

2   . . . then screw the slide-hammer into the hole and operate it. Tap with a hammer near the edge of the dent to help 'pop' the metal back to its original shape. When you're finished, the dent area should be close to its original contour and about 1/8-inch below the surface of the surrounding metal

3   Using coarse-grit sandpaper, remove the paint down to the bare metal. Hand sanding works fine, but the disc sander shown here makes the job faster. Use finer (about 320-grit) sandpaper to feather-edge the paint at least one inch around the dent area

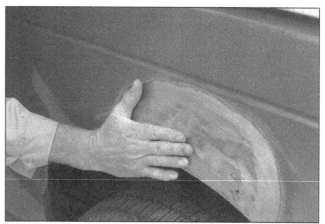

4   When the paint is removed, touch will probably be more helpful than sight for telling if the metal is straight. Hammer down the high spots or raise the low spots as necessary. Clean the repair area with wax/silicone remover

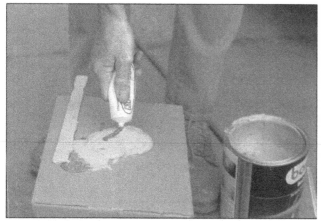

5   Following label instructions, mix up a batch of plastic filler and hardener. The ratio of filler to hardener is critical, and, if you mix it incorrectly, it will either not cure properly or cure too quickly (you won't have time to file and sand it into shape)

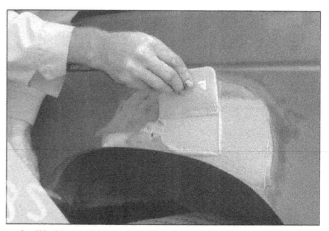

6   Working quickly so the filler doesn't harden, use a plastic applicator to press the body filler firmly into the metal, assuring it bonds completely. Work the filler until it matches the original contour and is slightly above the surrounding metal

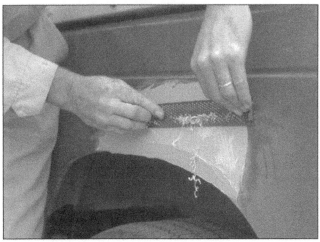

7 Let the filler harden until you can just dent it with your fingernail. Use a body file or Surform tool (shown here) to rough-shape the filler

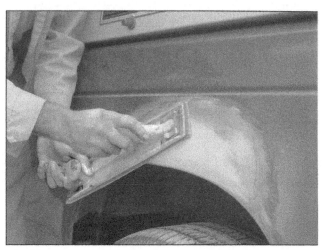

8 Use coarse-grit sandpaper and a sanding board or block to work the filler down until it's smooth and even. Work down to finer grits of sandpaper - always using a board or block - ending up with 360 or 400 grit

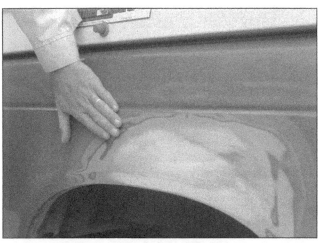

9 You shouldn't be able to feel any ridge at the transition from the filler to the bare metal or from the bare metal to the old paint. As soon as the repair is flat and uniform, remove the dust and mask off the adjacent panels or trim pieces

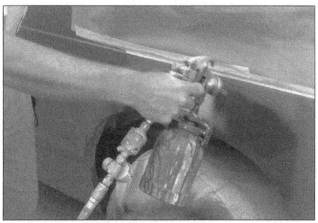

10 Apply several layers of primer to the area. Don't spray the primer on too heavy, so it sags or runs, and make sure each coat is dry before you spray on the next one. A professional-type spray gun is being used here, but aerosol spray primer is available inexpensively from auto parts stores

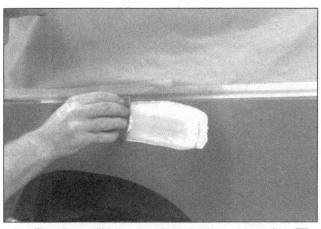

11 The primer will help reveal imperfections or scratches. Fill these with glazing compound. Follow the label instructions and sand it with 360 or 400-grit sandpaper until it's smooth. Repeat the glazing, sanding and respraying until the primer reveals a perfectly smooth surface

12 Finish sand the primer with very fine sandpaper (400 or 600-grit) to remove the primer overspray. Clean the area with water and allow it to dry. Use a tack rag to remove any dust, then apply the finish coat. Don't attempt to rub out or wax the repair area until the paint has dried completely (at least two weeks)

damp cloth. If they do become stained (which can be more apparent on light coloured upholstery) use a little liquid detergent and a soft nailbrush to scour the grime out of the grain of the material. Do not forget to keep the headlining clean in the same way as the upholstery. When using liquid cleaners inside the car do not over-wet the surfaces being cleaned. Excessive damp could get into the upholstery seams and padded interior, causing stains, offensive odours or even rot. If the inside of the car gets wet accidentally it is worthwhile taking some trouble to dry it out properly. DO NOT leave oil or electric heaters inside the car for this purpose. If, when removing mats for cleaning, there are signs of damp underneath, all the interior of the car floor should be uncovered and the point of water entry found. It may only be a missing grommet, but it could be a rusted through floor panel and this demands immediate attention as described in the previous Section. More often than not both sides of the panel will require treatment.

On cars fitted with the factory sunroof avoid touching the interior canvas. Keep it clean and rectify all tears immediately. Consult your local FIAT Agent as to the most suitable type of repair depending on the material used. Keep the stays and fixings very lightly but frequently oiled, particularly at the front of the roof, and periodically release and roll back the roof so that it does not become too stiff and weak.

4   Minor repairs to bodywork

See photographic sequence on pages 182 and 183

1   A car which does not suffer some minor damage to the bodywork from time to time is the exception rather than the rule. Even presuming the gate post is never scraped or the door opened against a wall or high kerb there is always the likelihood of gravel and grit being thrown up and chipping the surface, particularly at the lower edges of the doors and sills.

2   If the damage is merely a paint scrape which has not reached the metal base, delay is not critical but where bare metal is exposed action must be taken immediately before rust sets in.

3   The average owner will normally keep the following 'first aid' materials available which can give a professional finish for minor jobs.
a)   An anti-rust primer
b)   Cellulose stopper for minor scratch filling
c)   Resin filler paste for filling larger areas and depths
d)   Assorted grades of wet and dry abrasive paper
e)   Primer
f)   Matched finish paint for brush or aerosol application.

4   Where the damage is superficial (i.e. not down the the bare metal and not dented) fill the scratch or chip with sufficient filler to smooth the area, rub down with paper and apply the matching paint.

5   Where the bodywork is scratched down to the metal, but not dented, clean the metal surface thoroughly and apply a suitable metal primer first - such as red lead or zinc chromate. Fill up the scratch as necessary with filler and rub down with wet and dry paper. Apply the matching colour paint.

6   If more than one coat of colour is required rub down each coat with cutting paste before applying the next.

7   If the bodywork is dented, depending on how bad and where the dent is, think first in terms of beating it out but if it is felt that the original shape will not be retrieved, panel replacement may be viable. See the next Section. However if it is to be 'mended' proceed as follows: First beat out the dent to conform as near as possible to the original contour. Avoid using steel faced hammers - use hard wood mallets or similar and always support the panel being beaten with a hardwood or metal 'dolly'. In areas where severe creasing and buckling has occurred it will be virtually impossible to reform the metal to the original shape. In such instances a decision should be made whether or not to cut out the damaged piece or attempt to recontour over it with filler paste. In large areas where the metal panel is seriously damaged or rusted the repair is to be considered major and it is often better to replace a panel or sill section with the appropriate piece supplied as a spare. When using filler paste in largish

quantities make sure that the directions are carefully followed. It is a false economy to rush the job as the correct hardening time must be allowed between stages and before finishing. With thick applications the filler usually has to be applied in layers - allowing time for each layer to harden. Sometimes the original paint colour will have faded and it will be difficult to obtain an exact colour match. In such instances it is a good scheme to select a complete panel - such as a door or boot lid - and spray the whole panel. Differences will be less apparent where there are obvious divisions between the original and resprayed areas.

5   Major body repairs

1   Because the body is built on the monocoque principle, major damage must be repaired by a competant body repairer with the necessary jigs and equipment.

2   In the event of a crash that resulted in buckling of body panels, or damage to the road wheels the car must be taken to a FIAT Dealer or body repairer where the bodyshell and suspension alignment may be checked.

3   Bodyshell and/or suspension mis-alignment will cause excessive wear of the tyres, steering system and possibly transmission. The handling of the car also will be affected adversely.

6   Doors - tracing and silencing rattles

1   The commonest cause of door rattles in a misaligned, loose or worn striker plate, but other causes may be:-
a)   Loose door handles, window winder handles and door hinges.
b)   Loose, worn or misaligned door lock components.
c)   Loose or worn remote control mechanism.

2   It is quite possible for door rattles to be the result of a combination of these faults, so a careful examination must be made to determine the causes of the noise.

3   If the nose of the striker plate is worn and as a result the door rattles, renew it and adjust the plate.

4   If the nose of the door wedge is badly worn and the door rattles as a result, then fit a new door latch assembly.

5   Should the hinge be badly worn then the pivot pin may be replaced; however if with a new pivot pin the hinge is still slack, then it is a workshop task to repair the hinges since the halves are welded to the shell and door, and are not by themselves available as spares from FIAT.

7   Interior handles and trim panel - removal and replacement

1   The doors are the usual pressed steel panel construction specially strengthened to resist sideways impact into the car. It is retained on the bodyshell by two hinges which are welded to the door and screwed to the front door pillar. The exterior fittings are secured from the door interior; the interior handles clip onto the spindles which project through the trim from the respective mechanisms bolted to the door interior structure.

2   In order to gain access to any component attached to the door it will be necessary to begin by removing the interior facing fittings and trim.

3   Make a note of the positions of the various handles on the interior face, so that they can be refitted in their original positions.

4   Depress the bezels around the handle spindle and with a piece of piano wire fashioned into a small hook; pull the wire clip retaining the handle on the splined end of the spindle, out of its slots (photo).

5   The handle and bezel can now be removed from the spindle (photo).

6   The door pull handle - arm-rest is secured by a couple of screws and once these are undone and removed this handle may be lifted from the door.

7   Having removed all the handles the trim may be prised from the door structure. Use a wide bladed screwdriver or similar, and

insert it between the door panel and the trim panel. Lever tne trim panel very carefully away from the door. This action will release the trim panel retaining clips.

8   Collect the spacers that fit between the trim panel and the door panel around the handle spindles (photo).

9   Refitting follows the reversal of the removal procedure.

8   Doors, exterior fittings - removal and replacement

1   All the exterior fittings are secured in position by screws which pass through from the inside of the door.

2   Remove the interior handles and trim panel as detailed in Section 7 to gain access to the screws retaining the exterior fittings.

3   The nuts and bolts which retain the exterior handle, door lock and latch are now fully exposed and can be undone to enable the appropriate component to be removed. (photo).

4   Replacement of those components follows the reversal of the removal procedure.

9   Doors-interior mechanisms - removal and replacement

1   The interior handles and trim panel will need to be removed as detailed in Section 7 in order to gain access to the interior mechanisms.

2   **The door latch relay handle** and linkage are retained to the door panel with three bolts. Once these bolts have been undone and removed the handle mechanism can be extracted from the door structure.

3   **The window wind mechanism** is a wire rope regulator comprising a loop of wire rope which passes around four pulleys mounted in the door structure. The wind handle is geared into the top forward pulley and the window support strip is clamped onto the rear vertical length of cable; as the handle is turned the

wire rope is moved around the pulleys and moves the window up or down as appropriate (photo).

4   To remove the window wind mechanism, lower the window to its lower position and undo the two screws which retain the clamping plate and wire rope to the window support strip. Slacken the nut and bolt on which the forward lower pulley rotates, and move the pulley and bolt rearward in the slot in the door frame, to slacken the wire rope. (photo).

5   The wind pulley and handle spindle assembly is secured to the inside of the door panel by three nuts. Undo and remove these nuts and extract the wind mechanism with the wire rope from the door structure.

6   **The sliding window** the wound window may be extracted from the door once the quarter light and forward guide has been removed.

7   The Quarter Light is retained by screws passing through the front edge of the window frame, and the sliding window forward guide rail which is integral with the quarter light, is bolted to the door inner panel.

Once the quarter light retaining screws have been undone and removed, the light may be lifted from the window frame and door structure.

8   Now that the quarter light has been removed the sliding window may be extracted from the door. The draught seals on the window guides and window frame periphery can be lifted off when the window has been removed.

9   Replacing the windows follows the reversal of the removal procedure.

10 Replacement of the winding mechanism follows the reversal of its removal procedure, except that the wire rope should be terminated before the trim panel and handles are refitted. The tension of the wire with the window fitted and the winding mechansim secured, should be sufficient to limit the vertical movement midway between the lower pulleys and approximately 0.5 inch.

7.4 Removing the handle retaining clip, with a specially fashioned hook

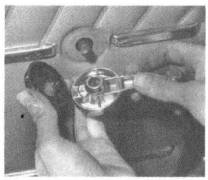

7.5 The interior handle and bezel removed. The splined end of the handle spindle and the handle retained clip are exposed

7.8 The spacer situated behind the trim panel around the handle spindle

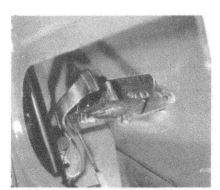

8.3 The view of the door latch mechanism, showing the handle link rods

9.3 The interior of the door exposed, showing the door latch mechanism and window wind mechanism

9.4 The clamping plate retaining the window support strip to the rear vertical stretch of wire rope

10 Door - removal and replacement

1 As mentioned earlier the door hinges are welded to the door, but are retained by three screws to the forward door pillar.
2 Removal is straightforward, necessitating only the removal of the screws securing the hinges, and the disengagement of the door movement stop.
3 Remember to support the door whilst it is being detached from the car bodyshell; the hinges will be severely distorted if the whole weight of the door acts on the single hinge.
4 As usual refitting follows the reversal of the removal procedure; except that the door catch should be checked for correct position.

11 Door - striker plate and catch - positioning, removal and replacement

1 The door catch is retained with screws to the door structure and its position cannot be adjusted. The movement of the catch pawl should be smooth and no sloppiness should be felt. The catch should be renewed once the pawls movement is found to slack.
2 The catch is removed from the inside of the door once the retaining screws have been undone. The exterior door handle and interior handle link rods need to be removed from the catch assembly before it can be extracted from the door. (photo).
3 When refitting the catch assembly ensure that the link rods from the interior and exterior handles are connected so that when the handles are moved the full catch pawl movement is achieved.
4 The striker plate is retained to the rear door pillar by screws; and the plates position is adjustable.
5 Removal and replacement of the striker plate is straightforward; and when adjusting the position the objective is to enable the door to be closed, without excessive effort, to a position when the outer surfaces of the door and coachwork are perfectly flush. The door should not rattle when the car is in motion.

12 Engine compartment hood and latch - removal and replacement

1 The hood is retained by two hinges and a latch, the latch is cable released from a handle fitted just below the dashboard. There is a spring loop hood stay, which locates in a slotted fitting attached to the underside of the hood.
2 Start to remove the hood by undoing the screws which retain the radiator grill in position. The grill may then be pulled clear of the car to reveal the bonnet hinge bolts underneath the top of the radiator aperture, (photo).
3 Remove the hinge bolts - 4 each hinge - detach the stay loop

and lift the hood from the front of the car, (photos).
4 The latch and its cable actuation is fairly straightforward comprising a hook to engage the striker on the underside of the hood. The hood is coupled to a cable which runs through fairleads to the actuating lever in the car. The latch mechanism is retained to the top of the rear engine compartment bulkhead by nuts on bolts which pass through the bulkhead.
5 If the latch is found to be worn, it should be replaced as a whole; there is no point trying to dismantle it.
6 The replacement of the engine hood and latch follows the reversal of the removal procedure.

13 Luggage compartment, bonnet and latch

1 The luggage compartment is hinged on the bodyshell in a like manner to the engine hood.
2 Mark the relative positions of the hinge and bonnet to ensure correct alignment of the components when refitted. Remove the nuts - 2 per hinge - which secure the hinge to the bonnet.
3 Lift the bonnet carefully away from the car.
4 The bonnet latch and lock is secured to the bonnet; and the striker on the boot sill is adjustable to provide sound operation of the latch and lock.
5 The latch and lock is retained by nuts and bolts and if worn do not try to affect a repair. Replacement of the latch and lock is the only reasonable course of action.
6 The striker position is reckoned to be correct when the top of the bonnet is flush with the adjacent coachwork; and when it does not require excessive effort to close the bonnet and lock it.

14 Windscreen Glass - removal and refitting

1 The methods of retaining the main windscreen in the Coupe is a departure from common practice. The screen is effectively bonded to the bodyshell with a thermally cured 'NEOPRENE' strip situated between the edge of the glass and the mounting flange around the inside of the screen aperture.
2 The screen location in the Spider cars follows the usual lines of employing a rubber extrusion which slips around the mounting flange in the screen aperture and around the periphery of the glass screen. The metal chromed trim is used to lock the screen and mounting flange in the rubber extrusion.
3 To begin with the procedure for removing the Spider windscreen:-
4 If the windscreen shatters, fitting a replacement screen is one of the few jobs the average owner is advised to leave to the experienced fitter. For the owner who wishes to do the job himself the following instructions are given:
5 Remove the wiper arms from their spindles, remove the interior rear view mirror and the visors. Cover the screen heating duct apertures with sticky tape.

12.2 Lifting the radiator grille from the car

12.3(i) Detaching the loop stay from the engine hood

12.3(ii) Lifting the hood clear of the car

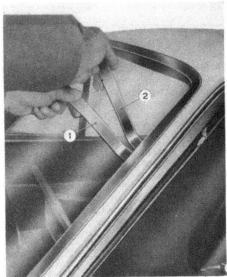

FIG.12.1. FIAT TOOLS NUMBER A78032 BEING USED TO REMOVE THE TRIM MOULDING FROM THE COUPE WIND-SCREEN

1   The tool used from the pronged end to catch the trim retaining clips
2   The tool used from the lugged end for locating the clips which retain the trim moulding

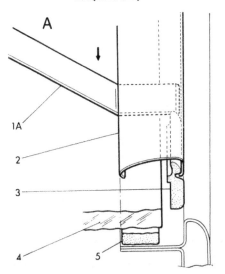

Fig.12.2. The use of piano wire to cut through the old bonding neoprene strip

6   Next extract the metal trim from the outside of the screen edge rubber. This trim locks the rubber extrusion around the glass screen edge.

7   Place a blanket or suitable protection on the car bonnet to prevent scoring the paintwork with the broken screen.

8   Move to the inside of the car and have an assistant outside the car ready to catch the screen as it is released.

9   Wearing leather gloves or similar hand protection push on the glass screen as near to the edge as possible, beginning at the top corners. The rubber extrusion should deform and allow the screen to move outwards out of the screen aperture. This óf course is not applicable if the screen has shattered.

10 Remove the rubber surround from the glass or alternatively carefully pick out the remains of the glass. Use a vacuum cleaner to extract as much of the screen debris as possible.

11 Carefully inspect the rubber extrusion surround for signs of pitting and deterioration. Offer up the new glass to the screen aperture and check that the shape and curvature of the screen conforms to that of the aperture. A screen will break quite soon again if the aperture and glass do not suit, typically if the vehicle has been involved in an accident during which the screen broke. A car bodyshell can be deformed easily in such instances to an extent when the aperture will need reshaping by a competant body repairer to ensure conformity with a new screen.

12 Position the new glass into the rubber extrusion surround, remember that the groove for the metal trim needs to be on the outer side of the screen assembly.

13 With the rubber now correctly positioned around the glass, a long piece of strong cord should be inserted in the slot in the rubber extrusion which is to accept the flange of the screen aperture in the bodyshell. The two free ends of the cord should finish at either the top or bottom centre and overlap each other.

14 The screen is now offered up to the aperture, and an assistant will be required to press the rubber surround hard against the bodyshell flange. Slowly pull one end of the cord, moving around the windscreen, thereby drawing the lip of the rubber extrusion screen surround over the flange of the screen aperture.

15 Finally ensure that the rubber surround is correctly seated around the screen and then press in the metal trim strip which locks the screen in the rubber. Once the glass has been fitted satisfactorily the windscreen wiper, visors and interior mirror may be replaced.

16 **The Coupe windscreen** FIAT supply the special kit which provides the neoprene strip, and the other special tools necessary

FIG.12.3. USE OF FIAT TOOL NUMBER A78032

1A Tool used from lugged end for locating trim moulding clips -
1B Tool used from pronged end for catching in trim moulding clips
2   Trim moulding
3   Trim moulding clip
4   Glass pane
5   Neoprene strip

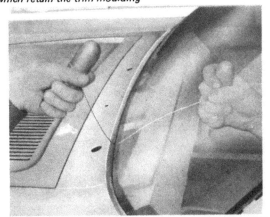

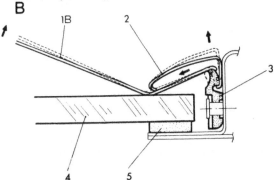

to fit a windscreen. The part number of the pack is 4173888. Begin removing the windscreen by taking those precautions, and completing those tasks outlined in paragraphs 4 to 7 in this Section.

17 Using the FIAT tools A78032 remove the trim from the periphery of the screen. A pair of those tools will be needed; and by inserting the pronged end of those tools underneath the inner lip of the trim and hooking up the trim retaining clips up, the trim may be removed.

18 If the screen is shattered remove as many fragments as possible before proceding to break the neoprene bondbond as described next. Note: a complete windscreen can be removed using the wire cutting technique described next quite satis- factorily.

19 Obtain a **length** of 0.0197'' (0.5 mm). Piano wire or wire supplied by FIAT in their windscreen fitting kit. Pierce a small hole through the existing neoprene bonding strip, and thread the piano wire through.

20 With the wire in position, you will need to enlist some help, because the next operation is to pull the wire, with a to-fro saw like motion, around the whole windscreen to cut the neoprene bonding strip.

21 If local stiff points are encountered in the neoprene, avoid using undue strain because this may damage the edge of the glass screen.

22 Then with all the glass removed (or screen) from the screen aperture, use a sharp knife to remove all the remaining neoprene from the mounting flange (and screen if it is intended to refit the old screen).

23 Thoroughly clean the mounting flange until no trace of old material can be found and until the flange bond surface is smooth and free from grease or dirt.

24 Check the condition of the clips around the screen frame used to retain the screen trim. Replace any found damaged.

25 Place the two rubber spacers, provided in the windscreen fitting pack mentioned in paragraph 16, on the bottom sill of the screen aperture about 8 inches in from the side pillars. Carefully, lower the windscreen into position in the aperture with its base edge resting on the spacers.

26 Make sure the glass screen is central in the aperture before fixing some lengths of masking tape at various points on the screen periphery and aperture.

27 Mark the types to ensure that when the screen is refitted with the neoprene bond strip in place, it can be centred straight away.

28 Note that the screen edge should be clear of the trim moulding clips, or else it will be difficult to fit the trim. When the screen has been finally fitted.

29 FIAT supply a gauge, tool no. 96801 which may be used to check the position of the screen.

30 Lift the screen away from the car; wipe the mounting flange clean again and then apply a thin coat of primer (included in the Windscreen Fitting Pack) to the mounting flange in the screen aperture. The coating should not be more than 3/8 inch wide, and apply it with a clean non-fluffy cloth.

31 If a new screen is being fitted the edges should be carefully cleaned with a cleansing solvent such as carbon tetrachloride. Wipe the edges dry.

32 The reel of neoprene strip may now be unpacked from the screen fitting kit. Connect the barred ends of the wire in the neoprene strip to a pair of 12 volt batteries arranged serially to produce 24 volts over the ends of the neoprene strip.

33 Allow the current to flow through the strip for ONE MINUTE this will dry up and preheat the compound.

34 Lift the prepared strip from the pack and place it around the inner edge of the glass screen. The strip should be positioned so that the tape side protrudes a little from the edge of the glass. Start this operation at the position of one of the spacers on the bottom edge of the glass. Once the strip is in position join the two ends by butting these together at right angles. and twist one end over the other. Smooth down the joint to avoid creating a leak point on the final product.

35 Clean the two wires and put any strip left over to one side for possible future use.

36 Refit the glass screen back to the aperture in the coachwork, and rest it on the two previously placed spacers. Adjust the windscreen to the previously marked centre position, indicated by the marks on the masking tape.

37 Reconnect the two wires from the strip to the 24 volts supply - take care that the clips and electrical connections do not scratch the paintwork.

38 Allow the current to pass through the strip for TWO MINUTES, - to soften the neoprene, then apply hand pressure to the edge of the screen to spread the softened strip until the wetted area on the glass is ¼ inch wide around the periphery of the screen.

39 FIAT tool A96802 is specially shaped to check that the position of the screen is correct. One edge of this tool is fashioned for the front windscreen and the other is fashioned for the rear screen.

40 There is just 5 to 6 minutes left to positon the screen correctly, before the curing stage of the neoprene strip begins.

41 As soon as the specified width of strip on the glass and mounting flange has been achieved, the hand pressure may be relaxed and the currrent left to flow through the strip for ONE HOUR to cure the neoprene.

42 Disconnect the batteries and allow the screen and bond to cool down.

43 Test the screen bond for leaks by pouring water over the screen. If any leaks are found they may be blocked with pieces of neoprene removed when the strip was fitted to the glass screen.

44 The metal trim may now be pressed back into position.

15 Coupe rear windscreen

1   The rear windscreen is attached in the same manner as the front windscreen. Paragraphs 16 to 44 in the previous section number 14 apply for the removal and refitting of the rear screen.

16 Exterior brightwork and trim

1   All bumpers and overriders are bolt on assemblies. Removal is straightforward, but the nuts and bolt threads will probably require penetrating oil because they are invariably rusted.

2   The body trim both internally and externally is of very simple construction. Its removal is obvious in each case: If a screw is not visible then it is either a push-on or slide-on fit. Check each part - and never force anything.

17 Interior trim and centre console - removal and refitting

1   The interior trim panels and fittings are usually screwed in position, although the rear seat panels are clipped to the bodywork in the same manner as similar panels are secured to the door.

2   **Centre console** The centre console consists of two main panel assemblies. The front upper panel assembly is used to hold the radio, if fitted, and on later models the air conditioning ducts. The lower panel assembly covers the transmission funnel to varying extents depending on the model of Coupe or Spider.

3   The upper panel assembly is retained by screws which pass through to the underside of the dashboard and the top of the lower panel assembly.

4   Once these screws are removed, the panel assembly may be lifted away. The electrical leads to the devices mounted on that panel may now be disconnected. (photo).

5   The lower panel assembly is now clear to be removed: begin by prising the cigarette ash box from the rear of the assembly to reveal the nut which secures the rear assembly to a stud on the transmission tunnel. Undo and remove the nut.

6   Unscrew the heater lever gate from the top of the panel assembly and then remove the gearlever boot and retaining plate.

7   Finally remove the screws securing the front end of the panel assembly to the transmission tunnel. (photo).

17.4 The upper centre console removed

17.7 Undoing the screws securing the
forward end of the lower centre console
to the transmission tunnel

17.8 Lifting the whole lower centre
console from the transmission tunnel

8   The whole of the lower centre console can now be lifted away to reveal the heater controls, handbrake and the heater itself below the dashboard. (photo).

9   Refitting both lower and upper centre consoles follows the reversal of the removal procedure.

18 Heater - removal, dismantling, reassembly and refitting

1   The heater radiator and motor unit is situated underneath the dashboard and is covered by the centre console. Flexible ducts take the air from the heater unit to the various outlets for derusting and air 'conditioning'.

2   It will be necessary to remove the whole of the centre console as directed in Section 17 of this chapter in order to gain access to the heater and controls.

3   The heater unit comprises a upper casing bolted to the bodywork behind the dashboard, a radiator and valve assembly which slots into the base of the upper casing. The motor fan and lower housing is clipped to the underside of the upper casing.

4   If you intend to work on the air deflectors, electric motor or fan, there is no need to disturb the radiator and therefore no need to drain the cooling system.

5   **Removal of heater motor and fan only**. Having removed the centre console; disconnect that single control cable attached to the lower air deflector. Then prise off the four spring clips which retain the upper and lower halves of the heater casing together. Remove the lower casing wheich will contain the motor and fan. There will be a third moulding between the two halves; this surrounds the fan and ducts the warm air down to the lower casing.

6   Once the lower casing has been removed the motor electrical connections can be detached, and the motor/fan unit unclipped from the inside of the lower casing.

7   If the motor is faulty there is no point trying to dismantle it, only complete motors are available as spares. The fan may be detached from the motor shaft after the single retaining nut has been undone, and the special hollow screw loosened in the centre of the fan boss.

8   Refitting the motor and fan follows the reversal of the removal procedure. Remember to fix the rubber spacers each side of the motor when it is reassembled into the lower casing.

9   **Removal of heater radiator**. It will be necessary to drain the whole cooling system as detailed in Chapter 2, remove the centre console as directed in Section 17, and then remove, the lower heating casing together with the motor and fan as described in paragraphs 5 to 8 of this section.

10 Continue by detaching the control cable from the water flow valve and the two water hoses from the ends of the metal radiator pipes.

11 The radiator may now be eased down out of the upper heater casing. There is a soft rubber strip fitted around the radiator between it and the inside surface of the upper heater casing.

12 With the radiator assembly free of the car, the pipes and valves may be detached as necessary. Do not try and dismantle the valve. FIAT supply whole valves only. It is possible to carry out minor repair work on the radiator with propietry sealing compounds; but it is advisable to entrust repair work to the local radiator specialist. They will have the expertise to tackle a soldering job on the radiator.

13 Refitting the radiator follows the reversal of the removal procedure, remember as always to use new gaskets at flange joints of the radiator supply pipes. Do not use sealing compounds instead of new gaskets, it will make it extremely difficult to break the joint next time it is to be dismantled.

14 **The heater controls.**The three heater controls are connected to the lower air vent, hot water supply and fresh air inlet control flap. It will be necessary to remove the whole of the centre console as directed in Section 17 to gain access to the cable controls and the lever assemblies.

15 The control cable sleeves are clipped to the transmission tunnel from the lever assembly between the front seats to the heater assembly.

On Spider cars the heater controls are situated in the forward central console, and therefore the control links are much shorter.

16 Once the consoles have been removed, the control cables may be detached from the bodywork and the lever bracket assembly unbolted from the transmission tunnel. On Spider cars the control levers are mounted on a bracket assembly on the top heater casing.

17 The cables may be detached from the operating levers by simply unhooking from the holes in the base of the levers. The forward ends of the cables, which attach to the valve and flap actuation levers, may be removed from those levers equally easily.

18 The control levers pivot on steel pins retained in the bracket assembly with split pins. Once those pins hav been removed the pivot pins can be extracted to free the operating levers. Retrieve the spacers and washers.

19 **The heater upper casing** the whole heater has been designed so that all maintenance and repair operations on the heater components may be carried out without needing to remove the upper casing. Study the exploded views of the heater components and the text in this chapter and then if the upper casing must be removed proceed as follows:

20 Remove the centre console, the lower heater casing together with the motor and fan, and finally remove the radiator itself.

21 The heater upper casing is secured to the bodywork behind the dashboard with four nuts and bolts; all of which are quite accessible. Retrieve the soft gasket between the casing and the mounting.

22 **Air ducting**. Air from the outside of the car is taken into a chamber between the engine compartment and dashboard through vents in the engine hood. The heater unit mounted beneath this chamber draws in the air as required through the inlet flap and deflectors. The purpose of the deflectors is to prevent rain water from being taken into the heater. At each end of this chamber, near the sides of the car, there are ducts which take cool fresh air directly to controllable vents at each

end of the dashboard in the car.
23 Spider cars have a slightly different ducting system: The cool air chamber is omitted and the heater takes fresh air directly from vents in front of the windscreen. The warm air is discharged downwards into the car and upwards to demist the windscreen.

24 Little is required in the way of maintenance to these ducts; the terminal vent flap assemblies are screwed to the reverse side of the fascia panel and flexible ducting connects the terminal vents to the feed ducts.

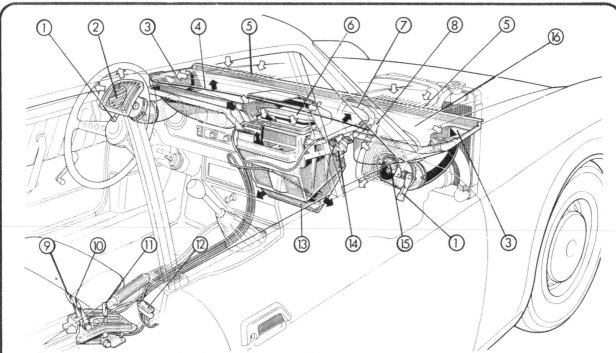

FIG.12.4. THE HEATING AND VENTILATING SYSTEM AS FITTED TO THE COUPE. THE SPIDER HEATER INSTALLATION DIFFERS ONLY BY THE EMISSION OF THE COOL AIR CHAMBER, BENEATH THE ENGINE HOOD

1 Levers controlling butterfly valves 15
2 Adjustable air outlets (fresh air delivery against windshield, side windows or onto front seat occupants)
3 Intakes of outlets 2
4 Fixed outlet, air against windshield
5 Air intake slots on hood
6 Shutter, external air admission into heater
7 Line, water delivery from engine to heater
8 Line, water return from heater to engine
9 Lever controlling shutter 6
10 Lever controlling cock 14
11 Lever controlling shutter 13
12 Three-position switch, heater fan
13 Shutter, heater air outlet to interior front compartment
14 Cock, hot water from engine radiator
15 Butterfly valves, air outlets 2
16 Cool air chamber

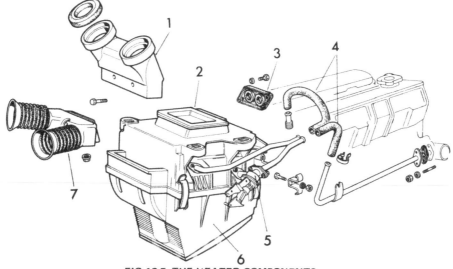

FIG.12.5. THE HEATER COMPONENTS

1 Warm air ducts to the windscreen
2 The upper heater casing
3 The seal around the water supply pipes fitted on the engine compartment rear bulkhead
4 Flexible water hoses to the heater
5 Inlet water flow control valve
6 Lower heater casing
7 Warm air ducts to the centre console

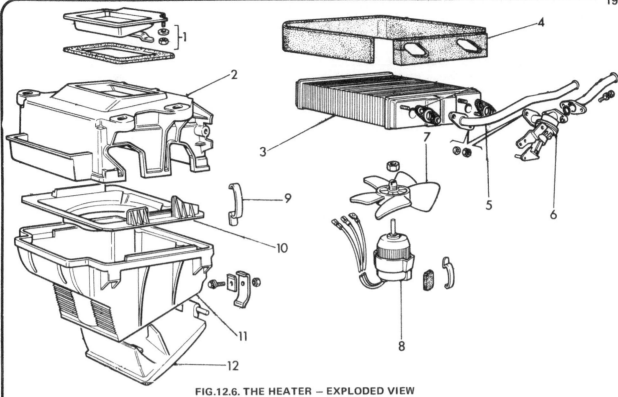

FIG.12.6. THE HEATER — EXPLODED VIEW

1 Air inlet control flap, and gasket
2 Heater upper casing
3 Heater radiator
4 Radiator periphery seal
5 Water outlet pipe
6 Water inlet control valve
7 Fan

8 Heater motor
9 Spring clip holding upper and lower casings together
10 Centre moulding which surrounds the fan to duct the warmed air down to the distribution passages
11 Heater lower casing
12 Warm air control flap

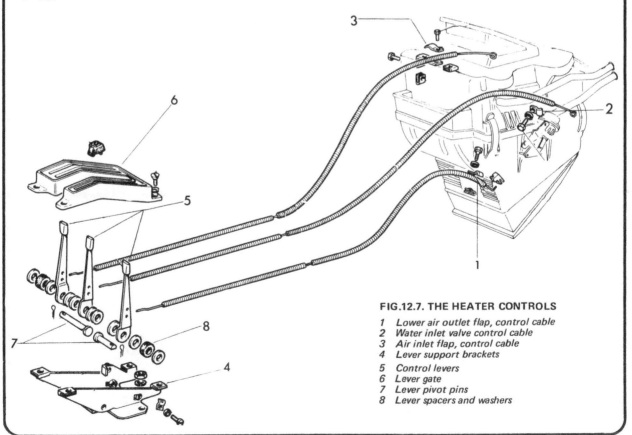

FIG.12.7. THE HEATER CONTROLS

1 Lower air outlet flap, control cable
2 Water inlet valve control cable
3 Air inlet flap, control cable
4 Lever support brackets

5 Control levers
6 Lever gate
7 Lever pivot pins
8 Lever spacers and washers

19 Petrol tank - removal and refitting

1   The petrol tank is located in the bottom left hand side of the luggage compartment next to the spare wheel. Early models will have a fairly simple tank arrangement, with only a few pipes and connections of various kinds. Later cars fitted with emission control devices will have more pipes and connections to the tank, to provide for the vapour emission control system described in Chapter 3.

2   Whenever dealing with the fuel system and particularly with the fuel tank work in a well ventilated workshop - or even better outside in the fresh air. Fuel vapours are very dangerous when allowed to accumulate in still air.

3   The tank exist as two halves joined at the horizontal middle; a gasket is fitted between the mating flanges. The fuel feed pipe enters the lower half, and a drain plug is also fitted to the lower half.

4   The fuel tank contents measuring transducer and vapour venting ports are mounted in the upper half.

5   If you intend to remove the tank begin by draining the contents, having hopefully used as much as possible beforehand - several gallons of petrol is not the easiest thing to store.

6   Once the tank has been emptied, the fuel feed pipes, vent pipes and engine supply pipes may be removed.

7   Disconnect both negative and positive terminals from the battery and stow them safely aside. You can now remove the electrical leads to the petrol tank contents sender unit.

8   Undo and remove the tank mounting nuts and bolts and lift the tank from the car.

9   The tank can be separated into its halves for cleaning before reassembly. It is essential to use a new gasket when reassembling the tank.

10 Refitting the tank follows the exact reversal of the removal procedure.

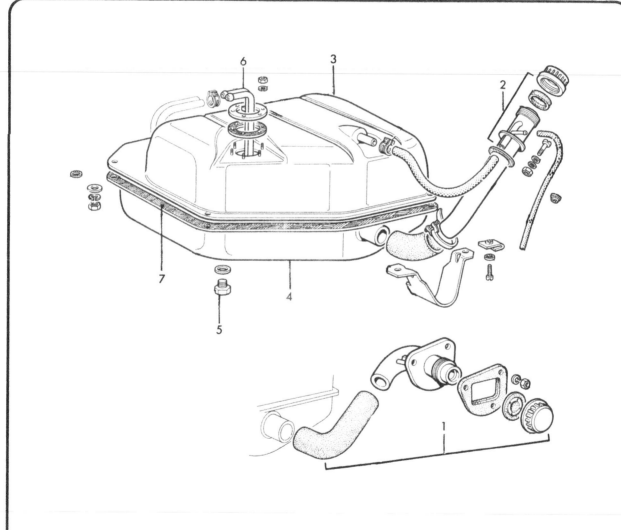

FIG.12.8. THE PETROL TANK INSTALLATION

1   Petrol fillter pipe assembly - Coupe
2   Petrol filler pipe assembly - Spider
3   Petrol tank upper half
4   Petrol tank lower half
5   The drain plug
6   Engine supply pipe assembly
7   Petrol tank joint gasket

Chapter 13 Supplement

Contents

1 Introduction

The purpose of this supplement to the Owner's Workshop Manual is to provide information on later models in the range. The author would like to thank Crouch's Garage, Ilminster, Somerset for assistance with its compilation.

2 Routine maintenance

In addition to the tasks listed on pages 7 to 11 of the Manual the following tasks should be added where applicable:

Every 6000 miles or 6 months
1 Check condition and tension of air conditioning compressor drivebelts.

Every 12 000 miles or 12 months
2 Remove, clean and examine the components of the crankcase

ventilation system.
3 Check fast idle for proper operation, including electrovalve, switches, lines and wires; renew components as necessary.
4 Check valves, lines and fittings of the fuel evaporative emission control system; renew components as necessary.
5 Check condition of wiring and connections of the ignition system.
6 Check the lines, manifolds, valves and air pump of the exhaust emission control system.

Every 25 000 miles
7 When the EGR service warning lamp is illuminated continuously arrange for your FIAT dealer to check the condition of the complete EGR system.
8 When the catalyst service warning lamp is illuminated continuously arrange for your FIAT dealer to renew the catalytic converter, and check the operation and condition of the converter system including the *Slow Down* indicator.
9 Check the operation of the switches and relays of the spark modulation device. Renew components as necessary.
10 Renew the activated carbon trap.

3 General data

The following engine types were fitted to the range

Coupe (UK)
1600	132 AC.000
1800	132 ACI.000

Coupe - Spider (USA)
1800	132 AC.040.3
1800	132 AI.040.4
1800	132 AI.040.5

Compression ratio
132 AC.000	9.8:1
132 ACI.000	9.8:1
132 AC.040.3	8.0:1
132 AI.040.4	8.0:1
132 AI.040.5	8.0:1

Maximum power output
132 AC.000 (4-speed transmission)	104hp
132 AC.000 (5-speed transmission)	108hp
132 ACI.000 (4-speed transmission)	114hp
132 ACI.000 (5-speed transmission)	118hp
132 AC.040.3	87hp
132 AI.040.4	78hp
132 AI.040.5	86hp

Valve timing
132 AC.000 engine
Inlet valve - Opens	12° BTDC
Closes	53° ABDC
Exhaust valve - Opens	52° BBDC
Closes	13° ATDC

132 ACI.000 engine
Inlet valve - Opens	15° BTDC
Closes	55° ABDC
Exhaust valve - Opens	55° BBDC
Closes	15° ATDC

132 AC.040.3 and 132 AI.040.4 engines
Inlet valve - Opens	22° BTDC
Closes	70° ABDC
Exhaust valve - Opens	70° BBDC
Closes	22° ATDC

132 AI.040.5 engine
Inlet valve - Opens	5° BTDC
Closes	53° ABDC
Exhaust valve - Opens	53° BBDC
Closes	5° ATDC

Valve clearances (cold)
	Inlet	Exhaust
UK models	0.18 in (0.45 mm)	0.24 in (0.60 mm)
USA models	0.18 in (0.45 mm)	0.20 in (0.50 mm)

Fuel system

Carburettor
Weber 28/36DHSA *(dimensions in inches)*
	Primary barrel		Secondary barrel
Venturi diameter	0.905		1.102
Main jet	0.049		0.061
Main air jet	0.076		0.059
Idling jet	0.019		0.027
Idling air jet	0.63		0.027
Float level from cover face to float		0.236	
Float travel		0.335	
Pump jet		0.043	

Weber 32ADFA 2/100 *(dimensions in inches)*
	Primary barrel		Secondary barrel
Venturi diameter	0.787		0.905
Main jet	0.041		0.051
Emulsion tube type	F74		F74
Choke calibration		77°F (25°C)	
Float level from cover face to float		0.21 to 0.29	

Exhaust gas CO analysis *(USA models with Weber 32ADFA carburettor)*
Catalytic converter models	3 ± 0.5%
Non-catalytic converter models	0.5 ± 0.2%

Weber 34DMS *(dimensions in inches)*

	Primary barrel	Secondary barrel
Venturi diameter	0.945	1.023
Main jet	0.049	0.061
Main air jet	0.071	0.071
Idling jet	0.019	0.027
Idling air jet	0.041	0.027
Emulsion tube	F61	F61
Pump jet	0.019	—
Power jet	0.43	—
Float level from cover face to float	0.275	

Solex C34EIES5 *(dimensions in inches)*

	Primary barrel	Secondary barrel
Venturi diameter	0.945	1.063
Main jet	0.049	0.059
Main air jet	0.059	0.059
Idling jet	0.018	0.031
Idling air jet	0.055	0.043
Emulsion tube	0.138	0.138
Pump jet	0.021	—
Fuel recirculation	0.039	
Float level from cover face to float	0.80	

Ignition system
Spark plugs
Later USA models	Champion N9Y, Marelli CW7LP or AC 42XLS

Coil
Make	Marelli BES 200A
Resistance - Primary windings	2.59 to 2.81 ohms
Secondary windings	6750 to 8250 ohms

Distributor (UK models)
Make (4-speed transmission)	Marelli S147L
Make (5-speed transmission)	Marelli S147H
Centrifugal advance	28° ± 2°
Static advance	10° BTDC
Contact gap	0.014 to 0.016 in (0.37 to 0.43 mm)
Dwell angle	55° ± 3°
Ignition cutout speed	6500 ± 100 (engine) rpm

Engine idle speed *(USA models with Weber 32ADFA carburettor)*
Catalytic converter models	800 to 850 rpm
Non-catalytic converter models	800 to 900 rpm

Fast Idle speed *(USA models with Weber 32ADFA carburettor)*
All models	1550 to 1650 rpm

Distributor (USA models)
Make	Marelli S144CAY
Static advance	0° (TDC)
Centrifugal advance	36° ± 1° 30'
Dwell angle	55° ± 3°
Additional points gap	0.012 to 0.019 in (0.31 to 0.49 mm)

Electrical system
Regulator
Type	RC2/12B
Resistance between terminal 15 and earth	27.7 ± 2 ohms
Resistance between terminals 15 and 67 open contacts	5.65 ± 0.3 ohms

Fuses (UK Coupe model)

	Circuits protected
A (16 amp)	Inspection lamp socket
	Electro-pneumatic horns
	Electric clock
	Cooling fan motor
	Courtesy lights
	Lamp on rear view mirror
B (8 amp)	Heater blower motor
	Windscreen wiper
	Windscreen washer electric pump
C (8 amp)	LH headlamp main beam
	Headlamp main beam warning light

D (8 amp)   RH headlamp main beam

E (8 amp)   LH headlamp dipped beam

F (8 amp)   RH headlamp dipped beam

G (8 amp)   LH front parking light
                                               RH tail light
                                               LH licence plate light
                                               Cigar lighter illumination
                                               Instrument panel lights

H (8 amp)   RH front parking lights
                                               LH tail light
                                               RH licence plate light
                                               Reversing lights

I (8 amp)   Tachometer (rev-counter)
                                               Oil pressure gauge and warning light
                                               Water temperature gauge
                                               Fuel gauge and low fuel level warning light
                                               Direction indicators warning light
                                               Stop lights

L (8 amp)   Voltage regulator
                                               Alternator relay coil

Separate line fuses   Fuel pump/relay/relay coil (8 amp)
                                               Cigar lighter (16 amp)
Fuses (USA Coupe model)

                                               Circuits protected
A (25 amp)   Electro-pneumatic horns
                                               Engine cooling fan motor

B (8 amp)   Windscreen wiper motor
                                               Heater fan motor
                                               Windscreen washer pump

C (8 amp)   LH headlamps main beams
                                               Headlamp main beam warning light

D (8 amp)   RH headlamps main beams

E (8 amp)   LH outer headlamp dipped beam

F (8 amp)   RH outer headlamp dipped beam

G (8 amp)   Front LH parking light
                                               Parking and tail light warning light
                                               Rear RH tail light
                                               Front left/rear right side marker lights
                                               Ideogram illumination optical fibre light source
                                               LH licence plate illumination lamp
                                               Cigar lighter illumination
                                               Instrument panel illumination
                                               Hazard warning light switch illumination

H (8 amp)   Front RH parking lamp
                                               RH rear tail lamp
                                               Front right/rear left side marker lamps
                                               RH licence plate illumination lamp

I (8 amp)   Reverse lamps
                                               Oil pressure gauge and warning light
                                               Water temperature gauge
                                               Fuel gauge and reverse indicator
                                               Tachometer (rev-counter)
                                               Brake system effectiveness and hand brake ON indicator
                                               Stop lights
                                               Turn signal lights and warning light
                                               Electrovalve for diverter valve
                                               Relay winding of electrovalve for diverter valve
                                               Electrovalve for EEC system

	'Fasten seat belts' indicator
	'Remove key' and 'fasten seat belts' buzzer winding
	Belt/starter interlock system electronic control unit
	EGR indicator relay winding
	EGR warning system (25 000 miles)
	EGR indicator (25 000 miles)
L (8 amp)	Voltage regulator
	Alternator field winding
Separate line fuses	Remove key and fasten seat belts buzzer also Belt/starter interlock electronic control unit (3 amp)
	Fuel pump and relay (8 amp)
	Cigar lighter, clock, courtesy lights, inspection lamp receptacle - hazard warning light system (not switch illumination) (16 amp)
	Rear window demister and warning light (where applicable) (16 amp)
	EGR indicator reset device (3 amp)

Fuses (USA Spider models)

A (25 amp)	Electro-pneumatic horns
	Engine cooling fan motor
B (8 amp)	Windscreen wiper motor
	Heater fan motor
	Windscreen washer pump
C (8 amp)	LH headlamp main beam
	Main beam warning light
D (8 amp)	RH headlamp main beam
E (8 amp)	RH headlamp dipped beam
F (8 amp)	LH headlamp dipped beam
G (8 amp)	Front LH parking lamp
	Parking and tail light warning light
	Rear RH tail light
	Front left/right rear side marker lamps
	LH licence plate lamp
	Cigar lighter illumination
	Trunk light
	Instruments illumination
	Ideogram illumination optical fibre light source
	Hazard warning lights switch illumination
H (8 amp)	Front RH parking lamp
	Rear LH tail light
	Front right/rear left side marker lamps
	RH licence plate lamp
I (8 amp)	Turn signal lights and warning light
	Stop lights
	Oil pressure gauge and oil warning light
	Water temperature gauge
	Fuel gauge and reserve indicator
	Tachometer (rev-counter)
	Brake system effectiveness and handbrake ON warning light
	Reversing lights
	Electrovalve for EEC system
	'Fasten seat belts indicator' and relay for buzzer
	Starter/belt interlock electronic control unit
	Idle stop solenoid
	Electrovalve for diverter valve
	Relay winding of electrovalve for diverter valve
	EGR indicator relay winding
	EGR warning system (25 000 miles)
	EGR indicator (25 000 miles)
L (8 amp)	Voltage regulator
	Alternator field winding

Separate line fuses

Remove key and fasten seat belts buzzer, starter/belt interlock electronic control unit (3 amp)
EGR indicator reset device (3 amp)
Cigar lighter, clock, courtesy light, hazard warning and indicator light, inspection lamp receptacle
Fuel pump and relay (8 amp)

4 Engine

General note
1   Engine removal is almost identical to the procedure described in Chapter 1 with exceptions as follows:
a   Later Spider and Coupe models are fitted with an electric fuel pump; therefore, ignore the instructions relating to the mechanical type (engine driven) fuel pump given in Chapter 1.
b   Later USA models are fitted with extra emission control equipment; therefore, refer to the Sections of this Chapter when disconnecting or removing any items not dealt with in Chapter 3.

5 Fuel system, carburation and emission control

Carburettors - general note
1   Refer to the General Data Section of this Chapter for information on those carburettors not covered in Chapter 3.
2   The Weber and Solex carburettors fitted to later models are very similar in appearance and function to the compound type described in Chapter 3.
3   It will be noted that those carburettors fitted to models destined for the USA have special features which help to reduce exhaust emissions to a minimum (eg the installation of a fuel shut-off valve to prevent the engine running-on after the ignition is switched off and to shut off the fuel supply during

deceleration with the throttle completely closed at an engine speed higher than 2650 ± 50 rpm). Shutting off the fuel during deceleration with the throttle completely closed is achieved through an electrical contact on the idle speed adjusting screw and through a tachymetric switch. This also helps to prevent the catalytic converter from overheating.

Carburettor (USA models) - idle CO setting procedure
4   First ensure that the valve clearances, ignition timing and contact breaker points gap (where applicable) have all been checked, and where necessary adjustments have been made. It is also essential to ensure that the engine has reached normal operating temperature before attempting to make any carburettor adjustments.
5   Connect a tachometer to the engine and insert the sampling probe of a CO tester into the exhaust tail-pipe.
6   Pinch off the air injection rubber tubing to the exhaust manifold, between the diverter valve and the check valve, using a pair of pliers or a suitable clamp.
7   Start the engine and check that the readings indicated on the two instruments are as indicated on the specification tag in the engine compartment.
8   If necessary, set the engine rpm by turning the idle speed adjusting screw clockwise to increase the rpm, and anti-clockwise to decrease the rpm.
9   The idle mixture can be set to the correct value by turning the mixture adjusting screw located at the base of the carburettor. Turning the screw in a clockwise direction will decrease the CO percentage (weaken the mixture), and turning in an anti-clockwise direction will increase the CO percentage

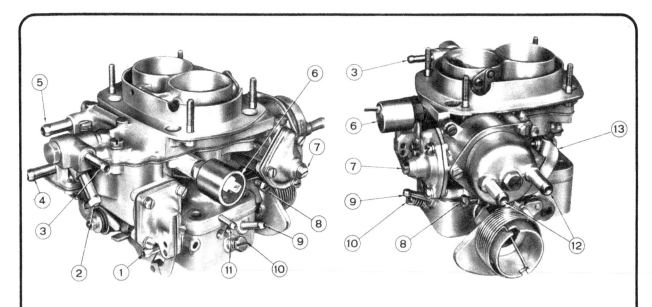

FIG. 13.1   WEBER 32 ADFA CARBURETTOR AS FITTED TO LATER USA SPIDER AND COUPE MODELS

1 *Blow-by connection*	6 *Idle stop solenoid*	9 *Canister tubing connection*
2 *Fuel inlet*	7 *Diaphragm device for partial*	10 *Idle mixture adjustment*
3 *Fuel recirculation outlet*	*opening of choke valves*	*screw*
4 *Fast idle connection*	8 *Idle speed adjustment screw*	11 *EGR tubing connection*
5 *Bowl vapour vent*		

12 *Automatic choke system water heating connections*
13 *Choke fast idle adjustment screw*

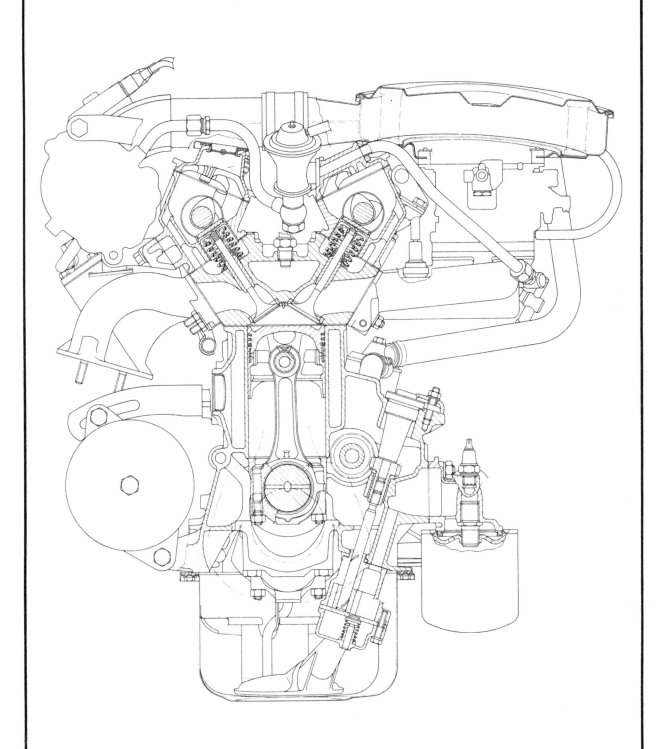

Fig. 13.2 Sectional end view of the 132 Series engine fitted to later USA Coupe and Spider models with emission control equipment

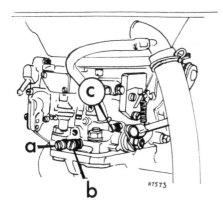

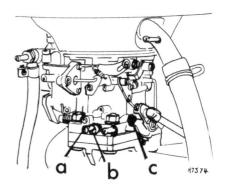

FIG. 13.3   WEBER 34 DMS CARBURETTOR AS FITTED TO SOME LATER UK COUPE MODELS

A *Slow running volume*
   adjusting screw   *adjusting screw*
B *Slow running speed*   C *Throttle stop screw*

FIG. 13.4   SOLEX C34EIES5 CARBURETTOR AS FITTED TO SOME LATER UK COUPE MODELS

A *Slow running volume*
   adjusting screw   *adjusting screw*
B *Slow running speed*   C *Throttle stop screw*

(enriches the mixture)

10 After adjusting the idle mixture it may be necessary to re-adjust the idle speed. Continue adjusting both the idle speed and mixture screws until both instruments read as indicated on the specification plate.

11 Remove the pliers or clamp from the air hose, then remove the tachometer and exhaust gas analyser.

Electric fuel pump - general description and location

12 Later UK and USA models are fitted with an electric fuel pump instead of the mechanical engine-driven type. The electric fuel pump is mounted on the rear bulkhead inside the luggage boot.

13 The two parts of the pump can be split to give access to the filter.

14 If the electrical operation or the diaphragm and valves fail, it is recommended that a new unit is fitted. The two separate halves of the pump are supplied as spares, but it is usual for retailers only to stock the complete unit as, if one part has failed, the other is likely to be near the end of its life.

Electric fuel pump - testing

15 If the electric fuel pump should fail it is first advisable to check that the line fuse protecting the fuel pump and its relay has not blown. This check can easily be carried out by either locating and inspecting the line fuse, or by disconnecting the live feed wire at the pump and substituting a 12 volt test lamp in place of the pump. Failure of the test lamp bulb to light-up will indicate either a blown fuse or a break in the wiring.

16 Assuming that the live feed is in order then it is reasonable to assume that the pump is faulty. Removal and refitting procedures are described in the next Section.

Electric fuel pump - removal and refitting

17 Disconnect the battery earth (negative) lead.

18 From within the luggage boot remove the cover panel from the top of the spare wheel.

19 Locate the fuel pump and disconnect the electrical feed wire at the plastic block connector.

20 Identify the inlet and outlet fuel pipes, and release their retaining clips. Carefully pull off the pipes from the fuel pump connections.

21 Undo the retaining bolts, and release the pump and earth wire.

22 Refitting the pump is the reverse of the removal procedure.

23 After installation of the pump, reconnect the battery earth

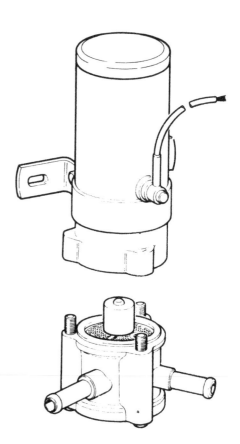

Fig. 13.5   The electric fuel pump

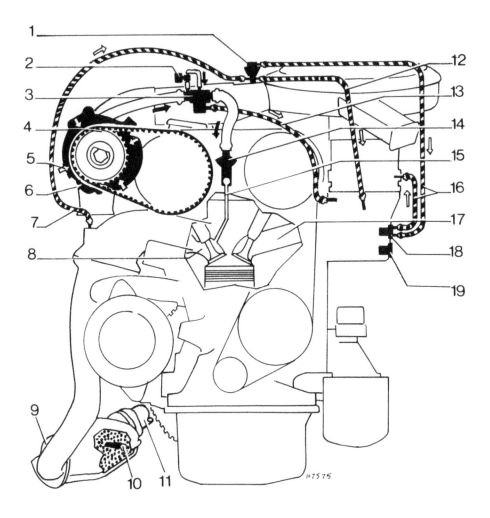

FIG. 13.6   TYPICAL LAYOUT OF THE EMISSION CONTROL SYSTEM (CALIFORNIAN MODEL SHOWN)

1 Exhaust gas recirculation
   (EGR) control valve
2 Electrovalve (normally
   closed) for diverter valve
3 Diverter valve
4 Air distribution line

5 Air intake
6 Air pump
7 EGR tapping line
8 Exhaust manifold
9 Catalytic converter
10 Thermo-couple

11 Exhaust pipe
12 Exhaust gas feedback line
13 Vacuum tapping line (intake
   manifold) for diverter valve
14 Air injection non-return valve
15 Air injector

16 Vacuum tapping line
   (carburettor/EGR valve)
17 Inlet manifold
18 EGR thermovalve
19 Thermostatic switch for
   electrovalve 2

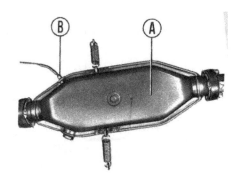

**FIG. 13.7   THE CATALYTIC CONVERTER INSTALLED
IN THE EXHAUST SYSTEM OF CALIFORNIAN MODELS**

A Catalytic converter
B Thermo-couple
   (heat sensor) unit

lead, switch on the ignition and check the pipe connections
for possible leakage.

*Emission control systems - general description and
changes*
24 To meet the latest emission control regulations has resulted
in the fitting of additional equipment. The function of each
part is briefly described below.

Exhaust gas recirculation (EGR)
25 In this system, a small part of the exhaust gas is introduced
into the combustion chamber to lower the spark flame temperature during combustion to reduce the nitrogen oxide content
of the exhaust gases. The system used for 1974 models consists
of a check and flow control valve activated by a vacuum signal
from the carburettor, an EGR valve control vacuum signal check
thermovalve actuated by the engine cooling water temperature,
and tapping points in the inlet and exhaust manifolds.
26 At 25 000 mile intervals an EGR warning light on the
dashboard illuminates to remind the driver that the EGR system
requires servicing.

202

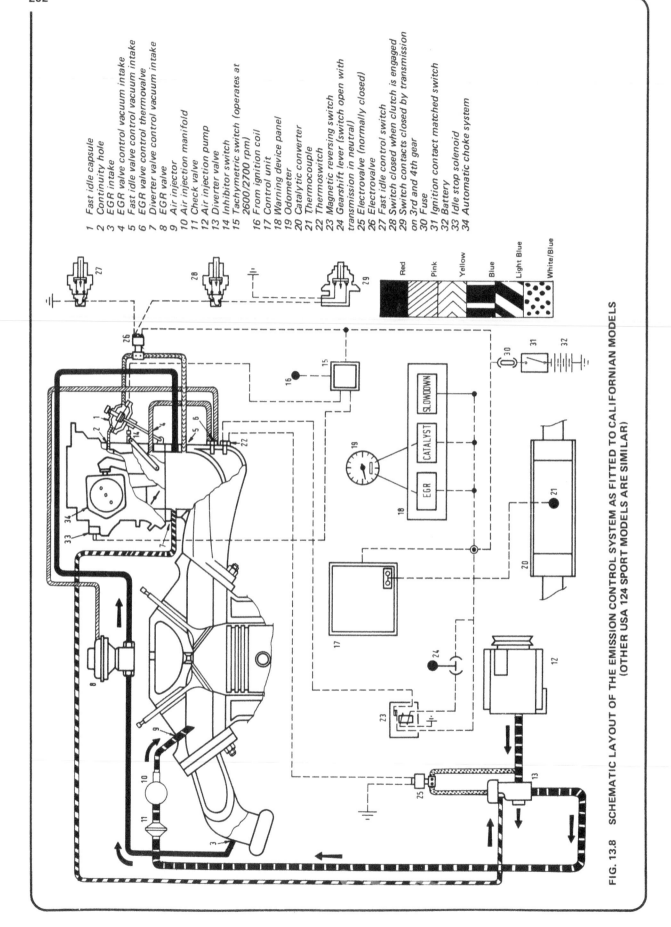

1 Fast idle capsule
2 Continuity hole
3 EGR intake
4 EGR valve control vacuum intake
5 Fast idle valve control vacuum intake
6 EGR valve control thermovalve
7 Diverter valve control vacuum intake
8 EGR valve
9 Air injector
10 Air injection manifold
11 Check valve
12 Air injection pump
13 Diverter valve
14 Inhibitor switch
15 Tachymetric switch (operates at 2600/2700 rpm)
16 From ignition coil
17 Control unit
18 Warning device panel
19 Odometer
20 Catalytic converter
21 Thermocouple
22 Thermoswitch
23 Magnetic reversing switch
24 Gearshift lever (switch open with transmission in neutral)
25 Electrovalve (normally closed)
26 Electrovalve
27 Fast idle control switch
28 Switch closed when clutch is engaged
29 Switch contacts closed by transmission on 3rd and 4th gear
30 Fuse
31 Ignition contact matched switch
32 Battery
33 Idle stop solenoid
34 Automatic choke system

Red
Pink
Yellow
Blue
Light Blue
White/Blue

FIG. 13.8   SCHEMATIC LAYOUT OF THE EMISSION CONTROL SYSTEM AS FITTED TO CALIFORNIAN MODELS (OTHER USA 124 SPORT MODELS ARE SIMILAR)

Air injection

27 This is a method of injecting air from an external compressor into the exhaust manifold in order to reduce hydrocarbons and carbon monoxide in the exhaust gas by providing conditions favourable for recombustion. The system comprises an engine driven air pump, diverter valve, check valve, air injection manifold air injectors and the associated hoses.

28 In this system, air is compressed by the air pump and directed through the check valve to the air injection manifold and air injectors. During high speed operation, excessive pump pressure is vented to the atmosphere through the diverter valve.

29 The check valve is screwed into the air injection manifold. The function of this valve is to prevent any exhaust gases passing into the air pump should the exhaust gas pressure in the exhaust manifold be greater than the air pump injection pressure. It is designed to close against the exhaust manifold pressure should the air pump fail as a result, for example, of a broken drivebelt.

Catalytic converter

30 Installed in the exhaust system of vehicles destined for California, this device speeds up the chemical reaction of the hydrocarbons and carbon monoxide present in the exhaust gases so that they change into harmless carbon dioxide and water. Air for the chemical process is supplied by the air injection pump.

31 In the event of the system overheating, a thermo-couple attached to the catalytic converter senses that the unit is overheating and relays a signal to the driver by activating the *Slow Down* warning light. During normal operating conditions, the warning lamp is illuminated during the engine start sequence

as an indication of its serviceability. It is not unusual for the warning lamp to come on during periods of hard driving, or climbing gradients for long periods in low gears.

32 At periods of 25 000 miles the catalytic converter must be renewed and the driver is warned that this mileage period has elapsed by a *Catalyst* warning lamp on the dashboard.

Emission control systems - checking and maintenance

Air pump drivebelt

33 Periodically check the air pump drivebelt tension and its condition.

34 The belt is of the non-adjustable type and, if found to be either slack or worn, must be renewed.

Diverter valve test

35 The valve can be checked by accelerating the engine and allowing the throttle to close quickly. A slight momentary rush of air should be heard at the diverter valve air outlet.

Check valve test

36 Disconnect the air supply hose at the check valve and start the engine.

37 Listen for exhaust leakage at the check valve. A slight fluttering of the valve is normal at idle speed.

Air pump test

38 Disconnect the output hose at the air pump and start the engine.

39 Check that a flow of air is being emitted from the pump outlet.

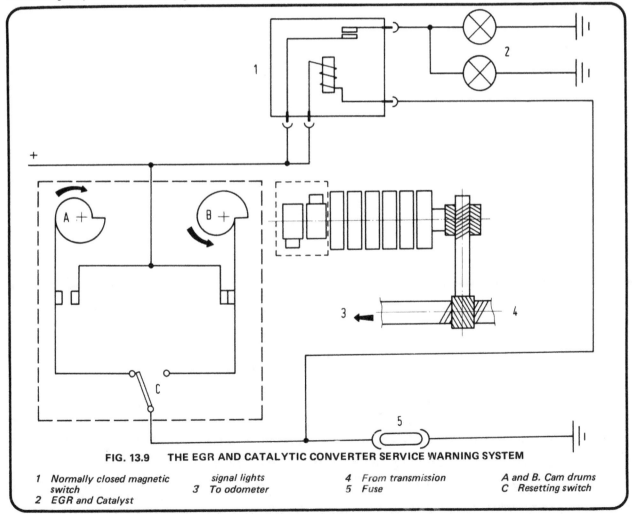

FIG. 13.9   THE EGR AND CATALYTIC CONVERTER SERVICE WARNING SYSTEM

1 Normally closed magnetic switch
2 EGR and Catalyst signal lights
3 To odometer
4 From transmission
5 Fuse
A and B. Cam drums
C Resetting switch

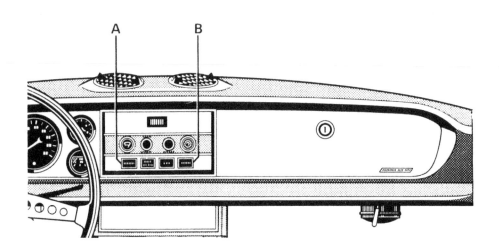

FIG. 13.10   THE CATALYTIC CONVERTER 'SLOW DOWN' AND 25 000 MILE SERVICE WARNING LAMPS

A Catalyst indicator   B 'Slow down' indicator

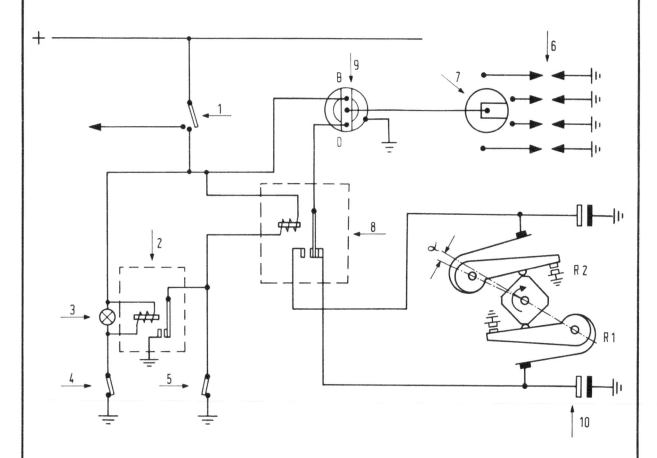

FIG. 13.11   CIRCUIT DIAGRAM FOR THE DUAL—POINT DISTRIBUTOR AS FITTED TO SOME USA MODELS

1 Ignition switch	5 Engine water-sensitive thermoswitch for dist-ributor	8 Ignition change-over switch
2 Cold start relay		9 Ignition coil
3 Oil pressure indicator light	6 Spark plugs	10 Capacitor
4 Oil pressure switch	7 Ignition distributor	R1 Breaker points

R2 Auxiliary breaker points operating by cold engine water and insufficient oil pressure
X = 10°

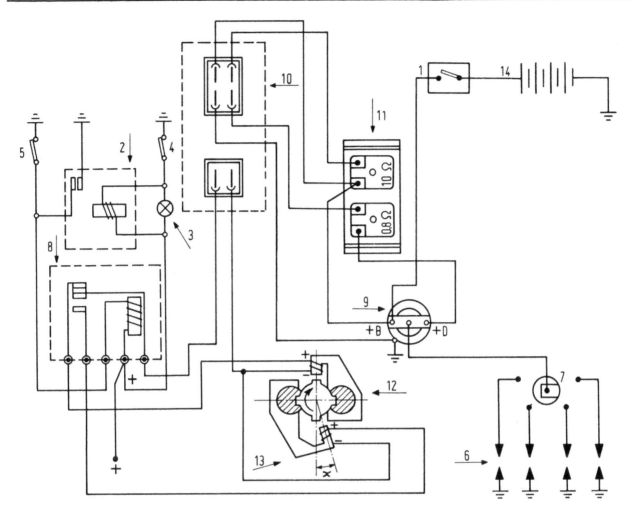

FIG. 13.12   CIRCUIT DIAGRAM FOR THE BREAKERLESS DISTRIBUTOR FITTED AS AN OPTION ON SOME USA MODELS

1 Ignition switch	thermoswitch for distributor	9 Ignition coil	12 Main detector
2 Cold start relay	6 Spark plugs	10 Inductive discharge	13 Auxiliary detector
3 Oil pressure indicator light	7 Ignition distributor	electronic control	14 Battery
4 Oil pressure switch	8 Ignition change-over switch	11 Dissipator with resistors	X = 10º
5 Engine water-sensitive			

6   Ignition system

Dual-point distributor - general description

1   Some later USA models are fitted with a distributor which has two sets of contact breaker points. The purpose of the additional set is to facilitate cold starting at cold temperatures. The additional contacts open at 10º BTDC and so give an advanced spark for easy starting.

2   Briefly the system functions as follows: When the engine coolant temperature is lower than 41 ± 5º F, as sensed by a water temperature thermoswitch and an insufficient oil pressure sensor, the engine is started and run through the additional set of contacts. When the coolant temperature reaches 59 ± 5º F the thermoswitch cuts out the additional contact set and the main contact set is brought into operation.

Ignition timing with distributor removed (dual-point type)

Note: *To accurately set the ignition timing requires the use of a stroboscopic timing light and a tachometer/dwell angle meter.*

3   Rotate the engine crankshaft until No 1 cylinder is on the compression stroke and the notch on the crankshaft pulley is lined up with the fixed reference which indicates 0º (TDC) basic ignition timing.

4   Remove the distributor cap and check that the contact breaker points gap is set to 0.012/0.014 in (0.31/0.49 mm). If necessary adjust both sets of points as described in Chapter 4.

5   Rotate the distributor spindle until the rotor arm is pointing to the insert in the distributor cap which is connected to the No 1 spark plug lead. .

6   Without removing the distributor shaft from this set position, insert it into the crankcase and secure it in position.

7   Now reconnect the wires to the ignition coil and distributor, and the HT leads to the spark plugs.

8   Connect the tachometer/dwell angle meter and the strobo- scopic timing light to the engine in accordance with the manufacturer's instructions.

9   Start the engine and after the warm-up period, check for a dwell angle of 55 ± 3º. If the reading is not within these limits, stop the engine and adjust the main contact breaker points as necessary to obtain the desired setting.

10 Now check the ignition timing using the stroboscopic

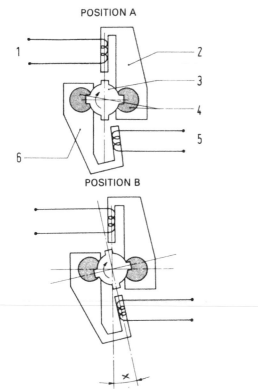

FIG. 13.13   THE NORMAL RUNNING POSITION A (0º/TDC)
AND THE COLD START POSITION B (10º BTDC) OF THE
BREAKERLESS DISTRIBUTOR

1   *Main detector*   5   *Auxiliary detector*
2   *Support*   6   *Support*
3   *Trigger wheel*   *X = 10º*
4   *Permanent magnet*

timing light; this should be 0º (TDC) at 850 engine rpm.
11 Adjustment of the timing is by rotating the distributor body,
after first loosening the clamp bolt, as described in Chapter 4.

Breakerless distributor - general description

12 This type of distributor, is offered as an alternative to the
dual point type, and is of the inductive discharge type.
Basically it is similar to the dual point distributor with a few
exceptions. The breaker points and cam used in the dual point
type are replaced by magnetic sensors and a trigger wheel.
13 The system consists of two permanent magnets and two
coils (one main and one auxiliary) as shown in Fig. 13.12.
14 The auxiliary detector is set in the 10º BTDC position and
has the function of advancing the spark when 'cold-starting'
the engine (shown as Position A in Fig. 13.13).
15 When the engine coolant temperature reaches 59 ± 5º F,
a thermoswitch will switch out the operation of the auxiliary
detector and the main detector will be brought into operation
(shown as Position B in Fig. 13.13).
16 When the trigger wheel is in Positions A or B, an electro-
motive force will be induced in the coil and transmitted to the
ignition electronic control to initiate a spark in the ignition coil.
17 The mechanical advance and distribution to the spark plugs
is identical to that of the conventional distributor.
18 The purpose of the electronic control unit is to process
the voltage pulse coming from the magnetic sensors and transmit
it to the primary winding of the ignition coil.
19 The main advantage of this system, when compared with
the conventional type, is that the breaking of the ignition coil
primary circuit is done by electronic means and not mechanical
so there are less adjustments to be made, and the amount of
maintenance is greatly reduced. Also the system is more reliable
and less prone to inaccuracy due to wear or contact point faults.

Ignition timing with distributor removed (breakerless type)

20 The procedure is identical to that described for the dual point
system, but ignore references to setting the contact points gap
and use of the dwell angle meter.

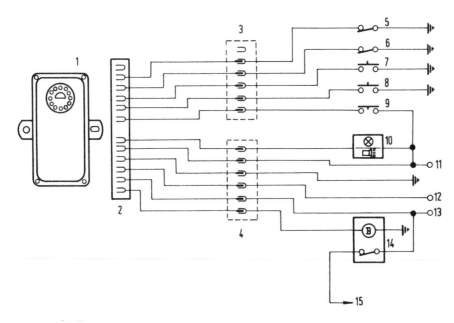

FIG. 13.14   SEAT BELT/STARTER INTERLOCK SYSTEM CIRCUIT DIAGRAM (AS FITTED TO SOME USA MODELS)

1   *Electronic control unit*   6   *Passenger belt retractor*   9   *Transmission switch*   12   *To battery positive terminal*
2   *12-way female connector*   *switch*   10   *Indicator and buzzer*   13   *Starting*
3   *Connector*   7   *Driver seat switch*   *warning system*   14   *Relay*
4   *Connector*   8   *Passenger seat switch*   11   *Ignition*   15   *To starter motor solenoid*
5   *Driver belt retractor switch*

FIG. 13.15   KEY TO WIRING DIAGRAM FOR LATER UK COUPE MODELS

1 Side lights/front direction indicators (5/21 watts)
2 Halogen headlamps - dipped beam
3 Halogen headlamps - main beam
4 Horn relay
5 Air horn compressor
6 Radiator fan motor
7 Ignition coil
8 Oil pressure gauge transmitter
9 Distributor
10 Spark plugs
11 Alternator
12 Radiator fan control switch
13 Repeater lights (4 watts)
14 Water temperature warning transmitter (moves gauge
     needle to full scale deflection over-riding temperature sensor)
15 Oil pressure warning transmitter
16 Headlamp relay
17 Water temperature gauge ballast resistor
18 Water temperature gauge transmitter
19 Voltage regulator
20 Radiator fan relay
21 Ignition warning relay
22 Starter motor
23 Battery
24 Fuel pump relay
25 Stop light switch
26 Fuel pump/relay fuse
27 Fuse unit
28 Heated backlight fuse (optional)
29 Heated backlight relay (optional)
30 Screen-washer pump
31 Cigar lighter fuse
32 Wiper motor
33 Wiper interruptor relay
34 Direction indicator flasher
35 Heated backlight switch and warning light
     (1.2 watts) (optional)
36 Power point
37 Side light w/l (3 watts) (green)

38 Speedometer lights (3 watts)
39 Direction indicator w/l (3 watts)
40 Headlamp w/l (3 watts)
41 Tachometer light (3 watts)
42 Ignition w/l (3 watts)
43 Tachometer
44 Fuel gauge
45 Fuel w/l
46 Fuel gauge light (3 watts)
47 Oil pressure w/l (3 watts)
48 Oil pressure gauge light (3 watts)
49 Oil pressure gauge
50 Water temperature gauge
51 Water temperature gauge light (3 watts)
52 Clock
53 Clock light (3 watts)
54 Light switch
55 Wiper speed selector switch
56 Instrument light dimmer
57 Panel light switch
58 Ignition switch
59 Door pillar switch
60 Headlamp switch
61 Direction indicator switch
62 Wiper/washer switch
63 Horn switch
64 Rear courtesy lights/switches (5 watts)
65 Heater fan switch
66 Cigar lighter/lamp (4 watts)
67 Heater fan motor
68 Front courtesy light/switch (5 watts)
69 Reversing light switch
70 Fuel pump
71 Fuel transmitter
72 Heated backlight (optional)
73 Rear direction indicators (21 watts)
74 Rear/stop lights (5/21 watts)
75 Reversing lights (21 watts)
76 Number plate light (5 watts)

Cable Colour Code

Arancio	=	Amber	Marrone	=	Brown
Azzurro	=	Light blue	Nero	=	Black
Bianco	=	White	Rosa	=	Pink
Blu	=	Dark blue	Rosso	=	Red
Giallo	=	Yellow	Verde	=	Green
Grigio	=	Grey	Viola	=	Mauve

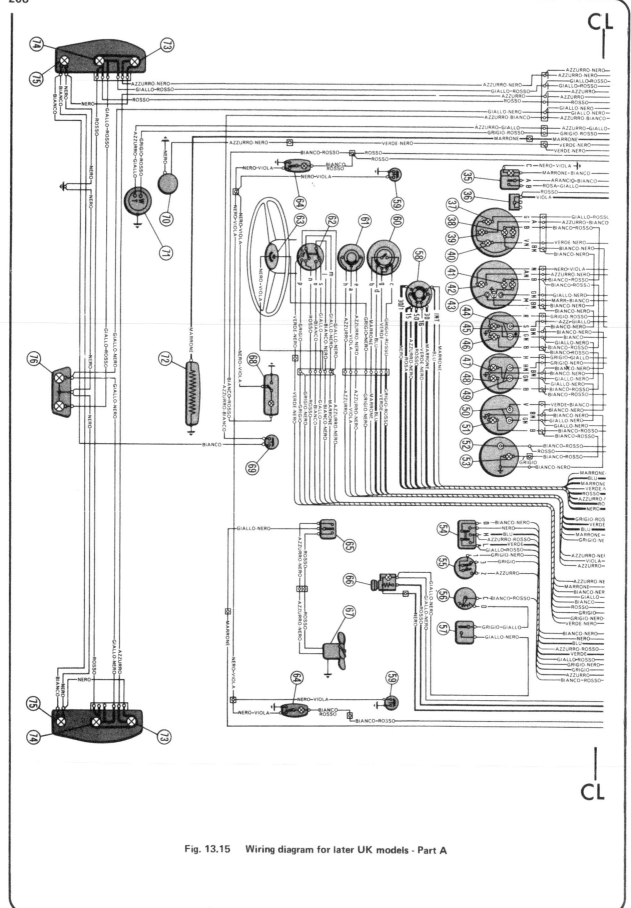

Fig. 13.15   Wiring diagram for later UK models - Part A

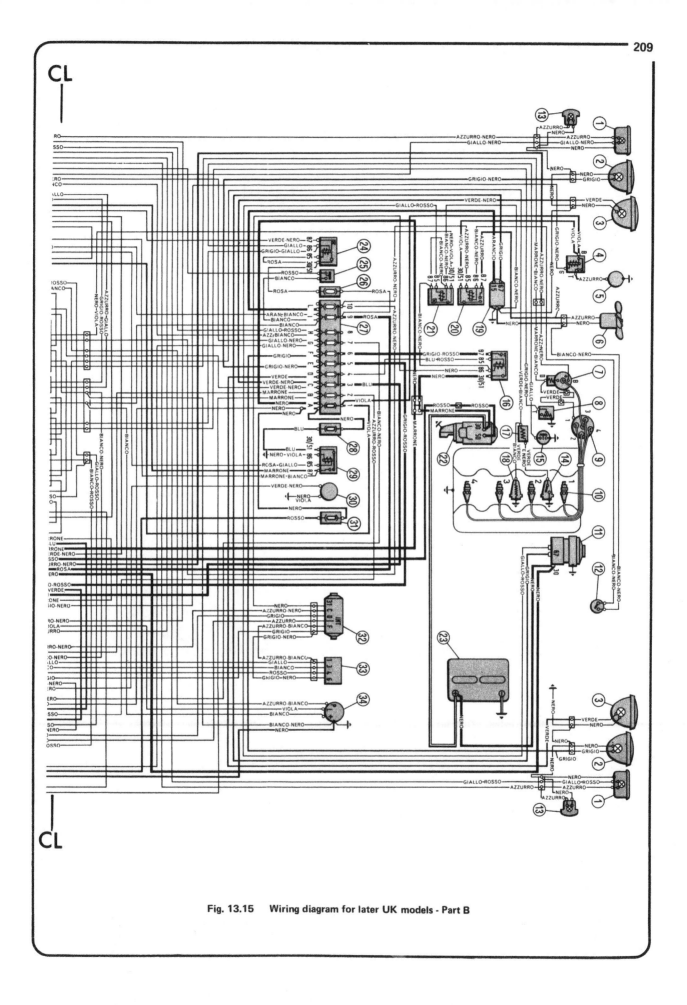

Fig. 13.15   Wiring diagram for later UK models - Part B

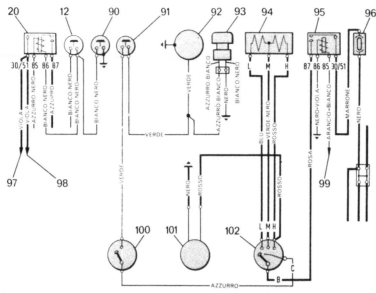

FIG. 13.16   TYPICAL CIRCUIT WIRING DIAGRAM FOR COUPE MODELS WITH OPTIONAL AIR CONDITIONING
EQUIPMENT (FOR COLOUR CODE SEE FIG. 13.15)

12 Temperature switch (on radiator)	92 Compressor	97 To terminal B on horn relay	100 Temperature and on/off control
20 Radiator fan relay	93 Fast idle solenoid valve	98 To fuse holder terminal 1 (see 27)	101 Fan motor (in place of heater motor - see 67)
90 Temperature switch (on condenser)	94 Fan speed selector resistor	99 To terminal + on direction indicator flasher	102 Fan speed selector switch
91 Low pressure switch	95 Master relay		
	96 In-line fuse		

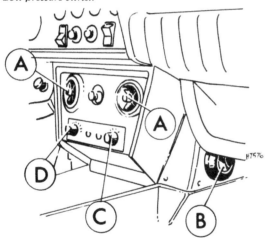

FIG. 13.17   AIR CONDITIONING EQUIPMENT DRIVER'S
CONTROL PANEL

A   Centre vents   C   Temperature control
B   Lower vents   D   Master switch/blower control

FIG. 13.18   THE AIR CONDITIONING COMPRESSOR UNIT
AND DRIVEBELTS

A   Retaining bolts   B   Compressor unit

---

7   Electrical system

Seat belt/starter interlock system - general description
1   Some USA models are fitted with a seat belt/starter interlock system as a safety feature. An electronic control unit ensures that the front seat occupants wear their seat belts by preventing operation of the starter, sounding a buzzer and operating a warning light.
2   On later models this system was modified to allow starting of the engine without fastening the seat belts; however, the *Fasten Belts* warning light and buzzer will operate for a few seconds.

---

8   Bodywork and fittings

Air conditioning equipment (where fitted) - maintenance
1   The air conditioning equipment fitted as an optional extra is generally trouble-free in operation. FIAT recommend that the owner inspects the compressor drivebelts at periods of 6000 miles (10 000 Km).
2   The drivebelts, if found to be split or frayed, should be renewed as a matched pair.
3   It is also necessary to check the belt tension at a point mid-way between the longest belt run. Apply a firm pressure of 22 lb (10 kg) at this point and, if the belt is correctly tensioned,

it will deflect 0.4 in (10 mm).

4 When it is necessary to either renew or adjust the belt tension, slacken the three mounting screws (marked A in Fig. 13.18), and move the compressor either inwards (towards the engine) to remove the belts, or outwards (away from the engine) to tension the belts.

5 In the event of the system becoming defective, it is essential that the repairs are entrusted to your FIAT dealer or a competent refrigeration engineer. Repairs on this type of system are beyond the scope of the average owner.

6 FIAT also recommend that the system is operated for a few minutes at regular intervals (even during the Winter) to adequately lubricate the compressor gland seal.

Heated rear window - general description

7 A heated rear window was offered as an optional extra on Coupe models only. The system consists of a control switch and warning light, and a special rear screen with heating element strips. A separate line fuse, rated at 16 amps protects the system.

8 Removal and refitting of the rear screen is identical to that described in Chapter 12, except that the wiring connections must be released.

General repair procedures

Whenever servicing, repair or overhaul work is carried out on the car or its components, it is necessary to observe the following procedures and instructions. This will assist in carrying out the operation efficiently and to a professional standard of workmanship.

Joint mating faces and gaskets

Where a gasket is used between the mating faces of two components, ensure that it is renewed on reassembly, and fit it dry unless otherwise stated in the repair procedure. Make sure that the mating faces are clean and dry with all traces of old gasket removed. When cleaning a joint face, use a tool which is not likely to score or damage the face, and remove any burrs or nicks with an oilstone or fine file.

Make sure that tapped holes are cleaned with a pipe cleaner, and keep them free of jointing compound if this is being used unless specifically instructed otherwise.

Ensure that all orifices, channels or pipes are clear and blow through them, preferably using compressed air.

Oil seals

Whenever an oil seal is removed from its working location, either individually or as part of an assembly, it should be renewed.

The very fine sealing lip of the seal is easily damaged and will not seal if the surface it contacts is not completely clean and free from scratches, nicks or grooves. If the original sealing surface of the component cannot be restored, the component should be renewed.

Protect the lips of the seal from any surface which may damage them in the course of fitting. Use tape or a conical sleeve where possible. Lubricate the seal lips with oil before fitting and, on dual lipped seals, fill the space between the lips with grease.

Unless otherwise stated, oil seals must be fitted with their sealing lips toward the lubricant to be sealed.

Use a tubular drift or block of wood of the appropriate size to install the seal and, if the seal housing is shouldered, drive the seal down to the shoulder. If the seal housing is unshouldered, the seal should be fitted with its face flush with the housing top face.

Screw threads and fastenings

Always ensure that a blind tapped hole is completely free from oil, grease, water or other fluid before installing the bolt or stud. Failure to do this could cause the housing to crack due to the hydraulic action of the bolt or stud as it is screwed in.

When tightening a castellated nut to accept a split pin, tighten the nut to the specified torque, where applicable, and then tighten further to the next split pin hole. Never slacken the nut to align a split pin hole unless stated in the repair procedure.

When checking or retightening a nut or bolt to a specified torque setting, slacken the nut or bolt by a quarter of a turn, and then retighten to the specified setting.

Locknuts, locktabs and washers

Any fastening which will rotate against a component or housing in the course of tightening should always have a washer between it and the relevant component or housing.

Spring or split washers should always be renewed when they are used to lock a critical component such as a big-end bearing retaining nut or bolt.

Locktabs which are folded over to retain a nut or bolt should always be renewed.

Self-locking nuts can be reused in non-critical areas, providing resistance can be felt when the locking portion passes over the bolt or stud thread.

Split pins must always be replaced with new ones of the correct size for the hole.

Special tools

Some repair procedures in this manual entail the use of special tools such as a press, two or three-legged pullers, spring compressors etc. Wherever possible, suitable readily available alternatives to the manufacturer's special tools are described, and are shown in use. In some instances, where no alternative is possible, it has been necessary to resort to the use of a manufacturer's tool and this has been done for reasons of safety as well as the efficient completion of the repair operation. Unless you are highly skilled and have a thorough understanding of the procedure described, never attempt to bypass the use of any special tool when the procedure described specifies its use. Not only is there a very great risk of personal injury, but expensive damage could be caused to the components involved.

Conversion factors

Length (distance)

Inches (in)	X	25.4	= Millimetres (mm)	X	0.0394	= Inches (in)
Feet (ft)	X	0.305	= Metres (m)	X	3.281	= Feet (ft)
Miles	X	1.609	= Kilometres (km)	X	0.621	= Miles

Volume (capacity)

Cubic inches (cu in; in³)	X	16.387	= Cubic centimetres (cc; cm³)	X	0.061	= Cubic inches (cu in; in³)
Imperial pints (Imp pt)	X	0.568	= Litres (l)	X	1.76	= Imperial pints (Imp pt)
Imperial quarts (Imp qt)	X	1.137	= Litres (l)	X	0.88	= Imperial quarts (Imp qt)
Imperial quarts (Imp qt)	X	1.201	= US quarts (US qt)	X	0.833	= Imperial quarts (Imp qt)
US quarts (US qt)	X	0.946	= Litres (l)	X	1.057	= US quarts (US qt)
Imperial gallons (Imp gal)	X	4.546	= Litres (l)	X	0.22	= Imperial gallons (Imp gal)
Imperial gallons (Imp gal)	X	1.201	= US gallons (US gal)	X	0.833	= Imperial gallons (Imp gal)
US gallons (US gal)	X	3.785	= Litres (l)	X	0.264	= US gallons (US gal)

Mass (weight)

Ounces (oz)	X	28.35	= Grams (g)	X	0.035	= Ounces (oz)
Pounds (lb)	X	0.454	= Kilograms (kg)	X	2.205	= Pounds (lb)

Force

Ounces-force (ozf; oz)	X	0.278	= Newtons (N)	X	3.6	= Ounces-force (ozf; oz)
Pounds-force (lbf; lb)	X	4.448	= Newtons (N)	X	0.225	= Pounds-force (lbf; lb)
Newtons (N)	X	0.1	= Kilograms-force (kgf; kg)	X	9.81	= Newtons (N)

Pressure

Pounds-force per square inch (psi; lbf/in²; lb/in²)	X	0.070	= Kilograms-force per square centimetre (kgf/cm²; kg/cm²)	X	14.223	= Pounds-force per square inch (psi; lbf/in²; lb/in²)
Pounds-force per square inch (psi; lbf/in²; lb/in²)	X	0.068	= Atmospheres (atm)	X	14.696	= Pounds-force per square inch (psi; lbf/in²; lb/in²)
Pounds-force per square inch (psi; lbf/in²; lb/in²)	X	0.069	= Bars	X	14.5	= Pounds-force per square inch (psi; lbf/in²; lb/in²)
Pounds-force per square inch (psi; lbf/in²; lb/in²)	X	6.895	= Kilopascals (kPa)	X	0.145	= Pounds-force per square inch (psi; lbf/in²; lb/in²)
Kilopascals (kPa)	X	0.01	= Kilograms-force per square centimetre (kgf/cm²; kg/cm²)	X	98.1	= Kilopascals (kPa)
Millibar (mbar)	X	100	= Pascals (Pa)	X	0.01	= Millibar (mbar)
Millibar (mbar)	X	0.0145	= Pounds-force per square inch (psi; lbf/in²; lb/in²)	X	68.947	= Millibar (mbar)
Millibar (mbar)	X	0.75	= Millimetres of mercury (mmHg)	X	1.333	= Millibar (mbar)
Millibar (mbar)	X	0.401	= Inches of water (inH₂O)	X	2.491	= Millibar (mbar)
Millimetres of mercury (mmHg)	X	0.535	= Inches of water (inH₂O)	X	1.868	= Millimetres of mercury (mmHg)
Inches of water (inH₂O)	X	0.036	= Pounds-force per square inch (psi; lbf/in²; lb/in²)	X	27.68	= Inches of water (inH₂O)

Torque (moment of force)

Pounds-force inches (lbf in; lb in)	X	1.152	= Kilograms-force centimetre (kgf cm; kg cm)	X	0.868	= Pounds-force inches (lbf in; lb in)
Pounds-force inches (lbf in; lb in)	X	0.113	= Newton metres (Nm)	X	8.85	= Pounds-force inches (lbf in; lb in)
Pounds-force inches (lbf in; lb in)	X	0.083	= Pounds-force feet (lbf ft; lb ft)	X	12	= Pounds-force inches (lbf in; lb in)
Pounds-force feet (lbf ft; lb ft)	X	0.138	= Kilograms-force metres (kgf m; kg m)	X	7.233	= Pounds-force feet (lbf ft; lb ft)
Pounds-force feet (lbf ft; lb ft)	X	1.356	= Newton metres (Nm)	X	0.738	= Pounds-force feet (lbf ft; lb ft)
Newton metres (Nm)	X	0.102	= Kilograms-force metres (kgf m; kg m)	X	9.804	= Newton metres (Nm)

Power

Horsepower (hp)	X	745.7	= Watts (W)	X	0.0013	= Horsepower (hp)

Velocity (speed)

Miles per hour (miles/hr; mph)	X	1.609	= Kilometres per hour (km/hr; kph)	X	0.621	= Miles per hour (miles/hr; mph)

Fuel consumption*

Miles per gallon, Imperial (mpg)	X	0.354	= Kilometres per litre (km/l)	X	2.825	= Miles per gallon, Imperial (mpg)
Miles per gallon, US (mpg)	X	0.425	= Kilometres per litre (km/l)	X	2.352	= Miles per gallon, US (mpg)

Temperature

Degrees Fahrenheit = (°C x 1.8) + 32

Degrees Celsius (Degrees Centigrade; °C) = (°F - 32) x 0.56

*It is common practice to convert from miles per gallon (mpg) to litres/100 kilometres (l/100km), where mpg (Imperial) x l/100 km = 282 and mpg (US) x l/100 km = 235

Index